AF564799

POLYMER CHEMISTRY

POLYMER CHEMISTRY

[For B.Sc. (Hons.) & M.Sc. Students]

By

M.G. ARORA

M. SINGH

M.S. YADAV

Edited by

SEEMA YADAV

ANMOL PUBLICATIONS PVT. LTD.

NEW DELHI - 110 002 (INDIA)

ANMOL PUBLICATIONS PVT. LTD.
Regd. Office: 4360/4, Ansari Road, Daryaganj,
New Delhi-110002 (India)
Tel.: 01123278000, 23261597, 23286875, 23255577
Email: anmolpub@gmail.com
Visit us at: www.anmolpublications.com

Polymer Chemistry

Reprint : 2017

ISBN: 81-261-1412-6

M.R.P.₹ 3000 (Inclusive of all Taxes)

PRINTED IN INDIA

Digitally Printed at Replika Press Pvt. Ltd.

Contents

Preface

In recent years very few books on 'Polymer Chemistry' have been published by Indian publishers. In all these books, the common drawback is that no book covers complete subject-matter. The present book is revised and enlarged edition and tries to cover the complete study of the subject. It has been written in such a way that students do not get confused while going through this book. Each topic included in this book is self-sufficient in itself and has been experienced in the light of modern development in a simple and elegant style. Throughout this book, it is assumed that the student understands the concepts completely in organic density. Every topic covered in this book can do full justification for most of the students. The latest syntheses of organic polymer have been included in this book. Tire of polymers analysis by various methods have been included in the oook.

We are thankful to our several immediate colleagues and friends for making useful suggestions during the preparation of manuscript of this book.

The authors express their sincere thanks to Dr. Kiran Pal and Dr. Ajay Kumar whose continuous inspirations have initiated the authors to bring out this book.

Suggestions for the further improvements towards this book shall be gratefully acknowledged and incorporated in the next edition.

Authors

1

Introduction to Polymers

1.1. Introduction

The word "*Polymer*" is derived from the Greek, "*poly*" or many, and '*meros*' or parts. However, *a polymer molecule may be defined as a number of repeating chemical units held together by covalent bonds*. This definition is necessary to distinguish polymers from crystals or liquids wherein repeating units are held together by ionic bonds or by hydrogen bonds or by even weaker forces such as dipole interactions.

The starting material from which the molecule is formed is known as the *monomer*. The repeating unit in the polymer is usually equivalent to, or nearly equivalent to, the monomer. However, some polymerisations involve the splitting off a small, usually water, during polymerisation; in these cases the repeating unit will not be exactly equivalent to the monomer. Thus, the repeating unit of poly (vinyl chloride) is —CH_2CHcl—; its monomer is vinyl chloride, $CH_2 = CHCl$. In Table 1.1, some linear high polymers, their monomers and their repeat units have been included.

In other cases, the chains are branched or interconnected to form three—dimensional networks.

High Polymers

A high polymer may be defined as one in which the number of repeating units is more than 100 or so. This number is known as the *degree of polymerisation* (DP). The lower limit 100 for the DP of high

Table 1.1: Some Linear High Polymers, Their Monomers, and Their Repeat Units

Polymer	*Monomer*	*Repeat Unit*
Polyethylene	$CH_2 = CH_2$	$—CH_2CH_2—$
Poly (vinyl chloride)	$CH_2 = CHCl$	$—CH_2CHCl$
Polyisobutylene	$CH_2{=}C(CH_3)_2$	$—CH_2—C(CH_3)_2—$
Polystyrene	$CH_2 = CH—C_6H_5$	$—CH_2—CH(C_6H_5)—$
Polycaprolactam (6-nylon)	$H—N(H)(CH_2)_5C(=O)—OH$	$—N(H)(CH_2)_5C(=O)—$
Polyisoprene (natural rubber)	$CH_2{=}CH—C(CH_3){=}CH_2$	$—CH_2CH{=}C(CH_3)—CH_2—$

polymers has been established because the physical properties needed for useful fibres, elastomers, plastics and coatings, are not characteristic of low molecular weight polymers. However, there is no upper limit for DP. Thus, the average molecular weight for most synthetic polymers is 10,000 to 100,000. However, in some cases, molecular weights of over 100 millions have been found.

1.2. Differences between Polymer and Macromolecule

By the term a 'macromolecule' means a large molecule. It is derived from the Greek word *macros* which means large. However, it is to be remembered that the term macromolecule and polymer are not equivalent. The polymers are composed of the repeating units while the macromolecules may not be composed of the repeating units.

1.3. Historical Outline

The development of Synthetic materials as substitutes for naturally occurring polymers has been largely responsible for the growth of polymer science.

From the earliest times man depended upon nature for polymeric materials like wood as fuel, furs and fibres as clothing, grain and flesh as food. Many polymeric materials behave as plastics, *i.e.*, in some stage of their fabrication they are soft and putty like and can be moulded into any desired shape and then set to retain that shape.

Many gums and resins that ooz out of the trees also behave like plastic. Sealing Wax when warmed becomes plastic cand an be impressed with a seal. Glass and clay are also plastic material of mineral origin. By early nineteenth century man had learned to mould many articles from natural gums and resins and make protective films of varnish from Shellac.

Rubber

During the early nineteenth century, Europe started getting rubber from Brazil. The latex that oozed from the bark of Hevea Bhaziliensis trees was coagulated by warming into rubber. For sometimes rubber remained interesting but useless commodity because it was tough and difficult to shape. In 1820, Thomas Hancock invented a method of masticating rubber so that it could be easily moulded under pressure; but it could not set into a shaft.

In 1839, vulcanisation process was discovered by Charles Goodyear. In this process the milled rubber is mixed with sulphur and then introduced in a mould and heated to 150°C for a few hours. The time and temperature of vulcanisation can be decreased by adding certain catalysts called **accelerators.** They generally contain nitrogen, sulphur or both.

On heating a chemical reaction took place between the rubber and sulphur which resulted in production of tough elastic rubber that retained its moulded shape. Natural rubber is plastic in nature while vulcanised rubber is elsatic. This vulcanisation process led to a rapid development of rubber industry.

The first synthetic rubbers to be commercially available in United States were **Thiokol** (1930) and **Neoprene** (1931). Both of these are still being produced commercially because they have special properties that are not matched by natural rubber. Various types of synthetic rubbers were introduced during (1939-43) World War II. After world war stereo rubbers have been made using stereo specific catalysts.

Plastics

In 1846, Schonbein nitrated cellulose to nitrocellulose. Parkes came to know about nitrocellulose and found out that it had plastic properties but it contracted on drying. He found out that nitrocellulose dissolves in molten camphor and this mixture could be moulded into any form.

He named this plastic **Parkesine.** The Parkesine was later named *celluloid* in U.S.A. In America during 1868, a prize of $10000 was offered for discovery of any substitute for Ivory. J.W. Hyatt made a billiard ball from nitrocellulose and camphor and called it celluloid and won this prize. Celluloid is highly inflammable as nitrocellulose is an explosive so attempts were made to discover other plastics.

In this attempt W. Krische coated paper with milk casein in 1897; but found that casein gets washed off with water. It was soon discovered that casein could be hardened and made waterproof by treatment with formaldehyde.

In early twentieth century plastics were largely obtained from natural products and the industry was a minor one. Most of the plastics were discovered by trial and error and no scientific explanation was available for their properties.

Resins

A Belgian chemist Leo Baekeland started investigation of tarry materials formed when phenol reacted with formaldehyde. He found out in 1909 that a resinous plastic substance is formed in this reaction. This substance could be heated and moulded into a shape. Further heating of a substance in the mould sets it in the find shape. Further heating does not resoften this substance. Baekeland patented this substance and called it **Bakelite.** Bakelite was the first synthetic polymer. The industrial of Bakelite led to a study and establishment success of polymer science.

James Swinburne also discovered phenol-formaldehyde resins at the same time but he was one day late in his application for patent than Baekeland.

Baekelite was a fore runner of many other modern synthetic polymers. For instance in 1912, Jacques Branden-burger introduced a famous transparent material, Cellophane.

The first phase of polymer chemistry started wth unlimited future prospects which encouraged over the years for the synthesis of synthetic polymers from available manomers making use of simple polymerisation techniques.

During 1926-1935 much progress was made in preparation and mechanism of polymerisation reactions and their characterization by various methods. Wallace Carothers from Willimington presented a huge account of the studies on polymerisation reactions carried out by him and his associates, at a meeting arranged by Faraday society in 1935, in Cambridge, England.

Another polymer symposia was organised by IUPAC in 1947 in Liege. At this conference the discussion included synthesis and technology of polymers like polyethylene, nylon, polyester. New characterisation methods such as X-ray scattering, x-rays, electron microscope, osmometry, nmr, IR, Raman spectroscopy etc. were now available for characterization of polymers. These methods become essential because of increasing complexity of new polymers.

The growth and development of vinyl polymers took place through several years.

Polyvinylchloride was first produced by Ostromislensky in 1912. However its commercial development occurred only when dibutylphthalate was used as plasticizer by Semor in early 1930's.

Reppe carried out researches in Germany and his work enabled large scale production of vinylpolymers from acetylene. In late thirties polystyrene and polymethacrylate were produced on large scale in Germany. ICI in England manufactured polymethyl methacrylate (Perspex) in 1936.

Polyethylene was discovered by ICI in 1933 while polytetra-fluoroethylene was discovered by Plunnekett. Urea-for-maldehyde resins which were made in 1920 were used for moulding. Melamine formal-dehyde resins were introduced by American Cyanamid Company in 1939.

Polyvinyl acetate was made commercially available in early 1930's. Poly ethers were developed by du Pont and produced on large scale in 1959. The poly carbonate were produced on large scale in 1960 by Bayer and General Electric.

Silicone polymers which were developed in 1930's became commercially important during and after the World War II. The starting material for producing silicone polymers are alkyl chlorosilanes, arychlorosilanes or substituted esters of orthosilicic acid.

Man-made Fibres

Audemars produced the first fibre in 1855 while experimenting with cellulose nitrate solution in either alcohol mixture. Hughes obtained a patent for making like thread from starch thickened with glue and resins, by about the some time.

J.W. Swan produced threads of cellulose nitrate in 1883 by forcing its solution through a spinning jet. This was called Swan Silk. Carbon filaments for early electric light bulbs were made from Swan Silk. Chardonnet manufactured artificial silk commercially in 1884 from cellulose nitrate.

In 1857, Schweizer discovered that ammoniacal solutions of copper hydroxide could dissolve cellulose. Cellulose is a versatile polymer which is found in plenty in nature in the form of cotton, hemp, jute, flax etc. In 1891, first attempts were made to spin a solution of cellulose. The cellulose produced like this is known as 'Bemberg Rayon' and is still being produced commercially in Germany, Italy, Japan and U.S.A.

In 1891 British chemistry Cross and Bevan discovered that wood pulp when treated with sodium hydroxide solution and carbon disulphide, gets converted into cellulose xanthate. When cellulose xanthate is dissolved in caustic soda solution it gives a viscous solution that could be spun to form fibres called 'Viscose Rayon'. Rayon is produced in large scale today in many countries including India. Rayon (also know as 'viscose') is in fact *regenerated cellulose*.

In 1894, Cross and Bevan acetylated cellulose to get cellulose acetate. In 1921 cellulose acetate fibres were marketed for the first time as 'Celanese'.

Regeneration of protein is used in food industry to make meat like products from plant proteins.

In 1927 Wallace Carothers started research at du Pont to synthesise polymer molecules. By 1938 a plant was set up for making polyamide fibres and in 1939 womens stocking were marketed with instant success. This fibre was called nylon.

The word nylon is now accepted as a generic term for the synthetic polyamides which are characterized by repeating —NHCO— amide linkage.

In U.K. in 1941 Whinfield and Dickson discovered the polyester fibre called terylene chemically it is polyethylene terephthalate. Teryene polymer can be melt spun into a fibre which is widely used in textile industry.

In 1945 du Pont prepared a synthetic fibre from acrylonitrile the first acrylic fibre was marketed under the name Orlon in 1950.

It is polyacrylonitrile and is made by polymerisation of vinyl cyanide.

Before World War II Germans developed a synthetic fibre from polyvinyl chloride. In America a copolymer of vinyl chloride and vinyl acetate was marketed as 'Vinyon'. Dow chemical marketed as 'Saran' a copolymer from vinylidene chloride and vinyl chloride. Saran fibre is characterized by its remarkable resistance of most of chemical reagents and fire.

In 1956 polyethylene fibres were prepared and later on polypropylene fibres were prepared. Fibres have been made from polytotrafloroethylene for special uses.

Polyurethane fibres were made in Germany during World War II. These were later improved. Today these fibres are sold as Spandex. They have properties like that of rubber.

Adhesives

Though animal glue was used as an adhesive for more than 3000 years but its commercial manufacture started only in 1808. 'Later on starch, casein and rubber based adhesives also came into use.' After 1940 several synthetic resin adhesives have been developed. Polyacrylates are used commercially. Copolymers of ethyl acrylate with methacrylate and small amounts of hydroxyl, carboxyl, amine, or amide comonomers are used to prepare high quality latex points for wood, wall board and masonry in homes.

Methyl cyanoacrylate is an extremely powerful adhesive.

Polymers

At present the word polymer is very popular but the acceptance of the concept of polymer took a long time. By 1850's the concept of atoms and molecules was largely accepted for inorganic compounds. The idea that large covalent bonded structure exist or could be prepared was not acceptable to the scientists in the nineteenth century. Kekule's work led to structure elucidation of organic compounds. Graham in 1861 found that solutions of rubber, cellulose and other such materials diffused through a membrane at a very slow rate. Attempts were unsuccessfully made to crystallic colloids.

German scientist Staudinger in 1920 stated that colloidal properties of organic materials are because of the large size of individual molecules. He also stated that macromolecules contain only primary covalent bonds.

He also showed that such materials can retain colloidal properties in all the solvents in which they dissolve. Additional support to this concept came from newer methods of molecular weight determination. By about 1930 the concept of polymer became established.

During the period 1920-1960 enormous polymers prepared in understanding the strucutre of native polymers and synthesising artificial counterparts.

Polymers nowadays are not difficult to prepare because of the easy availability of the raw materials. Most of the synthetic polymers are of recent origin but they have made an impact on our daily life. They have found extensive applications in textile fibres, rubber and rubber goods, building materials, packaging, fancy decorating articles and ion-exchange resins.

1.4. Degree of Polymerisation and Orientation in Polymers

If two identical molecules combine chemically a *dimer* is obtained. Acetylene for instance, is dimerised to vinylacetylene. If smaller molecules of a susbtance unite then a large molecule, a *polymer*, of high molecular weight is obtained. The individual small molecule from which a polymer is formed is called a *monomer*. The chemical process for the formation, of a polymer is called *polymerisation*. This is exemplified in the following equation.

$$\text{n monomer} \xrightarrow{\text{Polymerisation}} \underset{\text{A polymer}}{(\text{monomer})\ \text{n}}$$

The repeating atomic groupings in a polymer each of which is monomeric in character is termed *repeat unit*. The length of the polymer chain is specified by the repeat units. This called the *degree of polymerisation* (DP) and is denoted by n.

The length of the polymer chain is specified by the number of repeating units in two chain. This is termed as the degree of polymerisation and denoted by DP. The molecular weight of the polymer is obtained by the product of the molecular weight of the repeating unit and the degree of polymerisation (DP). For example, using poly (vinyl chloride), a polymer of DP 1000 has a molecular weight of $63 \times 1000 = 63000$. Most high polymers useful for plastics, rubbers or fibres have molecular weights between 10,000 and 1,000,000.

Orientation in Polymers

In a polymer made of long chain like molecules, molecular motion is curtailed and rotation even around single bonds is curtailed. As a result head to tail polymers can have varying orientation of the side groups which can be either orderly or disorderly fashion with respect to the chain. These different arrangements are **isotactic, syndiotactic** and **atactic** and are shown for propylene. In *isotactic* (or cis arrangement) all the methyl groups lie on one side of chain and the hydrogen atoms on the other

```
         R         H          R         R          H
        /    H    /     H    /    H    /     H    /
   /   C    /    C     /    C    /    C     /    C
 \ C  ≡  \ C   ≡   \ C    ≡  \ C   ≡  \ C    ≡
  ≡   H   ≡    R    ≡   H     ≡    H    ≡    R
  H       •H        H         H         H
```

Such polymers are obtained by Zieglar-Natta catalyst. *An example* of this class of arrangement is *natural rubber*. Isotactic polymers have high degree of crystallivity, are more denser, and strong because of presence of regularity of the chain structure.

If the methyl and hydrogen atoms fall alternatingly above and below the chain then this orientation is referred to as *sydiotactic* (for *trans* or alternating arrangement).

```
         R         H          R         H          R
  H     /    H    /     H    /    H    /     H    /
  /    C    /    C     /    C    /    C     /    C
 \C  ≡   \ C   ≡   \ C    ≡  \ C   ≡  \ C    ≡  \
  ≡   H   ≡    R    ≡   H     ≡    R    ≡    H
  H       H         H         H         H
```

Gutta Percha is an example of this arrangement. The third type *atactic* (Greek without order) the groups are arranged randomly above and below the chain. This type of radical is obtained by free radical polymerisation.

```
         R         R          R         R          R
  H     /    H    /     H    /    H    /     H    /
  /    C    /    C     /    C    /    C     /    C
 \C  ≡   \ C   ≡   \ C    ≡  \ C   ≡  \ C    ≡  \
  ≡   H   ≡    H    ≡   H     ≡    H    ≡    H
  H       H         H         H         H
```

1.5. Classification of Polymers

Polymer refers to a generic name which is assigned to a vast number of materials of high molecular weight. These materials are known to exist in numerous forms and numbers because of a very large number and types of atoms present in their molecules. Polymers can be having different chemical structures, physical properties, mechanical behaviour, thermal characteristics, etc., and would be classified on different ways, as now described.

1. Natural and Synthetic Polymers

Depending on their origin, it is possible to group polymers as natural or synthetic.

(a) ***Natural polymers.*** These are the polymers which are isolated from natural materials. These are also known as biological polymers. Examples of natural polymers are rubber, wool, cellulose, starch and proteins. Although these materials are extremely important in our daily life, they have the following two diadvantages:

(i) Their physical properties are fixed by the nature of the particular material, and cannot normally be varied.

(ii) The supplies are limited by agricultural considerations, and thus the materials are often expensive and subject to rapid fluctuations in price.

(b) ***Synthetic polymers.*** These are the polymers which are synthesized from low molecular weight compounds. Typical examples are: polyethylene, PVC, nylon and terylene.

These are manufactured from cheap and readily available petroleum fractions, and the physical properties may be "tailor made" for almost any desired application.

When we talk of polymers, conventionally it means the synthetic or man-made polymeric substances.

The first purely synthetic polymer was the phenolformaldehyde family of synthetic resins discovered by Baekeland in Germany and first produced commercially in 1907. In 1930, polystyrene was first manufactured in Germany.

Synthetic polymers are mainly of two types:

(i) *Linear polymer.* It is one in which the repeating units are similar to the links in a very long chain. A linear polymer usually has end groups which are different from the repeating unit because it is necessary in some way to terminate the polymerisation. Although the end groups have no effect on the mechanical properties of a polymer, they may have a profound effect on the stability, solubility and adhesive properties of the polymers. A linear polymer chain would invariably exhibit coiling due to chain flexibility, arising mainly due to internal rotation around a single bond.

(ii) *Branched polymer.* It is one in which some of the molecules are attached as side chains to the linear chains. However, in a branched polymer, the undivided molecules are still discrete. The branching may be random and at irregular intervals or may occur at regular intervals with equal lengths of branches. These branches may join to another adjacent chain. These types are shown in Figs. 1.1 (a) to 1.1 (d) as branched polymers, comb type polymers and ladder type polymers. More branching at random points connecting many chains give rise to net work or cross linked polymers [Fig. 1.1 (e)].

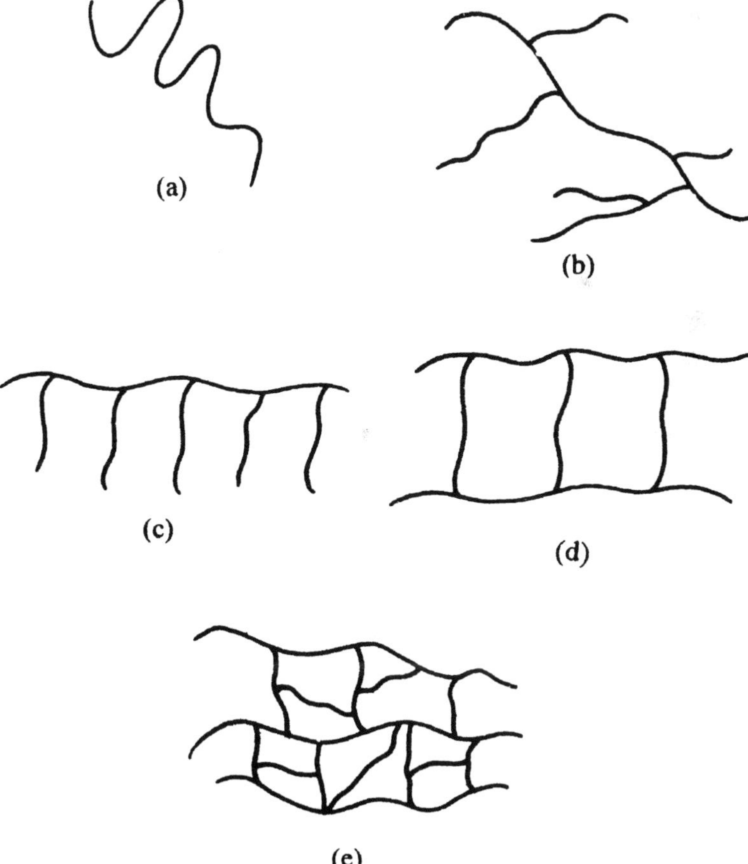

Fig. 1.1: Polymer chains. (a) linear, (b) branched, (c) comb type, (d) ladder type, (e) net work.

2. Organic and Inorganic Polymers

(a) ***Organic Polymers.*** An organic polymer may be defined as a polymer whose backbone chain is essentially made of carbon atoms. However the atoms attached to the side valencies of backbone carbon atoms include hydrogen, oxygen, nitrogen, sulphur, etc. The majority of synthetic polymers are organic polymers. They find valid appreiciations in every field and are very extensively studied.

In reality, the number and types of organic polymers have been to large that when we talk of polymers, we mean organic polymers.

(b) ***Inorganic Polymers.*** An inorganic polymer may be defined as a polymer whose backbone chain is not having carbon atom. Examples of inorganic polymers are glass, silicone, rubber, etc.

A few examples of organometallic and inorganic polymers are listed in Table 1.2.

Table. 1.2: Examples of Organometallic and Inorganic Polymers

Polymer	*Chain structure*
Polysiloxanes	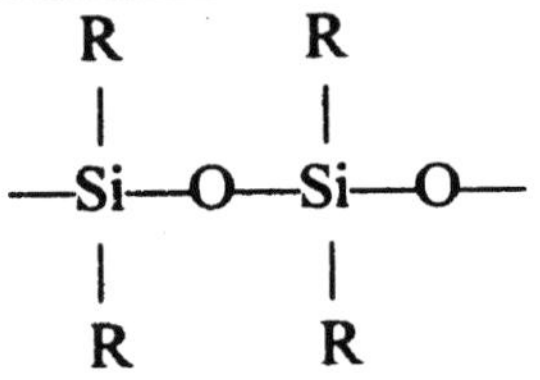 $\begin{array}{c}\text{R}\quad\quad\text{R}\\ \vert\quad\quad\;\vert\\ \text{—Si—O—Si—O—}\\ \vert\quad\quad\;\vert\\ \text{R}\quad\quad\text{R}\end{array}$
Polytitanoxanes	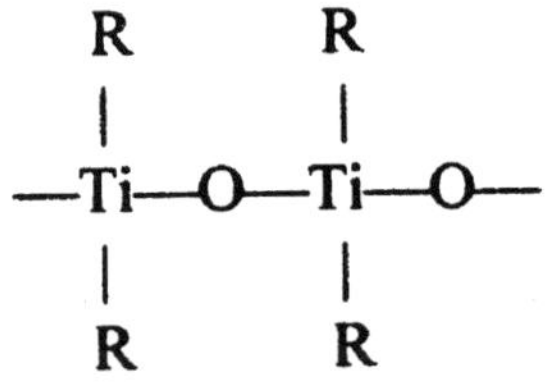 $\begin{array}{c}\text{R}\quad\quad\text{R}\\ \vert\quad\quad\;\vert\\ \text{—Ti—O—Ti—O—}\\ \vert\quad\quad\;\vert\\ \text{R}\quad\quad\text{R}\end{array}$
Polyphosphoric acid	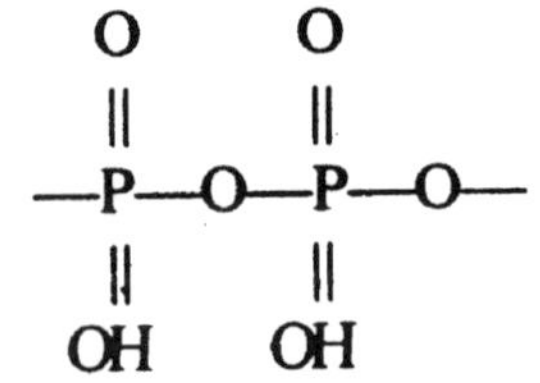 $\begin{array}{c}\text{O}\quad\quad\text{O}\\ \Vert\quad\quad\;\Vert\\ \text{—P—O—P—O—}\\ \Vert\quad\quad\;\Vert\\ \text{OH}\quad\text{OH}\end{array}$
Polyphosphates	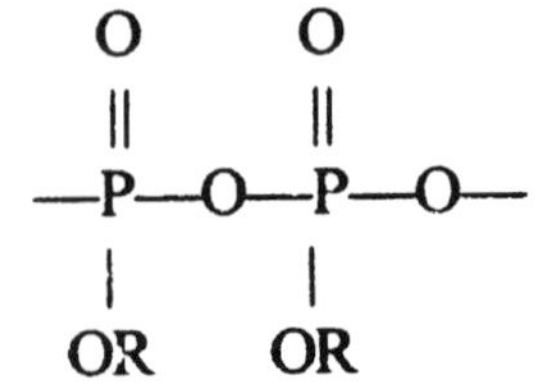 $\begin{array}{c}\text{O}\quad\quad\text{O}\\ \Vert\quad\quad\;\Vert\\ \text{—P—O—P—O—}\\ \vert\quad\quad\;\vert\\ \text{OR}\quad\text{OR}\end{array}$

Polyarsenates

In addition to these, there are many more polymers like phosphonitrichloride, stibnates, amorphous silicon dioxide, natural and synthetic silicates etc.

3. Thermoplastic and Thermosetting Polymers

(a) ***Thermoplastic polymers.*** These are the polymers which soften on heating and stiffen on cooling. Examples of such polymers are polyethylene, PVC, nylon and sealing wax.

It is to be remembered that thermoplastics soften on heating and can be converted into any shape that they are able to retain on cooling. However, the process of heating, reshaping and retaining the same on cooling can be repeated several times. In this type of polymers there is no cross linking between the chains. The intermolecular forces in thermoplastic polymers are intermediate to those of elastomers and fibres.

(b) ***Thermosetting polymers.*** These are the polymers which on heating undergo chemical change and convert themselves into an infusible and insoluble mass. These polymers are normally made from semifluid polymers with low molecular masses by heating in a mould. Bakelite is the common example of thermosetting polymer.

The thermosetting material undergoes a permanent change upon melting and thereafter sets to a solid which can not be remelted. This may be attributed to excessive cross linking between the chains forming three dimensional network of bonds.

4. Plastics, Elastomers, Fibres and Liquid Resins

Depending on their ultimate form and uses, polymers can be classified into plastics, elastomers, fibres or liquid resins. These are defined as follows:

(a) ***Plastics.*** These are the polymers which can be shaped into hard and tough utility articles by the application of heat and pressure. Example of plastics are polystyrene, PVC and polymethyl methacrylate.

Plasticizers are substances which can impart plasticity to a polymer i.e., facilitate the conversion of solid and brittle resins into a dough-like condition so that it can be moulded in a desired shape.

Esters of phthalic acid such as dimethyl phthalate and high boiling solvents usually are termed as plasticizers. It is added, for instance, to hard poly vinyl chloride (PVC) to soften it.

(b) ***Elastomers.*** These are the polymers which have rubber-like elastic properties, *i.e.*, they readily undergo reversible deformation and elongation. Examples of elastomers are natural rubber, synthetic rubber, silicon rubber etc. In elastomers, the polymer chains are held together by weakest intermolecular forces which permits the polymers to be stretched under stress but they regain shape when the stress is relieved. It is possible to modify the elasticity of such polymers. For example natural rubber which is a gummy material having poor elasticity is a polymer of isoprene, when heated with sulphur forms a material which has modified elasticity and is known as **Vulcanized rubber.** Elasticity of rubber is a special physical property of rubber that results from the structure of polymer. In vulcanized rubber the isoprene chains are cross linked by sulphur which prevents the slippage of chains on application of stress. It also helps to regain the shape after removal of stress. Thus, vulcanised rubber is more elastic than natural rubber.

(c) ***Fibres.*** If polymers could be drawn into long filament-like materials whose length is at least 100 times its diameter, polymers are said to have converted into fibres. Typical examples of fibres are nylon and terylene. These are the polymers which have quite strong interparticle forces such as hydrogen bonds.

(d) ***Liquid resins.*** Polymers used as adhesives, polting compounds, sealants, etc., in a liquid form are described as liquid resins. Typical examples of liquid resins are commercially available epoxy adhesives and polysulphide sealants.

The above classification is however, not rigid, since by suitable structural modifications it is possible to confer the desired properties on a high polymer. Thus, polyvinylchloride and polyurethane can function either as elastomers, or plastics. Polypropylene and polyamides can function as plastics or fibres.

5. Classification of Polymers based on Chemical Structure

This type of classification is very useful for understanding the properties and behaviour of these materials. From the known correlation betweent the structure and behaviours, one can synthesise a tailor-made molecule showing the desired properties. From the knowledge of structure, one can explain why polyvinylchloride is having a higher dielectric constant than that of polyethylene, why nylon is having a higher melting point than that of polystyrene, why nylon is less soluble in common solvents. In Table 1.3 the chemical structure of monomers repeat units and the derived names of polymers have been included.

Table 1.3: Chemical Classification of Polymers

Monomer structure	*Polymer*
(A) Rubber and related polymers	
$CH_2=C(CH_3)-CH=CH_2$	Polyisoprene, natural rubber
$CH_2=C(Cl)-CH=CH_2$	Polychloroprene
$CH_2=CH-CH=CH_2$	Polybutadiene
$HO-Si(CH_3)_2-OH$	Polydimethyl siloxane
(B) Vinyl polymers (X is shown below for a vinyl monomer $CH(X)=CH_2$)	
X	Plolymer
H	Polyethylene
Cl	Polyvinylchloride
CH_3	Polypropylene
CN	Polyacrylonitrile
CN	
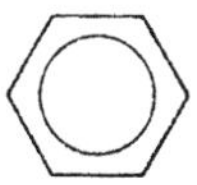 (benzene ring)	Polystyrene
OH	Polyvinylalcohol
$OCOCH_3$	Polyvinylacetate
COOH	Polyacrylic acid
$COOCH_3$	Polymethyl acrylate

COOR — Polyacrylic acid esters

OR — Polyvinyl ethers

$COCH_3$ — Polymethylvinylketone

$CONH_2$ — Polyacrylamide

N

H_2C CO

H_2C—CH_2 — Polyvinyl pyrrolidone

N — Polyvinyl carbozole

(C) Vinyl polymers in which two or more hydrogen atoms are substituted

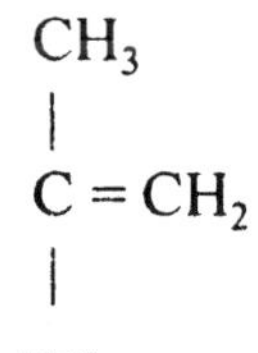

CH_3

|

$C = CH_2$ — Poly α-methyl styrene

|

CH_3

|

$C = CH_2$ — Poly methacrylic acid

|

COOH

CH_3

|

$C = CH_2$ — Poly methylmethacrylate

|

$COOCH_3$

CH_3

|

$C = CH_2$ — Poly methacrylic acid esters

|

COOR

$CF_2 = CF_2$	Polytetrafluoroethylene
$CH_2 = CCl_2$	Polyvinylidine chloride
$CF_2 = CFCl$	Polytriflurochloroethylene

(d) Polyaromatics

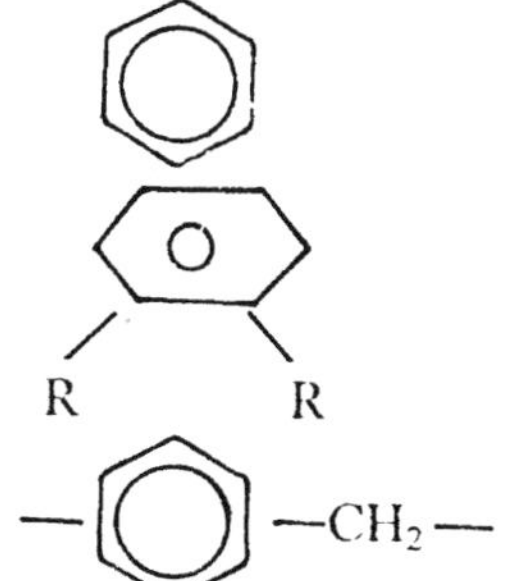

Polypnenylenes

Substituted polyphenylenes

Polymethylphenylenes

(E) Polymers with oxygen in the main chain

HCHO — Polyformaldehyde

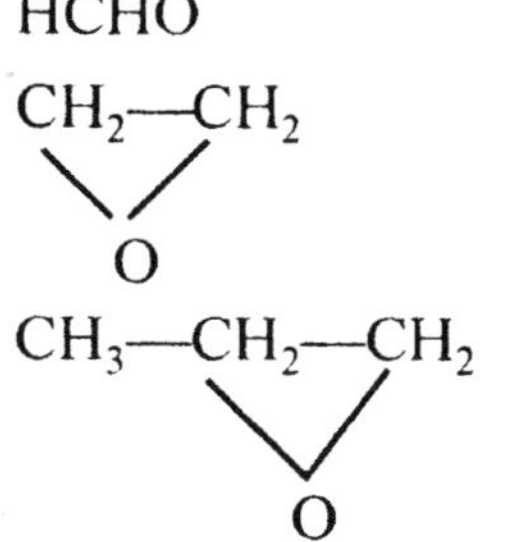

Polyethyleneoxide or Polyether

Polypropylene oxide

(E) Condensation polymers Repeat Units

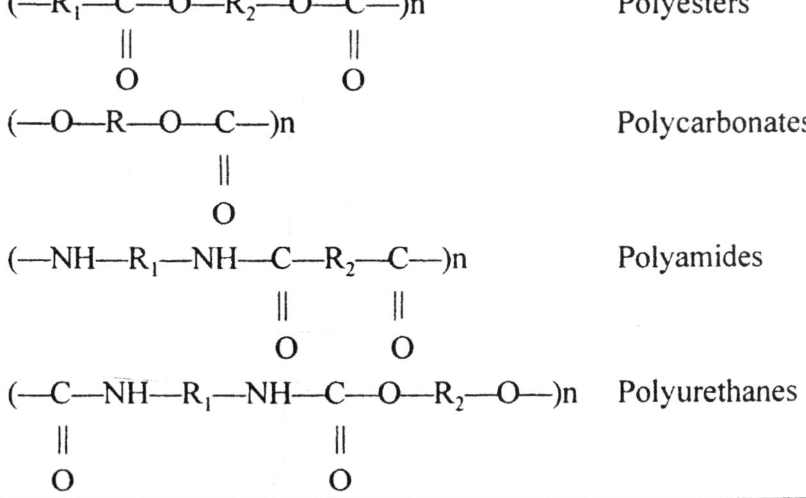

The tables are not exhaustive and include only typical and common examples of monomers/repeat units and polymers named after them. Many number of polymers are derived from vinyl unit. Besides this, a general classification as polyesters, polyamides, polyethers is based on the common chemical linkages found in these polymers.

6. Classification based on the Mode of Synthesis of Polymers

On this basis polymers may be classified as (i) **addition polymers and** (ii) **condensation polymers.**

(i) ***Addition polymers.*** These are the products formed when the monomer units are repeatedly added to form long chains without the elimination of any by product molecules. The monomer units are generally unsaturated compounds usually derivatives of alkenes. The molecular mass of such polymers is an integral multiple of the molecular mass of the monomer unit. Examples are polyethene (polyethylene), polypropylene, orlan, polyvinylchloride (PVC) etc.

This type of polymers are obtained by **addition polymerisation.** Such a polymerisation process involving addition of monomer units to the growing chain is known as **chain growth polymerisation.**

(ii) ***Condensation Polymers.*** This type of polymers are formed by the process of **condensation polymerisation** when a monomer contains active functional groups (generally two) which react together with the elimination of a simple molecule such as H_2O then the product formed is called condensation polymner. For example Nylon-66, Nylon-66 is a condensation polymer of hexamethylene, diamine and adipic acid.

Some other example of condensation polymers are Dacron (a polymer of ethylene glycol), polyester (a polymer of p-terephthallic acid), Bakelite (a polymer of phenol and formaldehyde).

The process of condensation takes place in a stepwise manner, ultimately resulting in the formation of polymer. This process is also known as **step growth polymerisation.**

7. Classification based on Composition

On this basis the polymers can be classified as (i) **Homopolymers** and (ii) **Copolymers.**

(i) ***Homopolymers.*** When a polymer is obtained from a single monomer unit it is called **homopolymer,** *e.g.*, polystyrene which is formed from monomer styrene.

Some other examples of homopolymers are polyethylene, polypropylene etc.

With some exceptions, polymers made in chain reactions often contain only carbon atoms in main chain **(homochain polymers).**

(ii) ***Copolymers.*** When a polymer contains two monomer units in the chain then it is known as a **copolymer.** In majority of commercial copolymers two kinds of repeat units are present but in a few of them three repeat units are also found. When a copolymer has three monomer units in the chain it is referred to as **terpolymer.** It is possible to designate copolymers as (a) random copolymers (b) alternating copolymers (c) block copolymers and (d) graft copolymers on the basis of the structural units as follows:

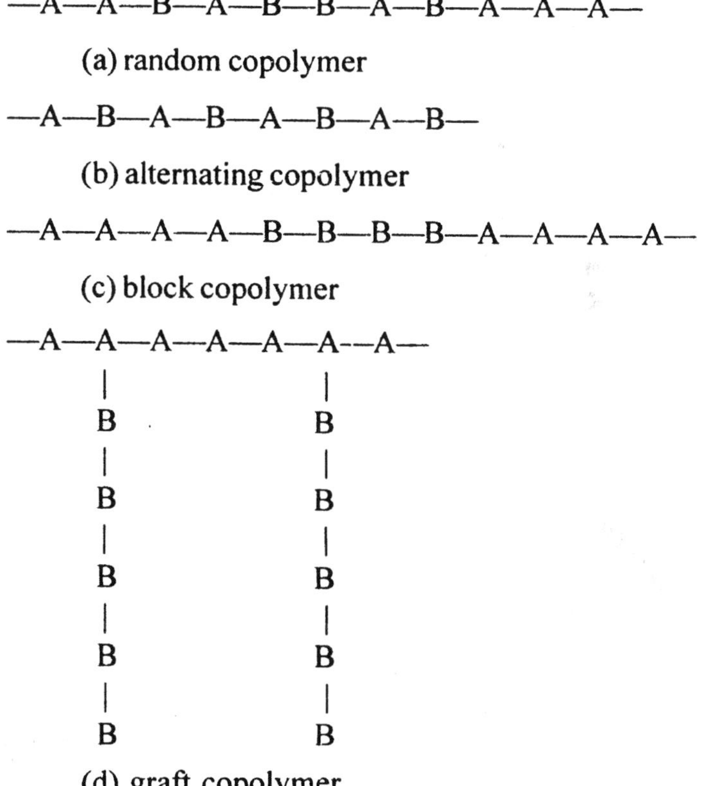

(a) random copolymer
(b) alternating copolymer
(c) block copolymer
(d) graft copolymer

A block copolymer consists of one in which blocks of repeating units of one type alternate with blocks of another type. Block copolymers are produced by the introduction of end groups which can be made to react under different conditions.

Block copolymers may be produced by the following four techniques:

(i) Utilise reactive end groups
(ii) Activate end groups
(iii) Fragment a chain to produce active end groups or
(vi) Transport reactive end groups into a different media.

A graft copolymer consists of a linear polymer chain of one type to which has been grafted side-chains of a different type of polymer. The formation of a graft copolymer can be brought about by either of two general methods:

(i) Initiation of chain growth of monomer B on an existing polymer molecule formed from monomer A or
(ii) Termination of chain growth polymer B by an existing polymer molecule formed from monomer B.

The importance of the block and graft copolymers is that the resultant material tends to exhibit the properties of each homopolymer. For example, pure polystyrene is quite brittle, whereas polymerisation in the presence of about 5 per cent of rubber produces a material which is strong and tough.

Polyacrylonitrile is an excellent textile fibre but is difficult to dye. However, by its copolymerisation or by grafting on a second polymer, it is possible to maintain the desirable properties of the fibre, yet produce a textile which can be processed in the usual way. Among the various factors that govern the copolymerization process, the concentration and reactivity of the monomer are quite important. At any given time, the chain may grow in four different ways as under. Here A* and B* are the radicals which are involved in propagating steps, whereas A and B are the respective monomers.

$$\text{———}A^* + A \longrightarrow \text{———}AA^* \quad \text{(I)}$$

$$\text{———}A^* + B \longrightarrow \text{———}AB^* \quad \text{(II)}$$

$$\text{———}B^* + A \longrightarrow \text{———}BA^* \quad \text{(III)}$$

$$\text{———}B^* + B \longrightarrow \text{———}BB^* \quad \text{(IV)}$$

If the monomer A is reactive as compared to B in reactions I and II, then reaction I will be preferred. If the monomer A add more quickly at the chain-ending of B (reactions III and IV), then III will be favoured. If the concentrations of A and B are equal, more of A will be incorporated into the chain.

A high concentration of B will help to compensate for its lesser reactivity, since the growing chain has a greater chance of meeting B molecules more often than A molecules. The relative reactivity of monomers depends on polar and steric effects of the substituents on the monomers.

The presence of electron-withdrawing substituents like —COOR, —CN, —$COCH_3$, etc. decrease the electron density at the site of the double bond in a vinyl monomer, relative to ethylene. Electron-donating groups like —CH_3, —OCH_3, CH_3COO— increase the electron density at the same site. The radical is electrophilic, since it contains an odd electron, and seeks electron-rich or nucleophilic centres.

The electron-withdrawing substituent when present in a free radical makes it more electrophilic, and electrophilic radical will seek a monomer containing an *electron-releasing* substituent and vice versa. Monomers with similar bond polarities like styrene and butadiene, are more likely to form random copolymers.

For example, by using 90 parts of vinyl chloride and 10 parts of vinyl acetate, the random copolymer formed has the toughness of poly (vinyl chloride), thermal stability of poly (vinyl acetate) and solubility akin to poly (vinyl acetate). These combination of properties makes it useful as a paint.

8. Classification based on Polarity

On the basis of polarity the polymers can be classified as (i) ionic polymers and (ii) non-ionic polymers.

9. Homochain and Heterochain Polymers

Homochain Polymers

Sulphur and selenium have a high tendency to form homochain polymeric compounds.

Ordinary rhombohedric sulphur has cyclic molecules containing eight sulphur atoms each. When sulphur is heated in the molten state, the ring appears into a linear polymer.

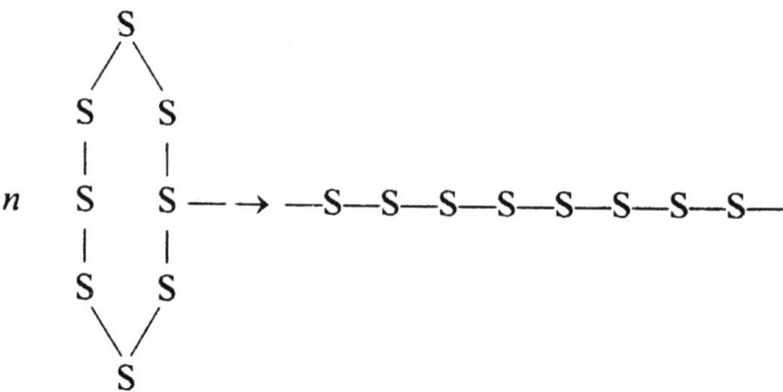

Heterochain polymers

Some of the heterochain polymers are given below. The polymeric structure of hydrogen boride can be represented as under:

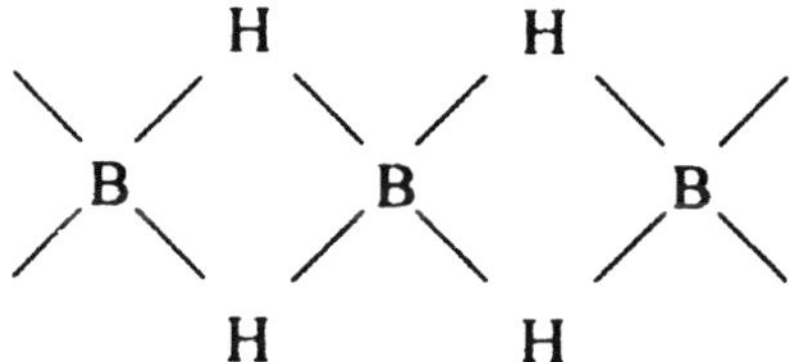

Silicone also forms the branched polymers as represented below:

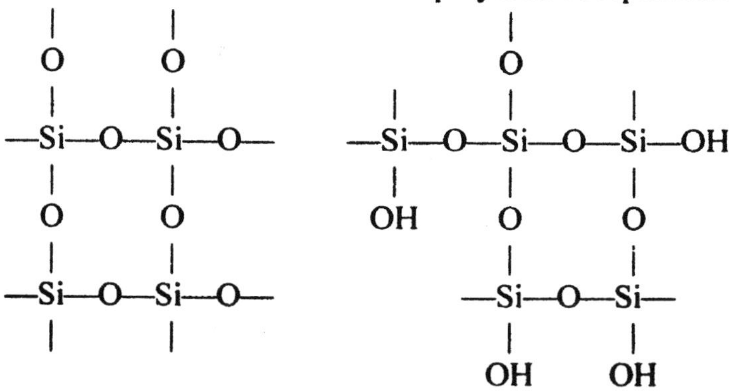

Silicone dioxide Polysilicic acid

Some of the polymers containing phosphorous, arsenic, selenium and tellurium are shown below:

```
 O     O     O
 ||    ||    ||
—P—O—P—O—P—O—
 |     |     |
 OH    OH    OH
```

Polyphosphoric acid

```
  O       O       O
  ||      ||      ||
—As—O—As—O—As—O—
  |       |       |
 OMe     OMe     OMe
```

Polyarsenates

```
  O       O       O
  ||      ||      ||
—Se—O—Se—O—Se—O—
```

Selenium containing polymers

```
  O       O       O
  ||      ||      ||
—Te—O—Te—O—Te—O—
```

Selenium containing polymers

1.6. Nomenclature of Polymers

Most of the polymers are better known by their trivial names or trade names. Polymers prepared from single polymers are denoted by prefixing *poly-* to the name of the monomer, *e.g.*, polyethylene, polypropylene, polyacrylonitrile, polystyrene, etc. If the monomer has substituents or has a multi-worded name, the name of the monomer is enclosed in parenthesis after the prefix *poly-*, *e.g.*, poly (methyl methacrylate), poly (vinyl alcohol) etc. Condensation polymers like that derived from ethylene glycol and terephthalic acid are named as poly(ethylene terephthalate).

1.7. Tacticity of Polymers

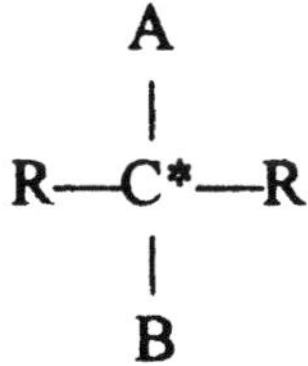

The variation in polymer structures is observed when the backone of the polymer molecule contains a carbon atom attached to two different side groups. Such polymers can have different configurational arrangements or **tacticity.**

Polymers wherein all adjacent asymmetric carbon atoms over the length of the polymer chain have the same space configuration are known as **isotatic polymers.**

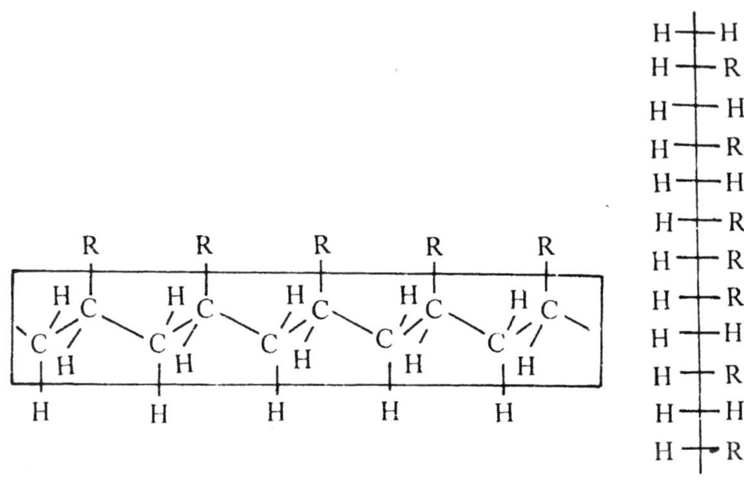

(a) Isotatic Polymer

Polymers in which chain is made up of units having opposite space configuration of each asymmetric carbon atom are known as **syndiotactic polymers.**

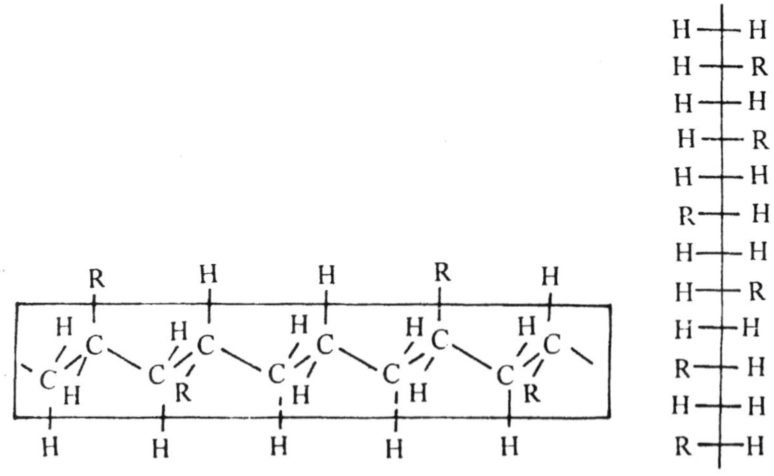

(b) Syndiotactic polymer

Polymers in which groups are arranged randomly in space over the chain are known as the **atatic polymer.**

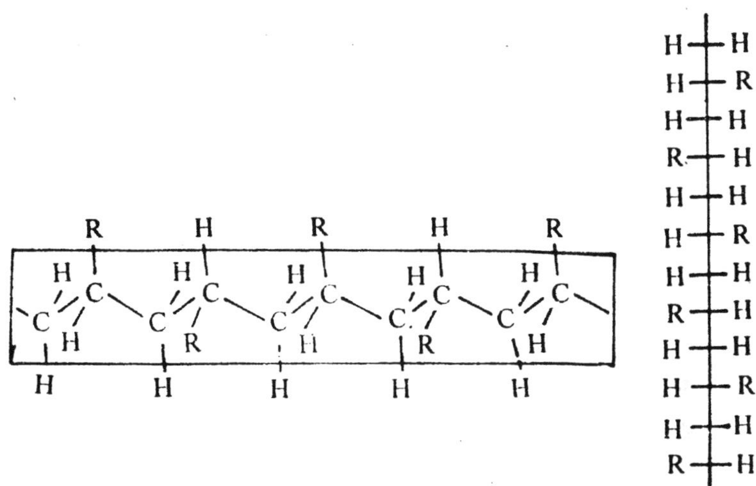

(c) Atactic polymer

1.8. Concept of Molecular Weights in Polymers

For low molecular weight compounds each compound has a definite molecular weight. For example, benzene has the molecular weight 78, methyl methacrylate has the molecular weight 100.12 and the molecular weight of vinyl chloride is 62.5. Thus a compound having molecular weight 100 can never be benzene. The molecular weight of these compounds are fixed and they can never be altered by any experimental methods. But with macromolecular compounds the case is totally different. It is a common observation that the macro-molecular compounds do not have a certain definite and permanent molecular weight. Polystyrene can have a molecular weight 50,000, it can have a molecular weight 10,000 or 100,000 or even as high as 10^6 and more.

Important Concepts on the Various Molecular Weights of High Polymers

Synthetic polymers are produced by the polymerisation or polycondensation of monomer molecules. Due to the particular mechanism of these reactions, the macromolecules so formed have varying degrees of polymerization. Even natural polymer compounds except certain proteins are not uniform with respect to their molecular weight. Because of this reason, any polymer sample has a **characteristic molecular weight distribution (MWD),** which can be represented by a distribution function or a distribution curve. At present, however, the determination of the molecular weight distribution curve is rather difficult. Hence, in polymer chemistry, the so called molecular weight of a polymer is merely a statistical average.

In case a polymer sample weighs W grams and contains n_1 monomer molecules each molecular wieght M_1, n_2 dimer molecules of molecular weight M_2, n_3 trimer molecules of molecular weight M_3....n_i i-mer molecules of molecular weight M_i etc., then the fraction of the molecules having the molecular weight M_i is given by

$$N_i = \frac{n_i}{\sum_i n_i}$$

their weight will be $W_i = n_i M_i$ and their weight fraction,

$$W_i = \frac{w_i}{\sum_i w_i}$$

The three common average molecular weights are

(i) Number average molecular weight $(\overline{M}n)$

(ii) Weight average molecular weight $(\overline{M}w)$

(iii) Average molecular weight (Z)

Number–average molecular weight is represented by,

$$\overline{M}n = \frac{\text{Total weight of sample}}{\text{Total moles of sample}}$$

$$= \frac{\sum_i n_i M_i}{\sum_i n_i} = \sum n_i M_i$$

Weight–average molecular weight is represented by,

$$\overline{M}w = \frac{\sum n_i M_{i^2}}{\sum n_i M_i}$$

Z–average molecular weight is represented by,

$$\overline{M}z = \frac{\sum_i n_i M_{i^3}}{\sum_i n_i M_{i^2}}$$

Hence, $\overline{M}n$ is the statistics average molecular weight calculated from the number molecular weight distribution function, $N(M)$. $\overline{M}n$ is the statistical average molecular weight calculated from the weight molecular weight distribution function, $W(M)$. $\overline{M}_2$ is the statistical average molecular weight calculated from the distribution function M W (M) or M^2 N (M).

For a polymer sample having uniform distribution, the molecular weight,

$$\overline{M}n = \overline{M}w = \overline{M}z$$

But for real polymer samples

$$\overline{M}n > \overline{M}w > \overline{M}z$$

and $\overline{M}n : \overline{M}w : \overline{M}z = 1 : 1.386 : 1.617$

Thus when,

$$\overline{M}w = 100{,}000,\ \overline{M}n = 72{,}000, \text{ and}$$

$$\overline{M}z = 116{,}700$$

It may be noted that the weight average molecular weight is higher than tne number average molecular weight.

The molecular weight of thé polymer is related to its chain length or to the degree of polymerization (DP).

Degree of polymerization (DP) is defined as the ratio between the molecular weight of the polymer and that of the monomer.

For growth of polymer chain it is essential that the radical encounters monomer molecules. The length of the chain formed is determined entirely by such random encounters. Some polymer chains may grow for a longer time than others. The product of polymerization process is a mixture of chains of differing lengths. Thus there is spread is molecular weight of the product.

Molecular Weight Distribution

The distribution of molecular mass is complicated in case of the radical polymerization, since there are various types of termination. Let us consider the simplest mode of termination, i.e., termination by disproportionation. In this case. $\overline{X}n$ is equal to v, the kinetic chain length. A parameter α can be assumed, α is defined as the probability that during an addition reaction, a chain will propagate rather than terminate. The parameter, α, is similar to the parameter, p, the extent of reaction in step-growth polymerization. Further, α can be represented as, under

$$\alpha = \frac{\text{rate of propagation}}{\text{rate of propagation + rate of all other reactions}}$$

If we consider a polymer consisting of i monomer units.

$$R\text{—}M\text{—}M\text{—}M\text{—}\ldots\ \ldots M$$

$$1 \quad 2 \quad 3 \qquad (i-1)$$

Then we find that the molecule M_i is formed by $(i-1)$ successive addition reactions. The probability, that one of these reactions has occurred is α and hence the probability that $(i-1)$ successive addition reaction have occurred is α^{i-1}. The probability, that the last reaction is termination rather than propagation is, $(1-\alpha)$.

Hence, the probability that M_i has been produced is,

$$P_{(i)} = \alpha^{i-1}(1-\alpha) \qquad \ldots(1.1)$$

The relationship between $\overline{M}w$, $\overline{M}n$ and α in case of condensation polymerization would be, $\overline{M}w / \overline{M}n = 1 + \alpha$

$$\text{If } \alpha \rightarrow 1, \text{ then } \overline{M}w\,\overline{M}n \sim 2 \qquad \ldots(1.2)$$

If the termination is by combination, then the relationship between $\overline{M}w$, $\overline{M}n$ and α would be

$$\overline{M}w / \overline{M}n = (2+\alpha)/2 \qquad \ldots(1.3)$$

and for long polymer molecules $\alpha \rightarrow 1$ and so,

$$\overline{M}w / \overline{M}n \sim 1.5 \qquad \ldots(1.4)$$

If we consider that equimolar quantities of reactants AA and BB react and that the reactivity of groups A or B remains unchanged at all stages of conversion. Then the degree of polymerisation changes with the extent of polymerisation that has occurred and the probability factor can give an idea about the molecular weight distribution at a particular conversion p as under.

If, p is the extent of reaction, which also gives the probability of reaction occurring between two functional groups to form a repeat unit of —AB— according to the following reaction.

$$n\text{A—A} + n\text{B—B} \rightarrow \text{A-[-AB-]}_{3n-1}\text{-B} + \text{by product}$$

The fraction of unreacted groups is $1-p$. If it is assumed that a tetramer is formed because of the reaction between 4 reactant molecules, 2 of A—A and 2 of B—B, in which 3 pairs of reactive groups have reacted. The probability of a tetramer formation is, p^3.

Similarly, the probability of formation of an n-mer, *i.e.*, a polymer molecule containing n number of repeat units, will be p^{n-1}. Hence if at a conversion of p, the total number of polymer molecules formed is N and that of the n-mers formed is N_n, then N and N_n are related as under:

$$N_n = N p^{n-1} (1-p) \qquad ...(1.5)$$

If N_0 is the number of reactant molecules initially present then

$$N = N_0 (1-p) \qquad ...(1.6)$$

Now, combining Eqns. (1.5) and (1.6), we get

$$N_n = N_0 (1-p) p^{n-1} (1-p)$$

or

$$N_n = N_0 (1-p) p^{n-1} \qquad ...(1.7)$$

Here N_n gives the number fraction of n-mer in the total number of polymer molecules present, N. The weight fraction W_n, similarly, is given by

$$W_n = \frac{N_n . n}{N_0} = n(1-p)^2 p^{n-1} \qquad ...(1.8)$$

For different values of p, the N_n and W_n distribution functions are shown in Figs. 1.2 and 1.3 respectively.

The curves in Fig. 1.2 and Fig. 1.3 show that if we take into consideration the number fraction, an appreciable quantity of very low molecular weight species will be present at all stages of conversion, in comparison to high molecular weight species. However, if we consider the weight fraction, the very low molecular weight species will be present in a negligible quantity, in comparison to high molecular weight species.

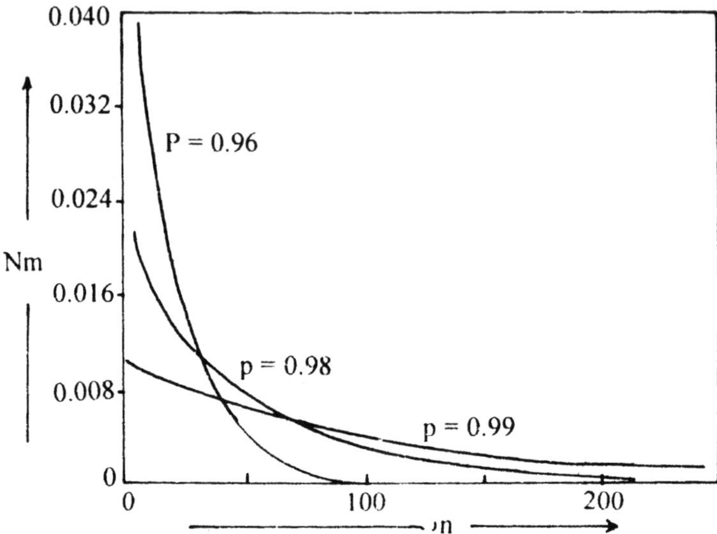

Fig. 1.2: Number-fraction distribution of chain molecules at different extents of reaction in polycondensation.

Equation 1.8 relating the weight fraction distribution with the extent of reaction can also be used for determining the extent of reaction that should be attained to get a maximum yield of a particular molecular weight species. Thus, from Eqn. (1.8) we can derive:

$$P_n = \frac{n-1}{n+1} \qquad \text{...(1.9)}$$

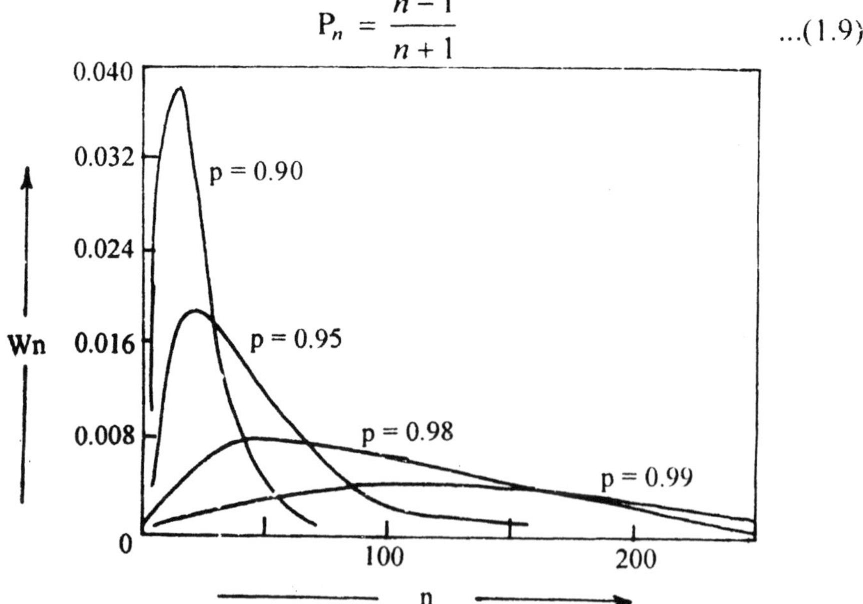

Fig. 1.3: Weight-fraction distribution of chain molecules at different extents of reaction in polycondensation.

where P_n represents the extent of reaction needed to obtain the n mer. Using Eqn. (1.9), we can calculate the extent of reaction essential to get a polymer with a maximum yield of, say, 9-mers as under:

$$P_a = \frac{9-1}{9+1} - \frac{8}{10} = 3.8$$

That is, at 80% conversion, the polymer produced will have a maximum percentage of 9-mers.

Ratio of Weight-average Molecular Weight to Number-average Molecular Weight

The breadth of the molecular weight distribution curve can be obtained as follows. The degree of polymerisation in given by

$$D_{yn} = \frac{1}{1+p}$$

Hence, $\overline{M}_n$ (the number-average molecular weight) can be expressed as

$$\overline{M}_n = \frac{M_0}{1-p} \qquad ...(1.10)$$

where M_0 is the molecular weight of the repeat unit.

Similarly the expression for $\overline{M}_n$ the weight-average molecular weight in the case of a condensation polymerisation is

$$\overline{M}_w = M_0 \frac{1+p}{1-p} \qquad ...(1.11)$$

Hence, the ratio

$$\frac{\overline{M}_w}{\overline{M}_n} = M_0 \frac{1+p}{1-p} \times \frac{1-p}{M_0} = 1+p \qquad ...(1.12)$$

From the above discussion we find that at very high conversions (that is, when p approaches unity), $\overline{M}_w / \overline{M}_n$ approaches a value of 2.

Extent of Reaction and Degree of Polymerisation

The relation between these are

$$\overline{D}_{pn} = \frac{1}{1-p}$$

For different values of p (1.4) the corresponding values of $\overline{D}_{pn}$, are listed in Table 1.4. To get a relatively high degree of polymerisation or a high molecular weight polymer, the extent of conversion should be very high.

From Table 1.4 we can see that even for a degree of polymerisation of 100 (which, in fact, corresponds to low molecular weight polymers), conversion has to be as high as 99%.

Table 1.4: Effect of extent of conversion on degree of polymeri-sation for polycondensation reaction

Values of P	*Extent of conversion (%)*	*Degree of polymerisation,* $\overline{D}_{pn} = \frac{1}{1-p}$
0.50	50	2
0.90	90	10
0.95	95	20
0.99	99	100
0.999	99.9	1,000
0.9999	99.999	100,000

2

Polymerisation and Polymerisation Techniques

2.1. Introduction

Polymers such as PVC, polyethylene, polypropylene or polystyrene are synthesised by using a process known as polymerisation.

Polymer synthesis is not difficult today. To synthesize a polymer we only need an appropriate quantity of the monomer and the catalyst and a suitable polymerisation reactor and we can obtain a polymer of our choice in terms of the required molecular weight, structure, crystallinity etc.

2.2. Classification of Polymerisation

Polymerisation reaction can be broadly classified into three catagories *i.e.*

(i) Addition polymerisation.
(ii) Condensation polymerisation.
(iii) Oxidative coupling.

(i) Addition or Chain Polymerisation

The term addition polymerisation was given by carothers (1929) was modified by H.F. Mark (1950) as chain polymerisation this type of polymerisation involve the self-addition of normal unsaturated molecules of one or two monomers without loss of any small molecule to give a single giant molecule. Reactions of this type involve the successive stages of initiation, propagation and termination common to chain reactions in general. An example of addition polymerisation is the polymerisation of propylene to polypropylene.

$$CH_3.CH = CH_2 \rightarrow -\left(-CH_2 - \underset{}{\overset{CH_3}{\overset{|}{CH}}} - \right)_n-$$

polypropylene

No by product is formed. The product has the same elemental composition as that of the monomer. The bifunctionality is provided by the double bond present in the monomer. Compounds containing reactive double bonds can undergo this type of reactions. Typical examples are vinyl compounds ($CH_2 = CHX$), allyl compounds ($CH_2 = CHCH_2X$), olefins ($CH_2 = CHR$) and dienes ($CH_2 = CR - CH = CH_2$) since most of these monomers can be classified as "vinyl", chain polymerisation is also known as *vinyl polymerisation.*

Monomers capable of undergoing chain polymerisation are listed in Table 2.1.

Table 2.1: Monomers capable of undergoing chain polymerisation

MONOMERS	
Compound	*Chemical Formula*
(i) *Vinyl compounds*	
Acrylamide	$CH_2 = CHCONH_2$
Acrylic acid	$CH_2 = CHCOOH$
Acrylonitrile	$CH_2 = CHCN$
Bromotrifluoroethylene	$BrFC = CF_2$
Butyl acrylate	$CH_2 = CHCOOC_4H_9$
α-cyanomethyl acrylate	$CH_2 = C(CN)COOCH_3$
Methacrylic acid	$CH_2 = C(CH_3)COOH$

Methacrylonitrile	$CH_2=C(CH_3)CN$
Methylmethacrylate	$CH_2=C(CH_3)COOCH_3$
α-methyl styrene	$CH_2=C(CH_3)C_6H_5$
Styrene	$CH_2=CHC_6H_5$
Nitro ethylene	$CH_2=CHNO_2$
p-nitro styrene	$CH_2=CHC_6H_4NO_2$
Tetrafluoro ethylene	$F_2C=CF_2$
Vinyl acetate	$CH_2=CHOCOCH_2$
Vinyl carbazole	$CH_2=CH(C_{12}H_8N)$
Vinyl chloride	$CH_2=CHCl$
Vinyl ethyl ether	$CH_2=CHOC_2H_5$
Vinyl ethyl ketone	$CH_2=CHCOC_2H_5$
2-Vinyl pyridine	$CH_2=CH(C_5H_4N)$
Vinylidene chloride	$CH_2=CCl_2$
(ii) *Allyl compounds*	
Allyl acetate	$CH_2=CHCH_2OCCH_3$
Allyl alcohol	$CH_2=CHCH_2OH$
Acryl chloride	$CH_2=CHCH_2Cl$
(iii) *Olefines*	
Ethylene	$CH_2=CH_2$
Propylene	$CH_2=CHCH_3$
Isobutylene	$CH_2=C(CH_3)_2$
(iv) *Dienes*	
1,3-Butadiene	$CH_2=CH—CH=CH_2$
Chloroprene	$CH_2=CH—CCl=CH_2$
Isoprene	$CH_2=CCH_3—CH=CH_2$

Most of these addition polymerisation are chain growth polymerisations as a particular polymer chain is formed in a single chain reaction.

The most important commercial polymers prepared by this method are obtained from monomers containing a carbon-carbon double bond. In such cases the reaction is called vinyl polymerisation. The addition polymerisation is also important for carbonyl compounds and 1,2-epoxides.

Chain polymerisation involves three major steps (*i.e.*, initiation, propagation and termination). This process of chain polymerisation can be brought about by a free radical, ionic or coordination mechanism.

(ii) Condensation Polymerisation

In the *condensation polymerisation* or *step growth polymerisation,* the polymer molecules are built up through many separate reaction of functional groups.

A condensation polymer may be defined as a polymer wherein the polymer unit contains fewer atoms less than the monomers from which the polymer is produced. In polymer formation, the condensation occurs between two polyfunctional molecules to produce a large molecule.

For example the condensation of ethylene glycol and terephthalic acid by the elimination of a molecule of water to yield polyethylene terephthalate (PET) as under:

$$n\ \mathrm{HO-CH_2-CH_2-OH} + n\ \mathrm{HOOC-C_6H_4-COOH} \longrightarrow$$

ethylene glycol   terephthalic acid

$$(\mathrm{-O-CH_2-CH_2-O-CO-C_6H_4-CO-})_n + 2n\ \mathrm{H_2O}$$

poly (ethylene terephthalate)

Functionality of a Compound

We have already seen that polymerisation is a process which allows simple low molecular weight compounds to combine and form a complex high molecular weight compound. For this to happen each molecule of the compound should possess the capability to react at least with two other molecules of the same or some other compound, *i.e.*, they should have a functionality of at least two. The functionality of a compound depends on the number of reactive sites it possesses. A compound assumes functionality due the presence of reactive functional groups such as —CH, —COOH, —NH_2, —SH, —NCO, etc. *The number of such functional groups per molecule of the compound gives its* **functionality.**

In some cases the functionality may arise due to the presence of double or triple bonds in them. The presence of double and triple bonds in a compound may make it polyfunctional *e.g.* the functionality of two in case of ethylene is due to the presence of double bond, ethylene can take up two hydrogen atoms or two halogen atoms as under:

$$CH_2 = CH_2 + H_2 \longrightarrow CH_3 - CH_3$$

$$CH_2 = CH_2 + Cl_2 \longrightarrow CH_2Cl - CH_2Cl$$

Other homologues of ethylene (*i.e.*, propylene, butylene, etc.), vinyl compounds (*e.g.* vinyl chloride, $CH_2 = CHCl$) styrene ($CH_2 = CH\ C_6H_5$), acetylene ($CH \equiv CH$) also are polyfunctional.

Phenol (C_6H_5OH) exhibits monofunctionality in the reaction given below:

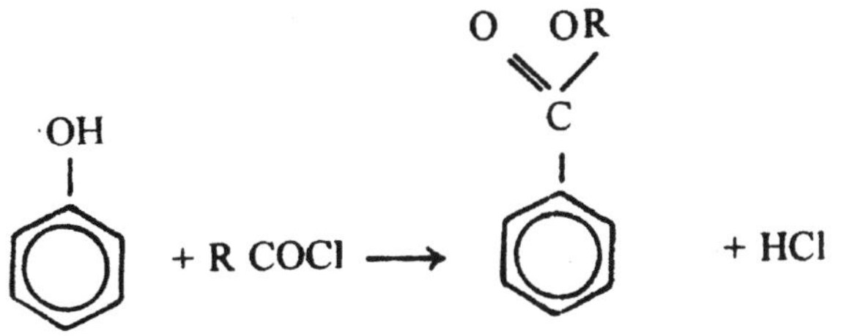

$$C_6H_5OH + R\,COCl \longrightarrow C_6H_5\text{–}C(=O)OR + HCl$$

However phenol shows trifunctionality in the following reaction:

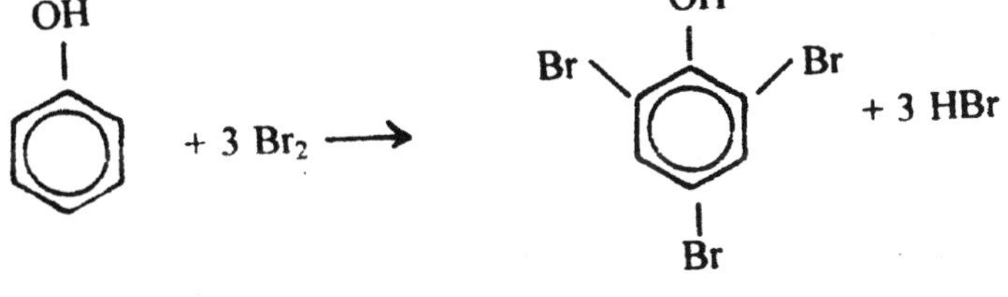

$$C_6H_5OH + 3\,Br_2 \longrightarrow \text{2,4,6-}Br_3C_6H_2OH + 3\,HBr$$

Low molecular weight compounds which have a functionality of two or more care called **monomers.**

Table 2.2. lists some compounds and their functionality.

Table 2.2: Functionality of some Compounds

Table 2.2:Functionality of some Compounds

Compound	*Formula of the Compound*	*Functional Groups Present Per Molecule*				*Functionality*
		—OH	*—COOH*	*—NH_2*	*—NCO*	
I. Carboxylic acids						
Acetic acid	CH_3COOH	—	1	—	—	1
Malonic acid	$HOOCCH_2COOH$	—	2	—	—	2
Tricarballylic acid	$HOOCCH_2CH(COOH)CH_2COOH$	—	3	—	—	3
Benzoic acid	C_6H_5COOH	—	1	—	—	1
Phthalic acid, isophthalic acid or terephthalic acid	$HOOCC_6H_4COOH$	—	2	—	—	2
II. Alcohols						
Ethyl alcohol	CH_3CH_2OH	1	—	—	—	1
Ethylene glycol	$HOCH_2CH_2OH$	2	—	—	—	2

Table

Trimethylol propane	$CH_3CH_2C(CH_2OH)_3$	3	—	—	—	3
Pentaerythritol	$C(CH_2OH)_4$	4	—	—	—	4
Benzyl alcohol	$C_6H_5CH_2OH$	1	—	—	—	1
Salicyl alcohol	$HOC_6H_4CH_2OH$	2	—	—	—	2
III. *Amines*						
Hexyl amine	$CH_3(CH_2)_4CH_2NH_2$	—	—	—	1	1
Hexamethylene diamine	$H_2NCH_2(CH_2)_4CH_2NH_2$	—	—	—	2	2
Aniline	$C_6H_3NH_2$	—	—	—	1	1
Phenylene diamines (ortho, meta and para)	$H_2NC_6H_4NH_2$	—	—	—	2	2
IV. *Isocyanates*						
Methyl isocyanate	CH_3NCO	—	—	—	1	1
Hexamethylene diisocyanate	$OCN(CH_2)_6NCO$	—	—	—	2	2
Toluene diisocyanate	$CH_3C_6H_3(NCO)_2$	—	—	—	2	2

Table

V. Antino acids						
Glycine	H_2NCH_2COOH	—	1	1	—	2
Glutamic acid	$HOOCCH_2CH_2$-$(NH_2)COOH$	—	2	1	—	3
Lysine	$H_2N(CH_2)_4CH(NH_2)COOH$	—	1	2	—	3
VI. Hydroxy acids						
Lactic acid	$CH_3CH(OH)COOH$	1	1	—	—	2
Malic acid	$HOOCCH_2CH(OH)COOH$	1	2	—	—	3
Tartaric acid	$HOOC(CHOH)_2COOH$	2	2	—	—	4
Gallic acid	$HOOCC_6H_2(OH)_3$	3	1	—	—	4
Hydroxy glutamic acid	$HOOCCH_2CH(OH)CH$-$(NH_2)COOH$	1	2	1	—	4

Difference between Condensation and Addition Polymerization

Condensation polymerization	*Addition polymerization*
1. The polymerization proceeds with the elimination of water, acid and alcohol.	1. No elimination takes place
2. The polymerization proceeds by the interaction of the functional groups. No initiator is required. Sometimes a catalyst is added to facilitate the process of polymerization.	2. An initiator is required in this type of polymerization. The polymerization proceeds through the interaction of the initiator with the π electron system of the vinyl monomers.
3. It involves a simple condensation reaction. The polymer is built up slowly by a sequence of discrete reactions, *i.e.*, the initiation, propagation and termination: $R_i = R_p = R_t$	3. It involves a typical chain reaction. The polymer molecule grows rapidly once initiation occurs. The initiation, propagation, and termination reactions are significantly different. $R_p > R_i \approx R_t$
4. A monomer molecule is capable of reacting with another monomer molecule or with a polymer molecule with equal facility. There is a rapid disappearance of monomer at an early stage of the reaction. The molecular weight grows with time.	4. In this case, a monomer molecule cannot react with another monomer molecule but only with an active endgroup or a polymer radical or ion. Several thousands structural units are added up in a second. The molecular weight of the polymer does not increase with time. The percentage of conversion changes with time.

Some other types of polymerisations are given below:

(i) Ionic Polymerisation

It is an addition polymerisation and is of commercial importance. Ionic polymerisations are classified as cationic or anionic depending on the nature of ions used to initiate the polymerisation process. In *anionic polymerisation* the active centres of the propagating chain is negatively charged where as *cationic polymerisation* involves carbonium ion as active species and propagation occurs by addition of monomer to carbonium ion.

(ii) Coordination Polymerisation

In this category are included the polymerisation reactions, of olefins and dienes which are catalysed by organometallic compounds. In this type of polymerization the first step is the formation of a monomer-catalyst complex which involves a coordinate bond between a carbon atom of the monomer and the metal of the catalyst. This coordinate bond formed acts as the active centre for chain propagation. This type of polymerisation depends on polarity of metal-carbon bond, solvent medium, the metal counter-ion placed in a particular spatial arrangement with respect to the anion. The stereo-regularity is imparted by the specific spatial arrangement of the monomeric unit inserted into the growing chain. It is possible to formulate lightly stereo-regular polymer by proper choice of catalyst/ solvent system. The well-known Ziegler-Natta catalysts belong to this category of polymerisation system.

2.3. Polymerisation Techniques

To produce a saleable polymer, the polymer produced should have a required molecular weight, molecular weight distribution and degree of branching. To obtain such a product various factors have to be taken into consideration. Factors like the nature of the monomer, the type of polymerisation mechanism chosen, the required physical form of the polymer and the viability of the process for industrial production dictate the physical conditions under which polymerisation is to be carried out.

The polymerisation reaction may be carried out in the solid phase, liquid phase and gas phase. Most of the commercial polymers are prepared in the liquid phase. More recently the solid phase polymerisation has also been used. The liquid phase polymerisation may be further subdivided into four categories depending on the nature of the physical system. i.e., (i) bulk polymerisation, (ii) solution, polymerisation, (iii) suspension polymerisation and (iv) emulsion polymerisation.

An attempt will be made to discuss these different polymerisation techniques in the paragraphs to follow.

Bulk Polymerisation

This is the simplest process and is widely used for synthesis of condensation polymers. The system is homogeneous and consists of monomer/polymer. In this process the monomer and initiator are kept in a reactor and heated to suitable temperature. The chain transfer agent whenever used for controlling the molecular weight is also dissolved in the monomer.

This process though simple has some draw backs associated with heat transfer. As polymer formation proceeds the medium becomes viscous and heat transfer becomes difficult. If polymerisation is exothermic, control becomes more difficult and sometimes it may lead to explosions. Because of the possibility of localised over heating leading to degradation and discolouration of polymer, this process is seldom used with large batches.

The product obtained by bulk polymerisation is of high purity because except initiator and chain transfer agent, no other additive which could contaminate the product is used.

This technique is used in free radical polymerisation of methyl methacrylate or styrene to obtaine transparent moulding powders etc.

Solution Polymerisation

In solution polymerisation, the reaction is carried out in presence of a solvent. The monomer is dissolved in a suitable inert solvent alongwith the chain transfer agent. A large number of initiators can be used in this process. The free radical initiator is also dissolved in the solvent. The ionic and coordination catalysts can either be dissolved or suspended in the medium. The solvent facilitates the contact of monomer and initiator and helps the process of dissipation of exothermic heat of reaction. It also helps to control viscocity increase.

The disadvantage of this process is the chain transfer due to the solvent and so it is difficult to get very high molecular weight products and generally the low molecular weight polymers are formed. To separate the polymer formed we have to use the process of evaporation of solvent for isolating the product. The product can also be isolated by precipitation in non-solvent.

This technique is advantageous when the polymer is to be used in its solution form *e.g.*, in case of certain adhesives and coatings. It is also advantageous to use this technique in systems where the polymer formed is insoluble in its monomer or solvent and precipitates out as a slurry and thus it can be easily isolated.

It is used in industry for preparing polyacrylonitrile by free radical polymerisation and polyisobutylene by cationic polymerisation. *Block copolymers are prepared exclusively by this technique.*

Suspension Polymerisation

By using this technique only water insoluble monomers can be polymerised. In this process the monomer is suspended as discrete droplets (0.1 to 1.0 mm diameter) in dilute aqueous solution containing protective colloids like polyvinyl alcohol and surfactants etc. The droplets have large surface area and can readily transfer heat to water. Suspension is brought about by agitating the suspension. Protective colloids prevent coalescence of the droplets. A monomer soluble initiator is used. The product is obtained by filteration or spray drying. This process cannot be carried out yet in a continuous process; hence batch processing has to be used.

Polymerisation proceeds to 100% conversion and the product formed is obtained as spherical beads or pearls, because of this it is sometimes known as *bead* or *pearl polymerisation.*

On commercial scale this technique is used to obtain polyvinyl chloride, polyvinyl acetate etc. In this the free radical initiators are used.

Emulsion Polymerisation

This is superficially similar to suspension polymerisation. But in this process a monomer dispersed in water, in presence of a surface active agent is polymerised to give a stable polymer latex. The surface active agents (surfactants) may be cationic, anionic or non-ionic. Surfactants commonly used are cetyltrimethyl ammonium bromide (CTABr), sodium lauryl sulphate (NaLS) and triton-X etc. The surfactants help to lower the surface tension at the monomer-water interface and also facilitate emulsification of the monomer in water. Because of their low solubility surfactants get fully dissolved or molecularly dispersed only at low concentrations and at higher concentrations 'micelles' are formed. The highest concentration where in all the molecules are in dispersed state is known as 'critical micelle concentration' (CMC). The CMC values of some surfactants are listed in Table 2.3.

We know that emulsifier molecules are made of two parts: a long non-polar hydrocarbon chain to which is attached a polar group which as —COONa, —SO_3Na, —NH_2HCl or —NBr. In micelle formation, the emulsifer molecules aggregate in such a manner that the polar end of the molecules gets aligned outward and the hydrocarbon ends come close to each other at the interior (Fig. 2.1).

Table 2.3: CMC values of some surfactants

Formulae of Surfactant	*CMC (g/l)*	*Temp. K*
Anionic		
$CH_3(CH_2)_6COONa$	6.5×10^1	293
$CH_3(CH_2)_{10}COONa$	5.6	293 – 343
$CH_3(CH_2)_7OSO_3Na$	3.0×10^1	298 – 323
$CH_3(CH_2)_{11}OSO_3Na$	2.6	298 – 333
$CH_3(CH_2)_5C_6H_4SO_3Na$	9.8	348
$CH_3(CH_2)_{11}C_6H_4SO_3Na$	4.0×10^{-1}	323 – 348
Cationic		
$CH_3(CH_2)_9NH_2.HCl$	8.5	298
$CH_3(CH_2)_{11}NH_2HCl$	2.7	303 – 323
$CH_3(CH_2)_7N(CH_3)_3Br$	7.8×10^1	298
$CH_3(CH_2)_{11}N(CH_3)_3Br$	5.4	298
Non-ionic		
$CH_3(CH_2)_7C_6H_{11}O_6$	7.3	298
$C_{12}H_{20}O_9(C_{16}H_{31}O_2)_2$	1.1×10^{-2}	293
$CH_3(CH_2)_{10}COOC_{12}H_{21}O_{10}$	7.1×10^{-3}	323

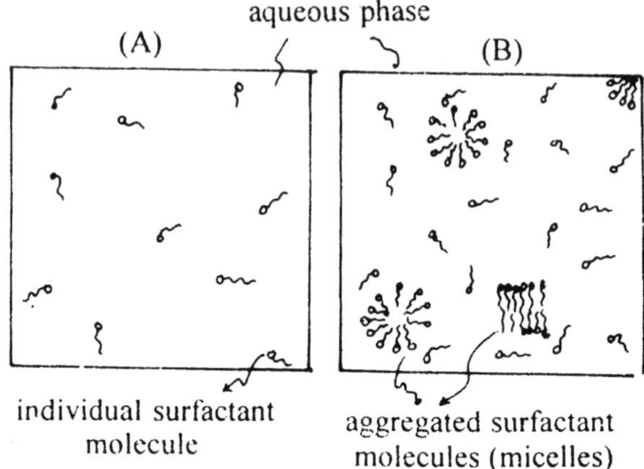

—: Non-polar hydrocarbon chain of the surfactant molecule
o: Polar head of the surfactant molecule

Fig. 2.1: Schematic representation of surfactant molecules distributed in water : (A) completely dissolved at low concentration and (B) dissolved as well as aggregated beyond critical micelle concentration (CMC).

When soap is dissolved in water, micelles are formed. These micelles are clusters of soap molecules with their hydrophobic end facing inwards and hydrophilic carboxyl groups facing outwards towards aqueous phase.

Due to the close proximity of the hydrocarbon ends of all emulsified molecules, the interior of micelle acts as a hydrocarbon phase where the monomer can be solubilised.

The monomers get absorbed in micelles resulting in their swelling. Water soluble initiators are used which form free radicals. Inorganic persulphates are commonly used as initiators. The initiator diffuses into a micelle and polymerisation proceeds. As more monomer is polymerised monomers from outside the micelle diffuse inside and the process continues; when another radical enters the micelle the polymerisation stops. This technique can give high molecular weight polymers.

At the end of the polymerisation, we have fine particles of the polymer, stabilised by the emulsifier layer and dispersed uniformly in the aqueous phase. This milky white dispersion is often called *'latex'*. The latex can either be used as such for making adhesives, water-soluble emulsion paints, etc., or the polymer can be isolated from the latex by destabilising the emulsion (using some electrolytes), by spray drying or by freezing.

The drawbacks are that latex has to be coagulated; filtered and dried to obtain a solid polymer. There is contamination with soap. But this is the preferred method for obtaining synthetic rubbers which requires a high molecular weight. By this method emulsion paints are also obtained.

This technique is extensively used for the free radical polymerisation of vinyl monomers containing water soluble initiators. The monomers like vinyl chloride, butadiene, chloroprene, vinyl acetate, acrylates and methacrylates are polymerised by this technique.

Melt Condensation

This process can be used for polymerisation of those monomers which do not decompose around their melting point. The process is generally carried out in an inert atmosphere of N_2 or CO_2 to avoid oxidation, decarboxylation, degradation etc. Some times the process may be carried out under reduced pressure so as to easily remove the byproduct formed and to obtain a product of high molecular weight. To avoid solidification of the melt inside the reactor the hot melt is passed directly to the processing equipments for extension, casting or spinning, this technique is used to produce polyethylene terephthalate from dimethyl terephthalate and ethylene glycol. Nylon –66 is also prepared using this technique.

Solution Polycondensation

In this technique the reactants are taken in solution form. The reactants are dissolved in a suitable inert solvent. Solvent also helps in removal of byproducts as it serves as an entraping agent for the byproducts formed. But due to the presence of solvent the probability of chain growth decreases and we get products with low degree of polymerisation.

This technique has been used to obtain many liquid polyester resins based on glycols and unsaturated dicar-boxylic acids. In their preparation generally high boiling aromatic hydrocarbons are used as solvents. Since water (formed as byproduct) forms an azeotropic mixture with high boiling aromatic hydrocarbons (solvent) so it can be removed easily.

Interfacial Condensation

In this technique polymerisation proceeds at the interface between an aqueous and organic medium. This technique is quite suitable for reactants that have highly reactive functional groups capable of reacting at ambient temperatures to produce condensation products.

This technique has been used to prepare fully aromatic polyamides from terephtholoyl chloride and paraphenylenediamine.

$$n\ ClOC-C_6H_4-COCl + n\ H_2N-C_6H_4-NH_2 \longrightarrow Cl\left[OC-C_6H_4-CONH-C_6H_4-NH\right]_n H + (2n-1)\ H_2O$$

For this preparation diamine is dissolved in water and acid chloride in an organic solvent (e.g., $CHCl_3$ or CCl_4). When the two solutions are mixed, at the interface, the diamine molecules diffuses into the organic phase and react with acid chloride to form polymer. The polymer formed precipitates out immediately and the byproduct (HCl) diffuses back into the aqueous phase in which it gets dissolved. When the precipitate of the product formed is removed it exposes a fresh surface of acid chloride to organic phase and more of the product is formed.

Since the process is diffusion controlled so a high molecule weight product can be obtained by this process.

Solid and Gas Phase Polymerisation

So far we have been discussing the processes that are carried out in liquid phase and are very popular and widely used for industrial preparation of polymers. However, the polymerisation process can also be carried out in solid and gaseous phases.

Solid phase polymerisation is used for chain polymerisation processes which are carried out at low temperatures. In such processes the thermal activation is difficult and so for activation of such processes radiation-activation technique is used. These processes are very slow. An example of such a solid phase polymerisation is the preparation of polyformaldehyde by the radiation polymerisation of solid trioxane.

```
      CH2 —O
     /       \                 hv
n   O         CH2      ------------->     [-CH2O-]n
     \       /
      CH2 —O
    (Trioxane)                          (Polyformaldehyde)
```

This polymerisation can also be brought about by making use at cationic initiators like BF_3.

Another example of solid-state polymerisation is polymerisation of diacetylene derivatives which results in the formation of highly crystalline polymer that also conducts electricity.

```
R— C≡C —C≡C —R
R— C≡C —C≡C —R
R —C≡C —C≡C —R \ hv                     //
                 \           — C —C≡C — C —R
R— C≡C —C≡C —R    ↘                      \\
                         R— C— C≡C — C — R
                              \\
                   R— C— C≡C — C — R
                            \\
             R— C— C≡C — C —R
                 \\
```

Gas phase polymerisation is known in case of a very few olefinic polymers. The methods used in gas phase polymerisation are (i) spraying the catalyst (generally Zeigler—Natta catalyst) into the gaseous monomer and (ii) feeding the gaseous monomer into a fluidised bed which is made up of the catalyst particles.

Examples of gas phase polymerisation are the polymerisation of ethylene and *p*-xylene.

Gas Phase Polymerisation of Ethylene

To carry out this process ethylene gas is passed through a fluidised bed column that is filled with Zeigler-Natta catalyst (a mixture of titanium chloride and alkyl aluminium in pentane medium) and is maintained at room temperature and a pressure of 4-5 atmospheres. The polymer formed is collected as a free-flowing powder.

CH_3 — C_6H_4 — CH_3 (p – xylene) $\xrightarrow[\text{Steam}]{950\ ^0C}$ $CH_2 - CH_2$ / $CH_2 - CH_2$ (DI – p – xylylene) $\xrightarrow[\text{1mm pressure}]{600\ ^0C}$

CH_2 / CH_2 (quinonoid) $\rightleftharpoons$ CH_2 / $\bullet CH_2$ (diradical) $\xrightarrow[\text{Quenched to } 50^0C]{\text{Metal Surface}}$ $-[\!-C_6H_4-CH_2\ CH_2-]\!-_n$ (DI – p – Xylylene)

Gas Phase Polymerisation of p-xylene

First of all *p*-xylene is dehydrogenated to obtain its dimer (*i.e.*, di-*p*-xylene). This is done by using superheated steam at 950°C. The dimer formed is a crystalline solid at room temperature and it is heated to 600°C at 1 mm pressure when it sublimes and forms and equilibrium mixture of diradical and a quinonoid. This equilibrium mixture when quenched to 50°C over metal surface results in the formation of a linear polymer known as poly-*p*-xylene. The polymer is formed on the metal surface.

3

Radical or Chain or Vinyl Polymerisation

3.1. Introduction

It is an addition polymerisation. This method is used for the polymerisation of substituted ethylenes. Each polymer molecule reaches its final stage rapidly through a series of stages. During each stage an unstable free radical intermediate is produced. Three distinct stages which are involved in the growth of the polymer molecule are (i) initiation (ii) propagation (iii) termination.

The over all rate of polymerisation and the size of the molecules formed are dependent upon the rate of these separate process.

Various vinyl monomers can be classified as—

(i) monomers which do not polymerise themselves *e.g.*, allyl alcohol, α-methyl styrene, allyl chloride etc.

(ii) monomers which polymerise rapidly only upon warming or an addition of suitable initiators, *e.g.*, ethylene, propylene, butadiene etc. and

(iii) monomers which show a strong tendency to polymerise *e.g.*, acrylic acid, nitroethylene, α-cyanoacrylate methyl malonate etc.

3.2. Initiation

Initiation of a chain reaction involves the formation of free radicals from an initiator. Benzoyl peroxide is a commonly used initiator.

$$C_6H_5\overset{\overset{O}{\|}}{C}O-O\overset{\overset{O}{\|}}{C}-C_6H_5 \dashrightarrow 2C_6H_5\overset{\overset{O}{\|}}{C}O \dashrightarrow 2C_6H_5 + 2CO_2$$

Benzoyl peroxide when heated formes a free radical. The free radical undergoes addition reaction with the monomer $CH_2 = CHX$ (X is an atom or group other than H) forming

$$R + CH_2 = \underset{\underset{X}{|}}{CH} \rightarrow RCH_2 - \underset{\underset{X}{|}}{CH}$$

a new free radical.

The double bond the site of initiation reaction, which gives rise to active centres and which may added to a monomer molecule without loss of reactivity.

Initiators

The compounds generally used as initiators are thermally unstable compounds which decomposse to produce the free radicals. The decomposition of initiators can be brought about by supplying them energy either in the form of heat or light. The molecule then cleaves homolytically producing free radicals e.g.

$$R - R \xrightarrow{\text{heat}} 2R.$$

Azoinitiators

The thermal decomposition of **azoinitiators** *e.g.*, azobis isobutyronitrile can be represented as under:

$$CH_3-\underset{\underset{CH_3}{|}}{\overset{\overset{CN}{|}}{C}}-N-N-\underset{\underset{CH_3}{|}}{\overset{\overset{CN}{|}}{C}}-CH_3 \xrightarrow{\text{Heat}} 2CH_3-\underset{\underset{CH_3}{|}}{\overset{\overset{CN}{|}}{C}} + N_3\uparrow$$

(AZOBIS ISOBUTYRO NITRILE) (CYANOPROPYL RADICAL)

Free Radical Initiators

Some of the compounds which are used as free radical initiators are listed in Table 3.1.

Table 3.1: Some free radical initiators

Common name	*Structure*	*Free radicls produced*
Azo – bis – cyano valeric acid	$HOOC—CH_2—CH_2—C(CN)(CH_3)—N=N—C(CN)(CH_3)—CH_2—CH_2—COOH$	$HOOC—CH_2—CH_2—C(CN)(CH_3)$
Azo-bis-diphenyl methane	$(C_6H_5)_2CH—N=N—CH(C_6H_5)_2$	$(C_6H_5)_2CH$
Azo-bis-methyl isobutyrate	$H_3C—C(CH_3)(COOCH_3)—N=N—C(CH_3)(COOCH_3)—CH_3$	$H_3C—C^{\bullet}(CH_3)(COOCH_3)$
Azo-bis-isobuty-ronitrile	$H_3C—C(CN)(CH_3)—N=N—C(CN)(CH_3)—CH_3$	$H_3C—C^{\bullet}(CN)(CH_3)$

(Cont.)

Table 3.1 : Cont.

Commercial name	Chemical structure	Free-radicals Formed
Benzdyl peroxide	$C_6H_5-\overset{O}{\overset{\parallel}{C}}-O-O-\overset{O}{\overset{\parallel}{C}}-C_6H_5$	$C_6H_5-\overset{O}{\overset{\parallel}{C}}-O^{\bullet}$ and C_6H_5
Text. butyl peroxide	$CH_3-C(CH_3)_2-O-O-C(CH_3)_2-CH_3$	$CH_3-C(CH_3)_2-O^{\bullet}$ and $^{\bullet}CH_3$
Cumyl peroxide	$C_6H_5-C(CH_3)_2-O-O-C(CH_3)_2-C_6H_5$	$C_6H_5-C(CH_3)_2-O^{\bullet}-C_6H_5^{\bullet}$ and $^{\bullet}CH_3$

Table 3.1: Cont.

Commercial name	Chemical structure	Free-radicals Formed
Cyclo heranone peroxide	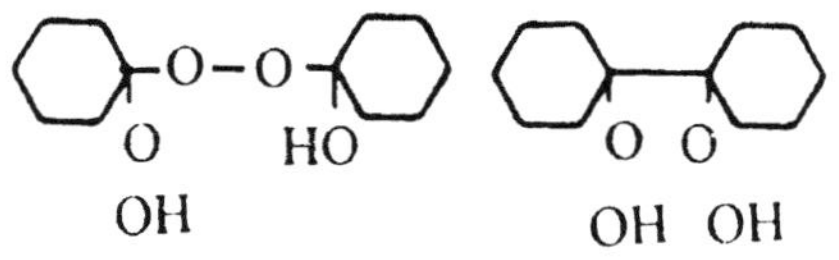 (cyclohexyl ring)–O–O–(cyclohexyl ring) with O, OH and HO substituents; (cyclohexyl ring)–(cyclohexyl ring) with O O, OH OH	OH and $CH_2{-}(CH_2)_4{-}COOH$
Glutaric acid peroxide	$COOH{-}(CH_2)_3{-}\overset{O}{\overset{\parallel}{C}}{-}O{-}O{-}\overset{O}{C}{-}(CH_2)_4\ COOH$	$HOOC{-}(CH_2)_3{-}\overset{O}{C}{-}O$ and $HOOC\ (CH_2)\ {}^{\bullet}CH_2$
Hydrogen peroxide	$HO{-}OH$	OH
Lauroyl peroxide	$CH_3{-}(CH_2)_{10}\ \underset{O}{\underset{\parallel}{C}}{-}O{-}O{-}\underset{O}{\underset{\parallel}{C}}{-}(CH_2)_{10}\ CH_3$	$CH_3{-}(CH_2)_{10}\ \overset{O}{\overset{\parallel}{C}}{-}O^{\bullet}$ and $CH_3\ (CH_2)_9\ {}^{\bullet}CH_2$

Table 3.1: Cont.

Commercial name	Chemical structure	Free- radicals formed
Text butyl hydroperoxide	$CH_3-C(CH_3)_2-O-OH$	$CH_3-C(CH_3)_2-\dot{O}$, $\dot{O}H$ and $\dot{C}H_3$
Cumene Hydroperoxide	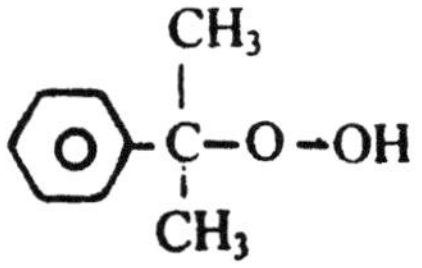 $C_6H_5-C(CH_3)_2-O-OH$	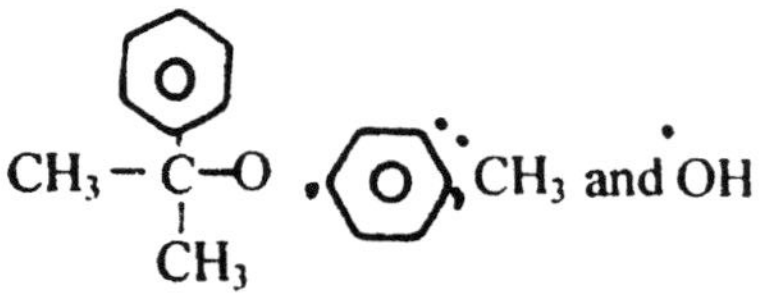 $CH_3-C(C_6H_5)(CH_3)-O\cdot$, $C_6H_5\cdot$, $\dot{C}H_3$ and $\dot{O}H$
Peracetic acid	$CH_3-C(=O)-O-OH$	$CH_3-CO\dot{O}$, $\dot{C}H_3$ and $\dot{O}H$
Perbenzoic acid	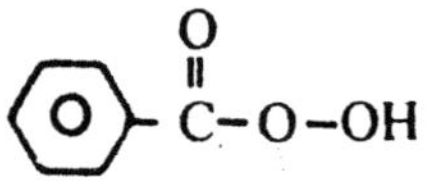 $C_6H_5-C(=O)-O-OH$	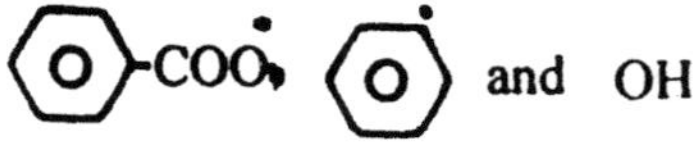 $C_6H_5-CO\dot{O}$, $\dot{C}_6H_5$ and OH
Text butyl Peracetate	$CH_3-C(=O)-O-O-C(CH_3)_2-CH_3$	$CH_3-CO\dot{O}$ $\dot{C}H_3$ and CH_3 C O (with CH_3 above and below C)

Table 3.1: Cont.

Commercial name	Chemical structure	Free-radicals formed
Text bytyl perpenzoate	$C_6H_5-C(=O)-O-O-C(CH_3)_2-CH_3$	$C_6H_5COO^{\bullet}$ $C_6H_5^{\bullet}$ CH_3 $O^{\bullet}C(CH_3)_2CH_3$
Diisopropyl percarbonate	$(CH_3)_2CH-O-C(=O)-O-O-C(=O)-O-CH(CH_3)_2$	$(CH_3)_2CH-O-C(=O)-O^{\bullet}$ $^{\bullet}CH_3$
Potassium persulphate	$KO-S(=O)_2-O-O-S(=O)_2-OK$	$K^{\oplus}\,{}^{\ominus}O-S(=O)_2-O^{\bullet}$

It has been found that the rate of decomposition of initiators depends on their chemical nature, temperature and solvent.

Some initiators can be decomposed by u.v. light e.g.

$$CH_3-\underset{CH_3}{\overset{CN}{\overset{|}{\underset{|}{C}}}}-N=N-\underset{CH_3}{\overset{CN}{\overset{|}{\underset{|}{C}}}}-CH_3 \xrightarrow{h\nu} 2CH_3-\underset{CH_3}{\overset{CN}{\overset{|}{\underset{|}{C}}}}+N_2\uparrow$$

where $h\nu$ represents light energy.

In such decompositions the rate of decomposition mainly depends on the intensity and wave length of radiation and not much on temperature. Solvent plays an important role in this type of decomposition as it controls the intensity of the radiation incident on the initiator molecule.

The polymerisation reaction initiated by u.v. light are known as "photoinitiated polymerisation."

Peroxide Initiators

Some times even catalysts can include initiators to decompose into free radicals. In such type of reaction an electron transfer mechanism is involved. Peroxides and hydroperoxides are decomposed in this way. For example, the decomposition of benzoyl peroxide by an aromatic tertiary amine at room temperature.

$$C_6H_5NR_2 + C_6H_5-\underset{\underset{O}{\|}}{C}-O-O-\underset{\underset{O}{\|}}{C}-C_6H_5 \longrightarrow [C_6H_5NR_2]^{\oplus}\;{}^{\ominus}O-\underset{C_6H_5}{C}{=}O + O{=}\underset{C_6H_5}{C}-O^{\cdot}$$

$$\longrightarrow C_6H_5NR_2 + C_6H_5-C(=O)OH$$

Similarly the decomposition of H_2O_2 by Fe^{2+} and that of hydroperoxide by CO^{2+} ion

$$H_2O_2 + Fe^{2+} \rightarrow Fe_3^{+} + HO^{-} + HO^{\cdot}$$

$$R—O—OH + Co^{2+} \rightarrow RO^{\cdot} + Co^{3+} + HO^{-}$$

$$R—O—OH + Co^{3+} \rightarrow ROO^{\cdot} + Co^{2+} + HO^{+}$$

Polymerisation reactions utilising such redox initiators are termed 'redox polymerisation'.

The formation of SO_4^{2-} radical is facilitated in presence of certain metal ions like Ag^{+}, Cu^{2+}, Zn^{2+}.

$$R—O—O—H + H—O—O—R \rightarrow R\dot{O} + R—O—\dot{O} + H_2O$$

Hydroperoxide

Hydroperoxides decompose in a bimolecular reaction with the formation of water. The activation energy of the peroxide decomposition reaction could be reduced by using some activators, *i.e.*, Fe^{2+}, Cu^{2+} and sodium hyposulphite etc. These initiators are generally used in case of emulsion polymerization.

The mechanism proposed for the formation of free radicals by the decomposition of peroxydiphosphate ion is as under:

$$\begin{array}{ccccccccccc} & & O & & & \vdots & & O & & \\ & & \| & & & \vdots & & \| & & \\ O & - & P & - & O & - & O & - & P & - & \bar{O} \\ & & | & & & \vdots & & | & & \\ & & \underline{O} & & & \vdots & & \underline{O} & & \end{array}$$

IV

Peroxydiphosphate ion

$$P_2O_8^{4-} \xrightarrow{\text{slow}} HPO_4 + \overset{2-\cdot}{P}O_4$$

$$\overset{2-\cdot}{P}O_4 + H_2O \longrightarrow H\bar{P}\bar{O}_4 + \dot{O}H$$

$$HP\dot{O}_4 + H_2O \longrightarrow H_2P\bar{O}_4 + \dot{O}H$$

$$\dot{O}H + HP\dot{O}_4 \longrightarrow H_2PO_4 + \tfrac{1}{2}O_2$$

$$\dot{O}H + \overset{2-}{\dot{P}}O_4 \longrightarrow HPO_4^{2-} + \tfrac{1}{2}O_2$$

Some other radicals formed by peroxide initiators can be represented as under:

$$C_6H_5-\overset{O}{\overset{\|}{C}}-O-O-\overset{O}{\overset{\|}{C}}-C_6H_5 \rightarrow 2\,C_6H_5-\overset{O}{\overset{\|}{C}}$$

benzoyl peroxide
I

Benzoyl radical

$$\downarrow$$

$$2\,\dot{C}_6H_5 + 2CO_2$$

Phenyl radical

$$C_6H_5-\underset{CH_4}{\overset{CH_3}{\overset{|}{\underset{|}{C}}}}-O-O-\underset{CH_3}{\overset{CH_3}{\overset{|}{\underset{|}{C}}}}-C_6H_5 \rightarrow 2C_6H_5-\underset{CH_3}{\overset{CH_3}{\overset{|}{\underset{|}{C}}}}-\dot{O}$$

Dimethyl peroxide
II

$$KO-\underset{O}{\overset{O}{\overset{\|}{\underset{\|}{S}}}}-O-O-\underset{O}{\overset{O}{\overset{\|}{\underset{\|}{S}}}}-OK \rightarrow 2\dot{S}O_4^-$$

Potassium Peroxydisulphate
III

Redox Initiators

We have already come across a few examples in which oxidation-reduction reaction can initiate vinyl polymerisation. Such a polymerisation reaction is known as **redox polymerisation.** In such reactions the *oxidant* is generally referred to as *initiator* and the *reductant* as *activator.*

The *redox polymerisation* shows the following features:

(i) They have very short induction period.
(ii) By this process high yield polymers having comparatively

high molecular weights are obtained in a short time.

(iii) They have a small activation energy, and

(iv) The process can occur at room temperature or even at lower temperatures.

This type of polymerisation can provide direct evidence for the existence of transient radical intermediates. Transition metals in higher valence state or such metals complexed with reducible organic substances are found to act as redox initiators. Some important redox initiators in use are Cr^{vi}, V^{v}, Ce^{iv}, Co^{iii}, Fe^{iii}, MnO_4^-, Mn^{iii} etc. Some of such initiators are discussed at length.

(i) Fenton's Reagent

Evans and coworkers carried out a detailed study of Fenton's reagents ($FeSO_4 - H_2O_2$) as a redox initiator, where $\dot{O}H$ are known to be the chain initiating species. The mechanism is as under:

$$Fe^{2+} + HO-OH \rightarrow Fe^{3+} + \overline{O}H + \dot{O}H$$

$$\dot{O}H - H_2O_2 \rightarrow H_2O + H\dot{O}_2$$

$$H\dot{O}_2 - H_2O_2 \rightarrow \dot{O}H + O_2 + H_2O$$

$$H\dot{O}_2 \rightleftharpoons \overline{O}_2 + H^+$$

$$\overline{O}_2 + H_2O_2 \rightleftharpoons O_2 + \overline{O}H + \dot{O}H$$

$$Fe^{2+} + \dot{O}H \rightleftharpoons Fe^{3+} + \overline{O}H$$

(ii) Metal Ions coupled with organic substrate(s)

Hexavalent Chromium Cr^{6+},

$$HCr\overline{O}_4 + S \div 2H^+ \xrightarrow{k_1} Cr^{4+} + \text{product}$$

$$S + Cr^{4+} \xrightarrow{k_2} \underset{\text{free radical}}{\dot{S}} + Cr^{3+} + H^+$$

Quinoequivalent Vanadium: V^{5+}

$$VO_2^+ + S \rightleftharpoons \text{Complex} \xrightarrow{\text{slow}} \underset{\text{free radical}}{S} + V^{4+}$$

Tetravalent Cerium: Ce^{4+}

$$Ce^{4+} + S \rightleftharpoons \text{Complex} \longrightarrow \dot{S} + Ce^{3+} + H^{+}$$

free radical

Permanganate—Oxalic acid

$$Mn^{4+} + C_2O_4^{2-} \xrightarrow{\text{measurable}} Mn^{3+} + CO_2 + \dot{C}O\bar{O}$$

$$Mn^{4+} + \dot{C}O\bar{O} \xrightarrow{\text{rapid}} Mn^{3+} + CO_2$$

$$Mn^{3+} + 2C_2\dot{O}O_4^{2-} \overset{\text{rapid}}{\rightleftharpoons} [Mn(C_2O_4)_2]^{-}$$

$$Mn^{3+} + C_2O_4^{2-} \dashrightarrow Mn^{2+} + \dot{C}O\bar{O} + CO_2$$

$$Mn^{3+} + \dot{C}O\bar{O} \dashrightarrow Mn^{2+} + CO_2$$

Formation of Active Centres-Initiation

The initiation of polymerisation takes place in two steps: The decomposition of initiatcr to form the free radical and the addition of the radical to the monomer with the generation of another radical. If the radical produced by decomposition of the initiator, I, is designated as R, then the process of initiation could be shown as under:

$$I \longrightarrow 2\dot{R}$$

$$R + CH_2 = CHX \rightarrow R - CH_2 - \underset{X}{\overset{H}{|}}C. \quad \text{(I)}$$

Vinyl monomer

$$\searrow R - \underset{H}{\overset{X}{|}}CH - CH_2$$

The production of the relative amounts of the two product raidcals is found to depend upon the difference between the activation energies of the two reactions. Theoretically, the activation energy for II is slightly greater than that for I, because the X group which is generally bulky such as Cl, CN etc. and hinders the approach of the R radical. Hence the lower activation energy of I favours the formation of the radical R—CH_2—CHX.

When free radical initiation is because of the addition of the radical to one end of the double bond, then we may expect that the radical would be attached to the end of the polymer chain. The presence of such end fragments of the initiator has been confirmed for radical polymerisations and with several monomers by end group analysis.

The rate of the initiation is found to depend on the rate of the initiator decomposition reaction, *i.e.*, on the half-life of the initiator, and it is, therefore, strongly temperature dependent. The course of the decomposition reaction and thus the constitution of the radicals that are generally formed, is found to be dependent on the nature of the solvent in which the decomposition takes place. Thus, dibenzoylperoxide decomposes in inert solvents, like benzene, into phenyl radicals under formation of CO_2. However, in the presence of styrene, where no CO_2 formation seems to take place the apparently intermediate benzoyl radicals starts the polymerisation.

Initiator Efficiencies

Many methods have been proposed for determining the efficiency, *f*, of the initiator. The most direct method depends on a quantitative assay of the polymer for initiator fragments, and its comparison with the amount of initiator decomposed. This is not difficult in those cases where the initiator leaves a reactive end group on the polymer or is radioactively tagged.

Direct quantitative determination of the number of initiator fragments combined with the polymer is feasible only under very exceptional circumstances. Another useful method depends on the determination of the molecular weight by a suitable method. The number of polymer molecules may then be calculated, and assuming termination by coupling, the number of combined, primary radicals may be considered to be twice the number of molecules, still another method for determining the efficiency depends on the reaction of the chain radicals stoichiometrically with certain inhibitors. Most initiators in typical vinyl polymerisations have efficiencies between 0.6 and 1 (*i.e.*, between 60 and 10% of all the radicals formed ultimately initiate polymer chains). A fraction of the radicals may disappear under certain circumstances also through different reactions, *i.e.*, through direct combination with atmospheric oxygen or other inhibiting substances present in the system.

3.3. Propagation

Propagation of the free radical take place with addition of each monomer to the growing chain and all these reactions are known as propagation. Because of propagation long chain radical is formed.

$$\underset{\substack{|\\X}}{RCH_3—CH.} + \underset{\substack{|\\X}}{CH_2=CH} \rightarrow RCH_2—\underset{\substack{|\\X}}{CH}—CH_2—\underset{\substack{|\\X}}{CH}$$

$$RCH_2\underset{\substack{|\\X}}{CH}CH_2\underset{\substack{|\\X}}{CH} + nCH_2=\underset{\substack{|\\X}}{CH} \rightarrow R—\{CH_2—\underset{\substack{|\\X}}{CH}\}_{n+1}—CH_2—\underset{\substack{|\\X}}{CH.}$$

The chain propagation occurs by a series of successive steps, all of which may be assumed to be governed by the same rate constant k_p, independent of chain length. The growth of the chain may be represented as follows:

$$PM^{\cdot} + M \xrightarrow{K\pi} PM_2$$

$$PM_2^{\cdot} + M \xrightarrow{K\pi} PM_3$$

$$\vdots$$

$$PM_{v-1}^{\cdot} + M \xrightarrow{K\pi} PM_v$$

(i) $\sim\sim CH_2-\underset{\substack{|\\X}}{\dot{C}H} + CH_2=\underset{\substack{|\\X}}{CH} \rightarrow \sim\sim CH_2-\underset{\substack{|\\X}}{CH}-CH_2-\underset{\substack{|\\X}}{\dot{C}H}$

Head-to-tail

(ii) $\sim\sim CH_2\ \underset{\substack{|\\X}}{\dot{C}H} + \underset{\substack{|\\X}}{CH}=CH_2 \rightarrow \sim\sim CH_2-\underset{\substack{|\\X}}{CH}-\underset{\substack{|\\X}}{CH}-\dot{C}H_2$

Tail-to-tail

(iii) $\sim\sim \underset{\substack{|\\X}}{CH}-\dot{C}H_2 + CH_2=\underset{\substack{|\\X}}{CH} \rightarrow \sim\sim \underset{\substack{|\\X}}{CH}-CH_2-CH_2-\underset{\substack{|\\X}}{\dot{C}H}$

Head-to-head

(iv) $\sim\sim \underset{\substack{|\\X}}{CH}-\dot{C}H_2 + \underset{\substack{|\\X}}{CH}=CH_2 \rightarrow \sim\sim \underset{\substack{|\\X}}{CH}-CH_2-\underset{\substack{|\\X}}{CH}\cdot CH_2$

Tail-to-head

The propagation continues till the chain growth is stopped by the free-radical site being 'killed' by some impurities or by a sheer termination process, or till there is no further monomer left for attack.

The structure of the growing chain can be represented as under:

$$R^{\ominus} \boxed{OR} \ R:+\dot{C}H_2 \div \dot{C}H(X) \longrightarrow R : CH_2-\ddot{C}H(X) \ \boxed{OR} \ R\text{-}CH_2\text{-}\overset{\ominus}{C}H(X)$$

where *n* gives the number of monomer units added up in the chain growth. The wavy line indicates the polymer chain made of *n* number of monomeric units. The mode of addition of the incoming monomer to the growing chain can be of the head-to-tail, tail-to-tail, head-to-head or tail-to-head type. If we consider the —CH_2— and the —CHX— parts of a monomeric unit its head and tail respectively, the four modes of addition can be represented as above:

Whenever a monomer unit iş added to the growing chain, the π-electrons come down to s-level and this process is accompanied by a release of about 20 K cal of energy. Chain propagation can be brought about without the help of any external energy.

The propagation steps are very fast because of the reactivity of the free radicals. In contrast to the initiation reaction, the propagation reaction much lower activation energy and, therefore, R_p, the rate of polymerization is less temperature-dependent. The average life-time, τ, of the growth process has been found directly in certain photoinitiation polymerisation reaction. For the gas phase polymerisation of vinyl acetate, vinyl chloride or methyl acrylate, and for the liquid phase polymerisation of vinyl acetate or styrene, τ, has a value of 10^{-3} to 10^{-2} sec for a chain length of about 1,000 molecules.

3.4. Termination

Termination can take place either by the combination of two free radical chains to form one molecule or by transfer of a hydrogen atom in a disproportionation reaction.

Coupling Termination

$$R\text{—}[CH_2\text{—}\underset{X}{\underset{|}{C}H}]_n\text{—}CH_2\text{—}\underset{X}{\underset{|}{C}H}.+.\underset{X}{\underset{|}{C}H}\text{—}CH_2\text{—}[\underset{X}{\underset{|}{C}H}\text{—}CH_2]_n'\text{—}R' \rightarrow$$

$$R\text{—}[CH_2\text{—}\underset{X}{\underset{|}{C}H}]_n\text{—}CH_2\text{—}\underset{X}{\underset{|}{C}H}\text{—}\underset{X}{\underset{|}{C}H}\text{—}CH_2\text{—}[\underset{X}{\underset{|}{C}H}\text{—}CH_2]_n\text{—}R$$

Disproportionation Termination

$$R{-}[CH_2{-}\underset{\underset{X}{|}}{CH}]_n{-}CH_2{-}\underset{\underset{X}{|}}{CH}. + \underset{\underset{X}{|}}{CH}{-}CH_2{-}[\underset{\underset{X}{|}}{CH}{-}CH_2]_n{-}R \rightarrow$$

$$R{-}[CH_2{-}\underset{\underset{X}{|}}{CH}]_n{-}CH = \underset{\underset{X}{|}}{CH} + \underset{\underset{X}{|}}{CH_2}{-}CH_2\,[\underset{\underset{X}{|}}{CH}{-}CH_2]_n{-}R$$

Thus initiator residues get incorporated into the polymer at one or both the ends of the polymer chain. These residues do not have any appreciable effect on properties of polymers.

It may be noted that the product molecules formed do not contain any free radical site and can not grow further. This process of termination brings about deactivation of growing chain and the polymer molecule so formed may be called a 'dead' polymer chain. The groups present at the two ends of the chain are known as '*end groups*'.

Linear Termination

More recently certain transitional metal ions have been used for initiation of vinyl polymerisation. It has been explained by Santappa *et al.* and Nayak *et. al.* that the termination in case of the metal ion initiation of polymerizztion to be linear, may be shown as below:

$$RM^{\cdot}_{n} + Me^{n+} \longrightarrow \text{Polymer} + Me^{n-1}$$

where Me represents the transitional metal ion and 'n' the valency of the metal ion. The termination reaction proceeds at the same rate as that of the initiation.

The molecular weight of polymer is found to depend on the relative rates of propagation and termination steps. For high molecular weight polymers propagation r.actoin rates must be much faster. Termination rate depends largely on concentration of free radicals in the system, *i.e.*, more free radicals lead to lower molecular weight polymers. Individual polymer chain grow to different lengths hence there is no definite molecular weight for polymeric material. The molecular weight is an average figure. For vinyl polymers average molecular weights is between 10^4 to 10^6.

Chain Transfer Reactions

In some cases *termination* may be brought about by 'transfer reactions'. In this type of reactions though the growth of one polymer chain is stopped due to formation of 'dead' polymer as in coupling or disproportionation reaction, yet there is a simultaneous generation of a new free radical that is capable of initiating a fresh polymer chain growth.

These reactions commonly take place in free radical polymerisation. In these reactions a growing polymer radical abstracts 2 hydrogen atom or some other atom from the initiator, monomer or a polymer or from any other species present in the system. This process results in an unpaired electron along the chain. Further propagation reaction results in formation of a side chain.

If chain transfer reactions occur to monomers it brings about reduction in molecular weight. This takes place in polymerisation of vinyl acetate.

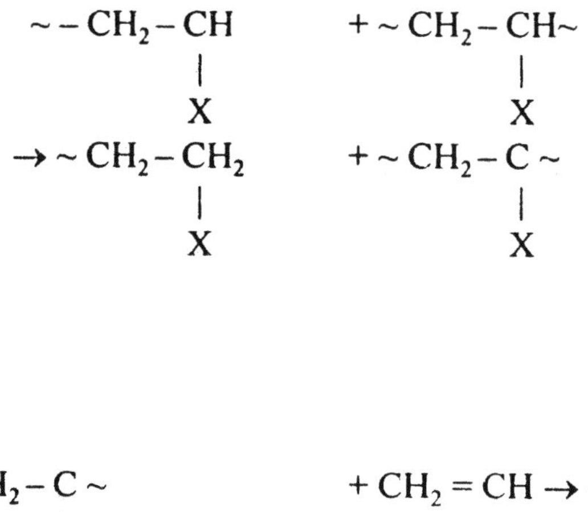

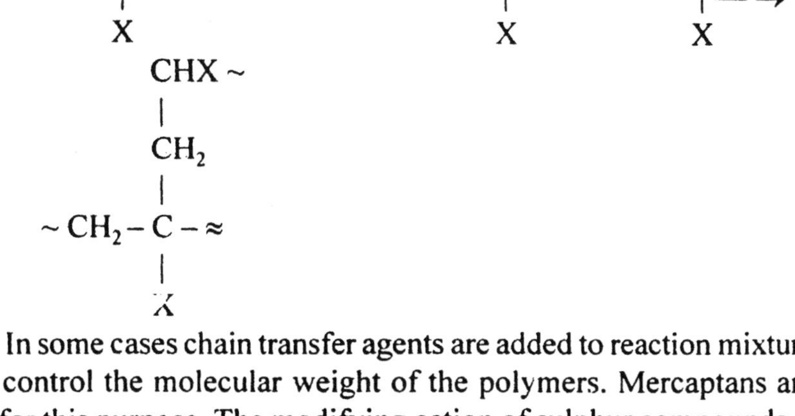

$$\sim\!-CH_2-\underset{\displaystyle X}{\underset{|}{CH}} \quad + \sim CH_2-\underset{\displaystyle X}{\underset{|}{CH}}\sim$$

$$\rightarrow \sim CH_2-\underset{\displaystyle X}{\underset{|}{CH_2}} \quad + \sim CH_2-\underset{\displaystyle X}{\underset{|}{C}}\sim$$

$$\sim CH_2-\underset{\displaystyle X}{\underset{|}{C}}\sim \quad + CH_2=\underset{\displaystyle X}{\underset{|}{CH}} \rightarrow \sim CH_2-\overset{\displaystyle CH-X}{\overset{|}{\underset{|}{\overset{\displaystyle CH_2}{\overset{|}{C}}}}}\sim \;\; \longrightarrow$$

$$\sim CH_2-\overset{\displaystyle CHX\sim}{\overset{|}{\overset{\displaystyle CH_2}{\overset{|}{\underset{\displaystyle X}{\underset{|}{C}}}}}}-\approx$$

In some cases chain transfer agents are added to reaction mixture or to control the molecular weight of the polymers. Mercaptans are used for this purpose. The modifying action of sulphur compounds is much greater than most of the solvents. The modifying action of diisopropylxanthate disulphide in the synthesis of SBR (styrene butadiene rubber) can be shown as under:

$$\sim CH_2-\underset{\underset{C_6H_5}{|}}{CH} + \begin{matrix} H_3C \\ H_3C \end{matrix}\!\!>\!CH-O-\overset{\overset{S}{\|}}{C}-S-S-\overset{\overset{S}{\|}}{C}-O-CH\!\!<\!\!\begin{matrix} CH_3 \\ CH_3 \end{matrix}$$

$$\rightarrow \sim CH_3-\underset{\underset{C_6H_5}{|}}{CH}-S-\overset{\overset{S}{\|}}{C}-O-\underset{\underset{CH_3}{|}}{\overset{\overset{CH_3}{|}}{CH}} + \underset{\underset{CH_3}{|}}{\overset{\overset{CH_3}{|}}{CH}}-O-\overset{\overset{S}{\|}}{C}-S\;CH_2-\underset{\underset{C_6H_5}{|}}{CH_5} \xrightarrow[\text{Polymerization}]{}$$

Similarly the chain transfer reactions with mercaptans can be shown as under:

$$\sim CH_2-\underset{\underset{C_6H_5}{|}}{\overset{\overset{H}{|}}{C}}\cdot + HS-R \rightarrow \sim CH_2-\underset{\underset{C_6H_5}{|}}{\overset{\overset{H}{|}}{C}}-H + R-\dot{S}$$

$$R-\dot{S} + \underset{\underset{C_6H_5}{|}}{CH}=CH_2 \rightarrow R-S-CH_2-\underset{\underset{C_6H_5}{|}}{\overset{\overset{H}{|}}{C}}\cdot$$

Inhibitors

Inhibitors are the substances that can kill or inhibit the chain growth by combining with free radicals and forming either stable products or inactive free radicals.

Phenols, quinones and aromatic amines reduce the rate of polymerisation by reacting with polymer radical. They lose a hydrogen readily but resultant radicals are not initiators. Inhibitors are added to monomers to prevent polymerisation during storage. Hydroquinone and t-butylcatechol in 0.001 to 0.1% concentration act as inhibitors.

Hydroquinone, nitrobenzene, dinitrobezene and benzothiazine are generally used as inhibitors in polymer industry. The inhibiting action of these can be shown as follows:

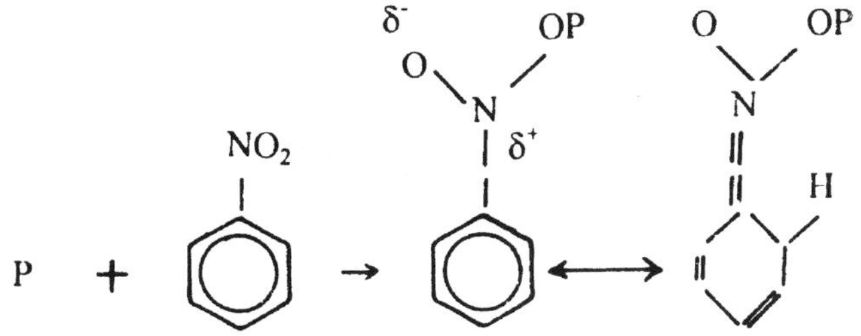

δ^- O, OP, N δ^+; O, OP, N, H

NO_2

P + (benzene ring) → (benzene ring) ⟷ (ring)

(Growing polymer chain) (inhibitor) Polymer chain with resonance-established end group)

In this case the inhibitor is nitrobenzene which adds on the growing chain P; yielding a polymer chain having a nitrobenzene end group carrying a radical site. The nitrocompound end of the chain is, resonance stabilised and the resonance-stabilised free-radical end is not active enough to attact a fresh monomer molecule and further propagate the chain. The free-radical nature of the end group is, however, powerful enough to recombine with the radical of another growing chain and terminate the growth of the latter:

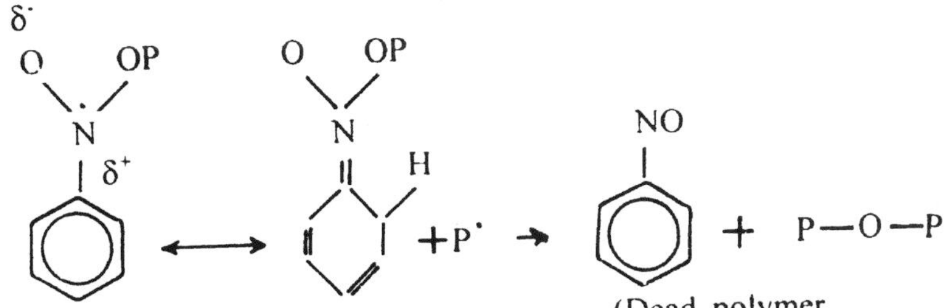

δ^- O, OP, N δ^+; O, OP, N, H; NO

(ring) ⟷ (ring) $+ P^{\cdot}$ → (ring) + P—O—P

(Dead polymer molecule)

In this way a single molecule of inhibitor has killed two growing chains.

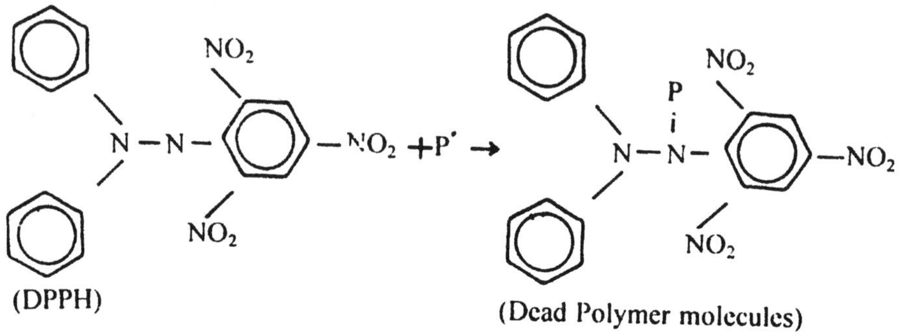

(DPPH) + P' → (Dead Polymer molecules)

In some cases, as in case of diphenyl picryl hydrazide (DPPH), the action of inhibitor may be due to the fact that it can exist in the form of stable free radicals which stops the chain growth by direct coupling.

Atmospheric oxygen also acts as an inhibitor. Its inhibiting action is due to its biradical nature

$$\dot{P} + \dot{O} - \dot{O} - \rightarrow P - O - \dot{O}$$

$$P - O - \dot{O} + \dot{P} - \rightarrow P - O - O - P$$

Majority of polymerisation reactions are carried out in an atmosphere of nitrogen to avoid contact with atmospheric oxygen.

Inhibitors are used for the storage of monomers but the inhibitor is destroyed and removed before use. For this the monomer is distilled or washed with an aqueous solution of sodium or potassium hydroxide. In actual practice in industries the inhibitors are killed by adding more quantities of initiators to the monomers.

Inhibitors are also used to arrest the polymerisation at some specified point in order to get a uniform product and to avoid cross-linking. For this purpose they are added near the end of polymerisation and are called "short stops".

Examples of important commercial products obtained by free radical polymerisation of substituted ethenes are polypropene (polypropylene). Polyphenylethene (polystyrene), poly-1chloroethene (poly vinylchloride) and poly 1-methoxy carbonyl-1 methylethene (polymethalmethacrylate).

Retarders

Retarders are the substances that reduce the rate of polymerisation the retarders reacts with the free radical and forms product which are incapable of adding monomer. A very effective retarder may act as inhibitor. Thus the distinction between retarder and inhibitor is merely of degree. Both retarder and inhibitor reduce the concentration of free radicals and shorten their average life and the length of the polymer chain.

The retarder may be a free radical which is too unreactive to initiate a polymer chain (*e.g.* triphenyl methyl or diphenyl-picrylhydrazyl). The mechanism of retardation is simply the combination or disproportionation of radicals.

4

Copolymerisation

4.1. Introduction

Copolymerisation is the process in which a mixture of two or more monomers gets polymerised to yield a product. The product obtained is known as a copolymer. A copolymer product contains some units of each type of monomer and is different from a physical mixture of individual polymer molecules formed by different monomers. It is not always possible to make a copolymer with any two or more monomers. When two monomers A and B are copolymerised the rate of polymerisation is determined by concentration of monomers. Four different propagation reaction can occur for copolymerisation of A and B. ~ A—A$^{\cdot}$, ~ A—B$^{\cdot}$, ~ B—B$^{\cdot}$, ~ B—A$^{\cdot}$. The monomers are randomly distributed in the polymer chain in most of cases but in case of copolymerisation of styrene and maleic anhydride, there is perfect alternate arrangement of monomers in the chain regardless of initial composition of monomers.

In general it is possible to represent copolymerisation in the following way:

$$xA + yB \rightarrow$$

—A—A—B—A—A—A—B—B—B—B—A—B—A—B—B—

The copolymer formed is composed of both the monomeric units A and B arranged at random. The A:B ratio and the degree of randomness is found to depend on the qunatity of the individual monomers taken, their amenability for copolymerisation and the polymerisation mechanism used.

Copolymerisation can be brought about by many types of polymerisation reactions. The majority of the commercially important copolymers, however, are made by free-radical, ionic or poly-condensation polymerisation.

4.2. Classification of Copolymer

Copolymers are classified into four categories depending upon the nature of the distribution of different monomers in the polymer chain. (i) Random copolymers, (ii) Alternating copolymers, (iii) Block copolymers, and (iv) Graft copolymers.

Random Copolymers

Random copolymers get formed by the random arrangement of monomer units in the chain as shown below:

—A—B—A—A—B—A—B—B—B—A—A—B—

Random copolymers are produced under specific polymerisation conditions.

Alternating Copolymers

In alternating polymers the monomers are produced alternately along the copolymer chain.

—A—B—A—B—A—B—A—B—A—B—

Alternating copolymers are produced during condensation polymerisation when two different types of monomers like diacid and diols are used.

Block Copolymers

Block copolymer is a linear copolymer that contains long chain of one monomer with another monomer in the polymer chain.

—A—A—A—A—B—B—B—B—A—A—A—A—A—

The block copolymers sometimes are made up of just two or three long blocks of each type of unit giving a material with interesting propertries.

In block copolymerisation, the macromolecule is made up of blocks of considerable length consisting entirely of one type of monomer.

~A—A—A~A~A—A A—B—B—B~B—B—B—C—C—C

~C—C—C~

The individual chain segements are generally not equally long. It is possible to obtain copolymers with different combinations of blocks.

~A~A—B~B—A~A—B~B—A~A~

A~A—B~B—A~A

~A—B—A—B—A—B~A—B—A—A—A—A~

A—A—A—A—B—A—B—A—B~

The blocks do not always have to be homopolymers, but might in turn be copolymers.

4.3. Preparation of Block Copolymers

Following two methods are used for preparation of block copolymers. In the first method;

1. A second monomer can be polymerised onto an existing polymer chain

2. Performed polymer chain could be combined with other polymeric chains with the aid of functional group.

3. The former procedure can be carried out simply by means of stoichiometric polymerisation where except for internal chain termination there is no termination. For example, we first polymerize α-methyl styrene with the aid of phenyl lithium. The chain length can be predetermined arbitrarily by means of the ratio of monomer to the initiator. When styrene is added the chain continues to grow without recognising the change in the monomer. Instead of styrene; methylmethacrylate, isoprene, or butadiene can also be used. These monomers could be added one after the other.

A certain sequence has to be followed so that the basicity of the chain end is never less than that of the new one which is produced by the addition of the next monomer.

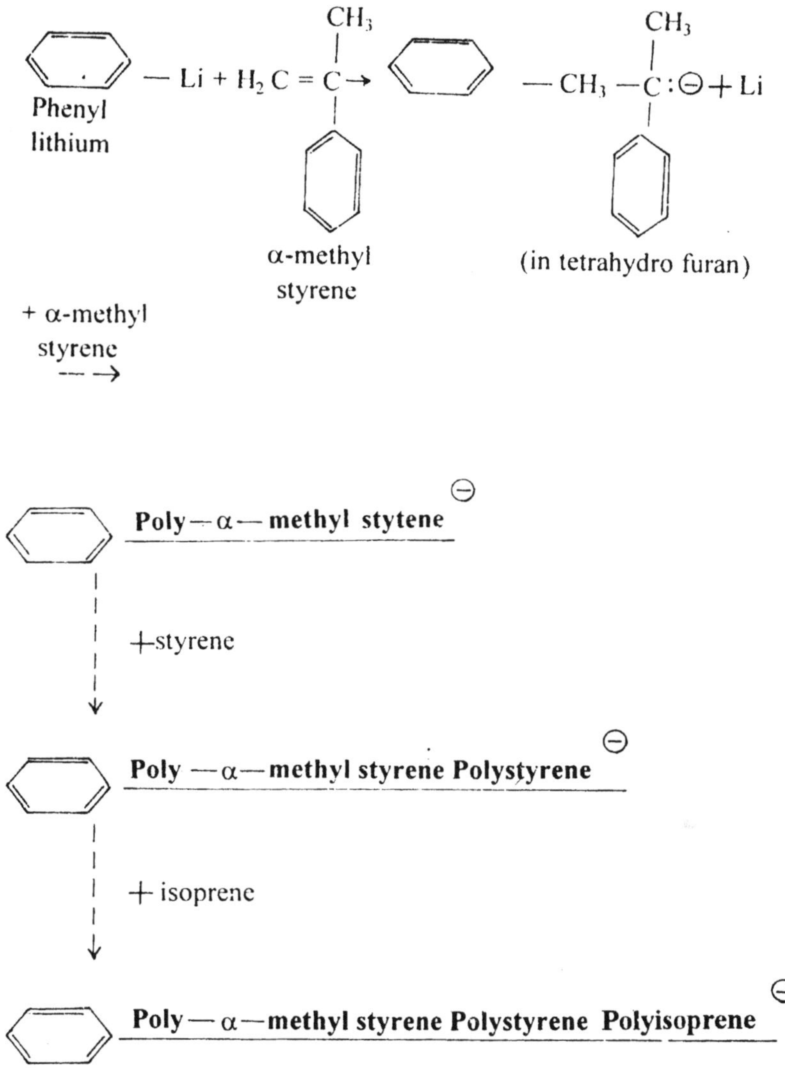

$$C_6H_5-Li + H_2C = \underset{\underset{C_6H_5}{|}}{\overset{\overset{CH_3}{|}}{C}} \rightarrow C_6H_5-CH_2-\underset{\underset{C_6H_5}{|}}{\overset{\overset{CH_3}{|}}{C}}:^{\ominus} + Li$$

Phenyl lithium α-methyl styrene (in tetrahydro furan)

+ α-methyl styrene ---→

C_6H_5 Poly—α—methyl stytene $^{\ominus}$

+styrene ↓

C_6H_5 Poly—α—methyl styrene Polystyrene $^{\ominus}$

+ isoprene ↓

C_6H_5 Poly—α—methyl styrene Polystyrene Polyisoprene $^{\ominus}$

In place of phenyl lithium, if we use sodium naphthalene as the initiator, then the chains may grow in both the direction and a block copolymer is formed with the primary polymer in the middle. By this method a block polymer such as polyisoprene—, Polystyrene—, poly—α—methyl styrene—, polystyrene—, isoprene can be obtained.

In the second procedure, we start with preformed polymer chain that is combined by the aid of functional groups. The reaction is of polyaddition of polymers and we obtain chains with an undetermined number of segements.

Polystyrene Polystyrene Polyacrilic ester

—————OH + HO————— + HOOC... ... ...COOH

esterification Polystyrene Polyacrylic ester Polystyrene

$$\xrightarrow{\text{esterification (Dicyclohexyl carbodi imide)}} \text{———O—}\underset{\underset{O}{\|}}{C}\text{... ... ...}\underset{\underset{O}{\|}}{C}\text{—O———}$$

(Dicyclohexyl carbodi imide)

Polystyrene having hydroxyl end groups can easily be obtained through stoichiometric polymerisation with sodium naphthalene and subsequent chain termination with ethylene oxide. Polyacrylates having carboxyl end groups can be obtained by polymerisation with an azocatalysed containing carboxyl groups as under:

$$2CH_3\text{—}\underset{\underset{O}{\|}}{C}\text{—}CH_2\text{—}CH_2\text{—}COOH \xrightarrow[-2H_2O]{+H_2N\text{—}NH_2,\ +2HCN}$$

Levulinic acid —2H

$$HOOC\text{—}(CH_2)_2\text{—}\underset{\underset{CN}{|}}{\overset{\overset{CH_3}{|}}{C}}\text{—}N = \underset{\underset{CN}{|}}{\overset{\overset{CH_3}{|}}{N}}\text{—}(CH_2)_2\text{—}COOH$$

$$\xrightarrow{60^\circ} 2\ HOOC\text{—}(CH_2)_2\text{—}\underset{\underset{CN}{|}}{\overset{\overset{CH_3}{|}}{C}} + N_2 + CH_2 = CH\text{—}COOR \longrightarrow$$

Polyacrylate with COOH end groups

The two polymers that are to be combined do not essentially have different end groups. It is also possible to combine polymers having the same end groups by addition of functional cross linking agents. For example polyesters with OH end groups and polyethers with OH end groups can be combined by a reaction with a diisocyanate.

4.4. Graft Copolymers

The phenomenon of graft copolymerisation dates back to the year 1933 but graft copolymer was officially defined in 1952 by IUPAC.

Generally the graft copolymers are made up of a main homopolymers chain with branches of another type of homopolymer.

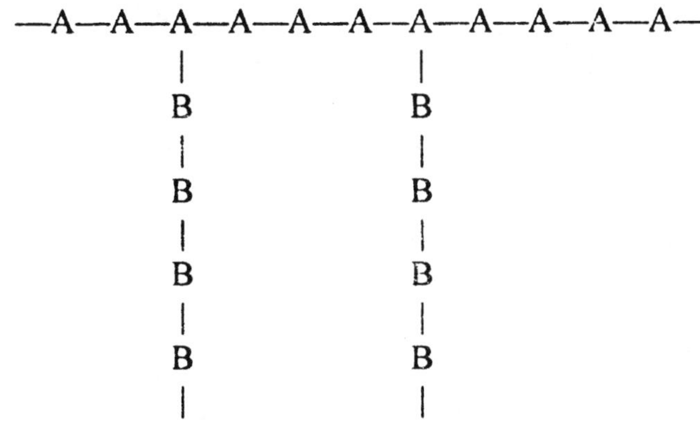

Graft copolymers are branched polymers in which monomers on the segments of the branches and the backbone are different. A typical example is copolymer of butadiene and acrylonitrile is grafted on a copolymer of styrene and acrylonitrile.

The main chain is chemically different from the branches in case of graft copolymers with branched molecules. The chemial nature, length and concentration of side chains may be different in each case. A graft copolymer may be schematically represented as in (Fig. 4.1.).

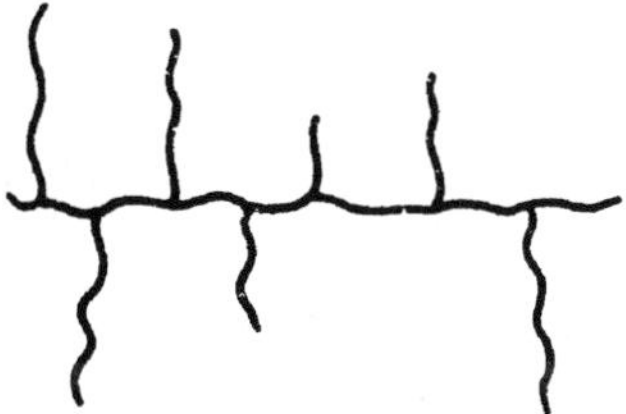

Fig. 4.1: General structure of graft copolymer.

Looking at the structure it can be realised that, the schematic representation has nothing to do with the real form of the macromolecules. And the structural diagram is only a substitute for the structural formula and concerns the chemical constitution only, not the form of the macromolecules. The form of the molecule in dilute solution is not influenced by the presence of branches, *i.e.*, whether the chain is linear or branches, *i.e.*, whether the chain is linear or branched it is always present in the form of a random coil. Only the coil density is large with branched molecule than with linear ones.

Using the process of graft copolymerisation, we can attach a large number of base polymers to each other. Due to impartibility of the components, however, it has to be assumed that even in a graft copolymer molecule the two components are spatially separated from each other. Thus, the structure of the graft copolymer molecule may be different depending on the ratio of the components, the density of the grafting points, the length of the side chains and the method by which it was formed. Figures 4.1 and 4.2. show some graft copolymer structure in a schematic fashion.

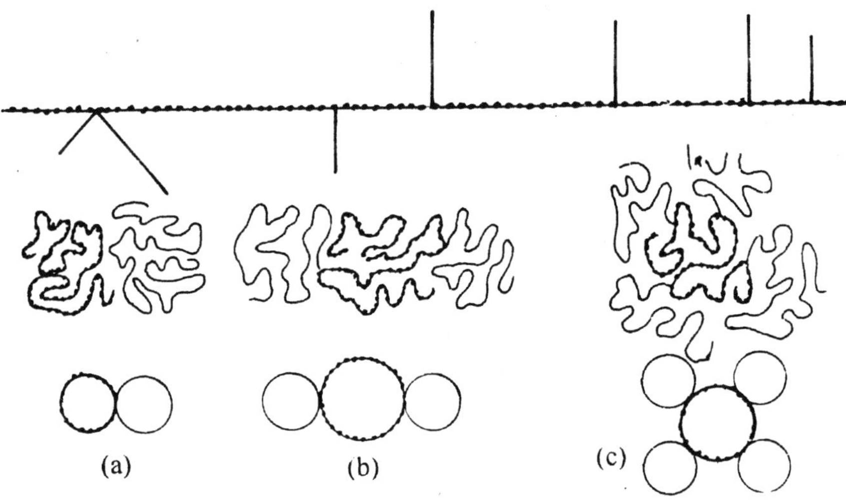

Fig. 4.2: Coil structure of graft—Copolymer molecules.

The formation of graft copolymers leads to the possibility of combining incompatible polymers in such a way that the components may be distributed homogeneously or they are atleast firmly fixed to each other at the phase boundaries, for example, with high-impact polystyrene, where the two components, polystyrene, (~ 90%) and rubber (~ 10%), are combined through grafting bonds in such a manner that a new material is produced which has inherited from polystyrene its hardness and from rubber its toughness.

4.5. Method of Grafting

Graft copolymerisation is a unique method which is used for modifying the properties of the base polymer. A varieties of new properties can be imparted to the base polymer by implanting a number of suitable polymers. Various methods used for grafting are: (i) Chemical method, (ii) Radiation method, and (iii) Photochemical method.

Chemical Method

Grafting through chain-transfer:

Fist the base polymer (for example polyacrylonitrile) is dissolved in a monomer such as styrene and then an initiator is added. The two possible ways of polymerisation are:

(i) Formation of polystyrene homopolymer

(ii) Grafting polystyrene onto polyacrylonitrile.

In the latter case, a free radical may be created by the interaction of the initiator with polyacrylonitrile producing a free radical on the backbone of the polyacylonitrile as under:

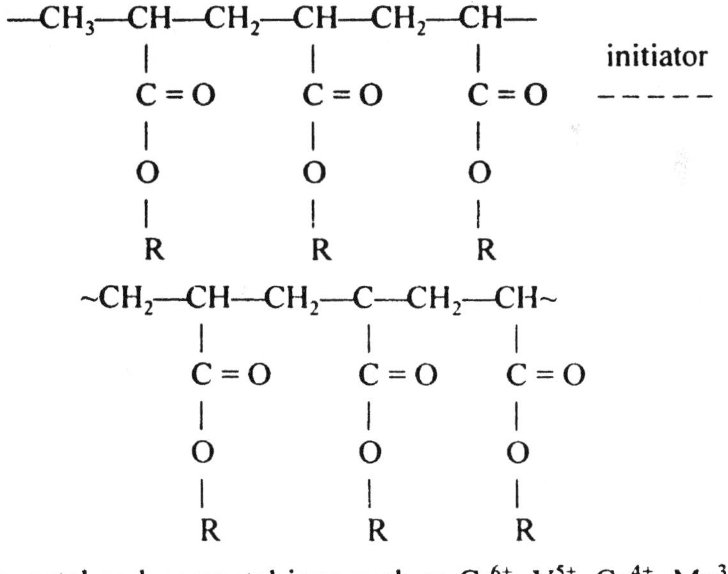

—CH_3—CH—CH_2—CH—CH_2—CH—

(each CH bearing | C = O | O | R)

initiator
- - - - - -

~CH_2—CH—CH_2—C—CH_2—CH~

(each CH / C bearing | C = O | O | R)

Some metal and nonmetal ions such as Cr^{6+}, V^{5+}, Ce^{4+}, Mn^{3+}, Fe^{3+}, peroxydisulphate ion and peroxydiphosphate ion are commonly used as the initiators.

In the second stage of the reaction, the free radical produced on the backbone of the base polymer initiates polymerisation which results in the formation of graft copolymerisation as under:

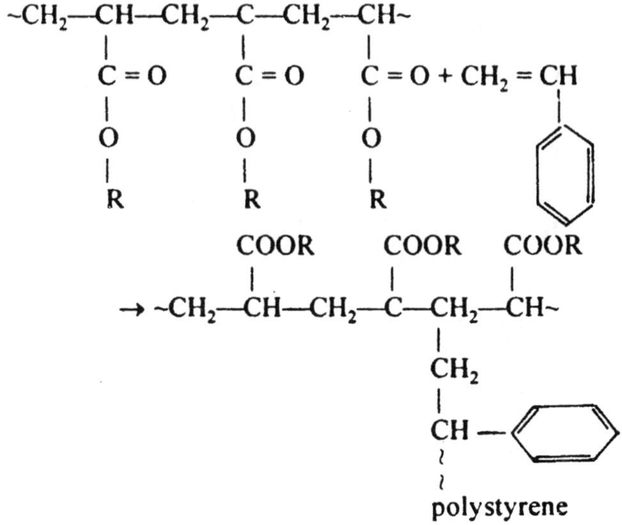

~CH$_2$—CH—CH$_2$—C—CH$_2$—CH~

C = O　C = O　C = O + CH$_2$ = CH

O　O　O

R　R　R

COOR　COOR　COOR

→ ~CH$_2$—CH—CH$_2$—C—CH$_2$—CH~

CH$_2$

CH —

polystyrene

The effect is found to be more pronounced with polybutadienes, *i.e.*, natural and synthetic rubbers. In the polybutadiene chain, the hydrogen atom of the—CH$_2$ group is more liable due to the neighbouring double bond and is thus easily attacked by a radical chain.

The tendency of polybutadiene to undergo chain transfer reactions can be used for the preparation of impact-resistance polymer where polystyrene has been grafted to polybutadiene as under:

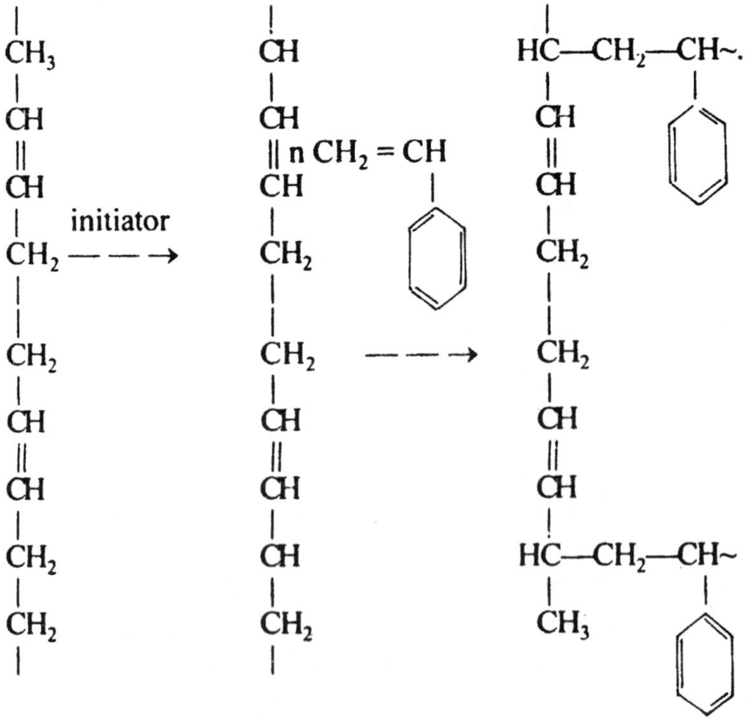

CH$_3$　CH　HC—CH$_2$—CH~.

CH　CH　CH

|| n CH$_2$ = CH

CH　CH　CH

initiator

CH$_2$ ——→　CH$_2$　CH$_2$

CH$_2$　CH$_2$ ——→　CH$_2$

CH　CH　CH

CH　CH　CH

CH$_2$　CH　HC—CH$_2$—CH~

CH$_2$　CH$_2$　CH$_3$

4.6. Preparation of Graft-Copolymers with the Acid Functional Group

Graft copolymers can be prepared by attaching performed polymers onto a base polymer with the aid of certain functional groups. The most suitable method of obtaining polymers with suitable functional group is copolymerisation with monomers which contain desired functional groups. The following list contains a number of relatively easily obtainable monomers with functional groups:

Acryloyl Chloride $CH_2=CH—COCl$

Acrylic acid $CH_2=CH—COOH$

Monoacrylate from glycols $CH_2=CH—COO(CH_2)_n—OH$

Glycidylacrylate $CH_2=CH—COO—CH_2—CH—CH_2$ (CH and CH_2 bridged by O)

Isocyanatoacrylate

$CH_2=CH—COO—CH_2—CH_2—N=C=O$

Dimethylamino ethyl-acrylate

$CH_2=CH—COO—CH_2—CH_2—N(CH_3)(CH_5)$

Maleicanhydride

CH = CH
| |
O = C C = O
(both C bridged by O)

Acrolein $CH_2=CH—CHO$

Ethylenecarbonate CH = CH
| |
O O

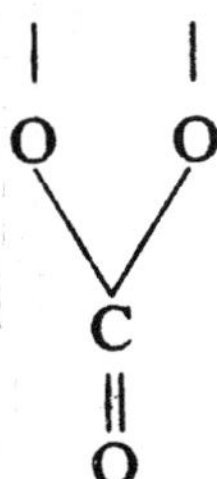

It is also possible to prepare polymers with functional groups by partial saponification of polyvinyl esters to give —OH groups or polyacrylic esters or polyacrylonitrile to give —COOH groups. Polymer with such functional groups at the chain ends are prepared by making use of initiators with functional groups *i.e.*, azocatalyst containing carboxylic groups as under:

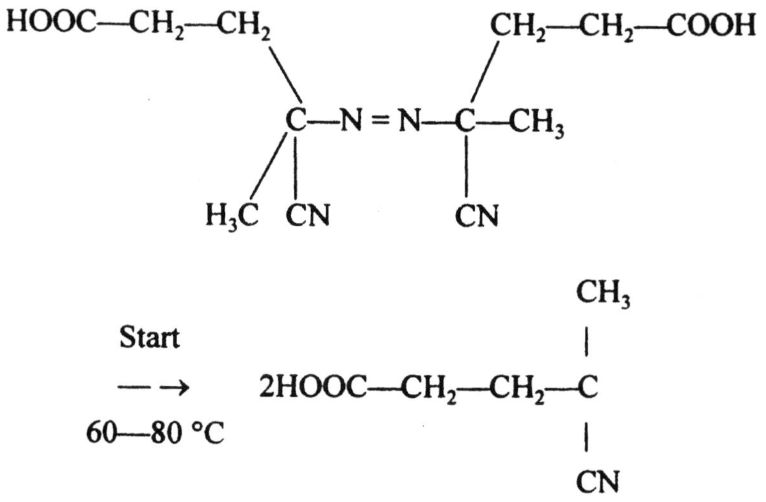

$$\begin{array}{c} HOOC{-}CH_2{-}CH_2 \qquad\qquad CH_2{-}CH_2{-}COOH \\ \diagdown \qquad\qquad \diagup \\ C{-}N{=}N{-}C{-}CH_3 \\ \diagup\ | \qquad\qquad | \\ H_3C\ \ CN \qquad\qquad CN \end{array}$$

$$\xrightarrow[60-80\ ^\circ C]{\text{Start}} 2HOOC{-}CH_2{-}CH_2{-}\overset{\displaystyle CH_3}{\underset{\displaystyle CN}{\overset{|}{\underset{|}{C}}}}$$

Polymers having one or two carboxyl end groups are prepared by anionic polymerisation and subsequent chain termination with Co^{2+} and further reaction with mineral acids.

4.7. Graft-Copolymers through Copolymerisation

Cross-linked graft-copolymers are prepared in a very simple manner by copolymerisation. In this an unsaturated polymer is dissolved in the monomer that is to be grafted on and the monomer is allowed to polymerize. The double bond of the polymer is incorporated into the growing chain of the polymerizing monomer.

An interesting example of this type of copolymer is the copolymerisation of butadiene and styrene in which first butadiene polymerizes giving rise to polybutadiene. The polybutadiene formed then reacts with styrene to yield a graft copolymer as under:

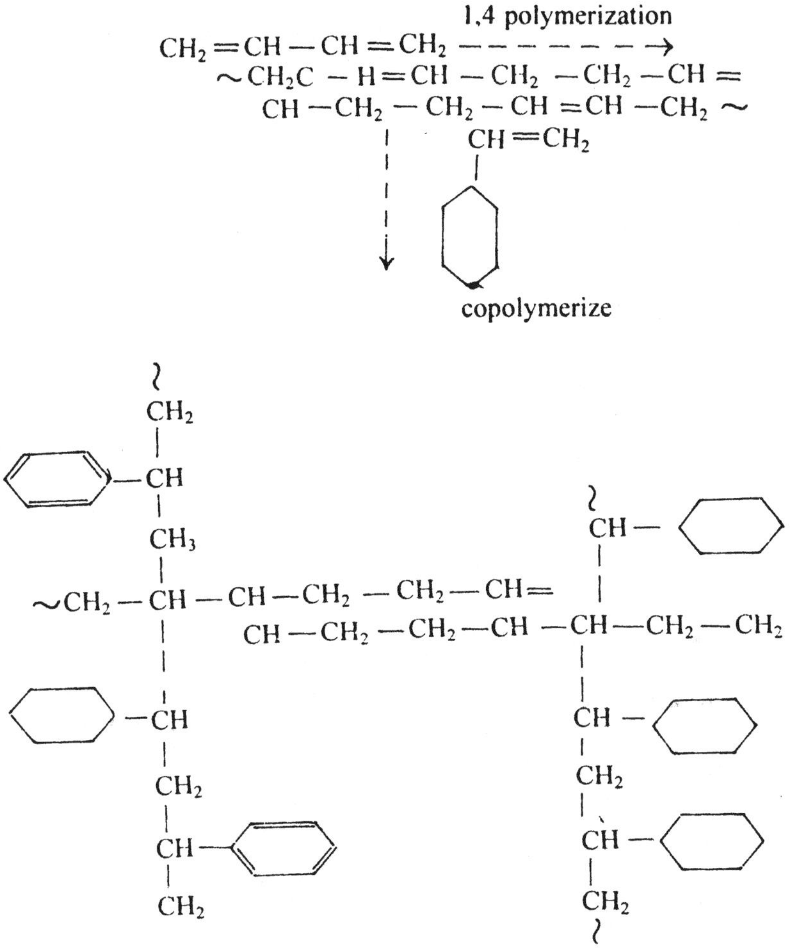

1,4 polymerization

$CH_2=CH-CH=CH_2$ - - - - - - - → $\sim CH_2C-H=CH-CH_2-CH_2-CH=$
$CH-CH_2-CH_2-CH=CH-CH_2\sim$

$CH=CH_2$ (styrene)

↓ copolymerize

$\sim CH_2-CH-CH-CH_2-CH_2-CH=CH-CH_2-CH_2-CH-CH-CH_2-CH_2$ (with phenyl-substituted CH and CH_2 side chains)

4.8. Reactivity Ratios

A copolymer of butadiene and styrene is known as poly (butadiene-costyrene). A copolymer obtained from two monomers is a bipolymer, a copolymer from three monomers is a terpolymer and so on. *Run numbers* R is defined as average number of monomer sequences in 100 monomer units. In a bipolymer from monomers M_1 and M_2, fraction $[M_1]/[M_2]$ in the copolymer formed is denoted by f and fraction $[M_1]/[M_2]$ in the reactant mixture is denoted by X. When f = X, it is called an azeotropic copolymer *i.e.*, a copolymer whose composition is same as that of the reactants. **Wall's equation** gives the general relation between copolymer composition f and composition of reactants (feed) X as—

$$f = rX \qquad ...(4.1)$$

where r is the *reactivity* ratio. Wall's equation is a simplified relation between copolymer composition (f) and the *feed composition* (X). However, on kinetic considerations it can be shown that the monomers M_1 and M_2 can react in following four ways:

homopolymerisation

$$M_1^{\cdot} + M_1 \xrightarrow{k_{11}} M_1 M_1^{\cdot}$$

kinetics of Polymerisation

$$M_2^{\cdot} + M_2 \xrightarrow{k_{22}} M_2 M_2^{\cdot}$$

heteropolymerisation

$$M_1^{\cdot} + M_2 \xrightarrow{k_{12}} M_1 M_2^{\cdot}$$

$$M_2^{\cdot} + M_1 \xrightarrow{k_{21}} M_2 M_1^{\cdot}$$

where k_{11} and k_{22} are propagation rate constants for the homopolymerisation and k_{12} and k_{21} are the propagation rate constants for heteropolymerisation reactions leading to copolymer formation. If F_1 and F_2 are mole fractions of monomers M_1 and M_2 in the copolymer, these will be related to the rates of disappearance of monomers from the feed as under:

$$F_1 = \frac{d[M_1]}{d[M_1]+d[M_2]}; \qquad \text{...(4.2)}$$

$$F_2 = \frac{d[M_2]}{d[M_1]+d[M_2]};$$

and hence,

$$\frac{F_1}{F_2} = \frac{d[M_1]}{d[M_2]}$$

$$\therefore \quad \frac{F_1}{F_2} = \frac{k_{11}[M_1^{\cdot}][M_1]+k_{21}[M_1][M_2^{\cdot}]}{k_{22}[M_2][M_2^{\cdot}]+k_{12}[M_2][M_1^{\cdot}]} \qquad \text{...(4.3)}$$

For an ideal case F_1 should be equal to X_1. If we assume steady state for free radical concentrations, *i.e.*, rate of conversion of $\sim M_1$ to

$\sim M_1 M_2$ is equal to rate of conversion of $\sim M_2$ to $\sim M_2 M_1$. Then we have,

$$k_{21}[M_1][M_2^{\cdot}] = k_{12}[M_2][M_1^{\cdot}] \qquad ...(4.4)$$

Hence using eq. 4.3, eq. 4.2 can be rewritten as under

$$\frac{F_1}{F_2} = \frac{k_{11}[M_1][M_1^{\cdot}] + k_{12}[M_2][M_1^{\cdot}]}{k_{22}[M_2][M_2^{\cdot}] + k_{21}[M_1][M_2^{\cdot}]}$$

$$= \left(\frac{k_{11}[M_1]}{k_{12}[M_2]} + 1\right) \Bigg| \left(\frac{k_{22}[M_2]}{k_{21}[M_1]} + 1\right)$$

$$= \frac{r_1 X + 1}{(r_2 / X) + 1}$$

$$\therefore \qquad f = \frac{r_1 X + 1}{(r_2 / X) + 1} \qquad ...(4.5)$$

where $r_1 = k_{11} / k_{12}$ and $r_2 = k_{22} / k_{22}$. Equation (4.4) is known as copolymer equation and correlates the feed composition X with copolymer composition f through a two paramerter equation *viz.* through two unknowns r_1 and r_2. In spite of its limitations and empirical development, the correlation is more useful than Wall's equation.

For different magnitudes of r_1, r_2 in a given copolymerisation reaction of two monomers M_1 and M_2. Tendency of randomness is greatest when $r_1 r_2 = 1$. In this case $r = 1/r_2$ *i.e.* $k_{11}/ k_{12} = k_{21}/ k_{22}$ and there is no preference of a growing radical to add to a particular monomer. Hence, this is an ideal system giving a random copolymer. In this case the copolymer equation (eq. 4.4) becomes identical to Wall's equation.

However, when $r_1 = r_2 = 0$, each growing radical has a preference to add to another monomer unit yielding a perfectly alternating copolymer of the type—

$$\sim M_1 M_2 M_1 M_2 M_1 M_2 \sim$$

In most cases the copolymer compositions lie between these two extreme situations of $r_1 \; r_2 = 1$ (random copolymers) or $r_1 = r_2 = 0$ (alternate copolymers), and correspond to $0 < r_1$

$r_2 < 1$. When $r_1 r_2 > 1$, there is tendency to form homopolymers and if both $r_1 > 1$, $r_2 > 1$, a long sequence of each monomer will be found in the chain, giving block copolymer. Such examples are only very few.

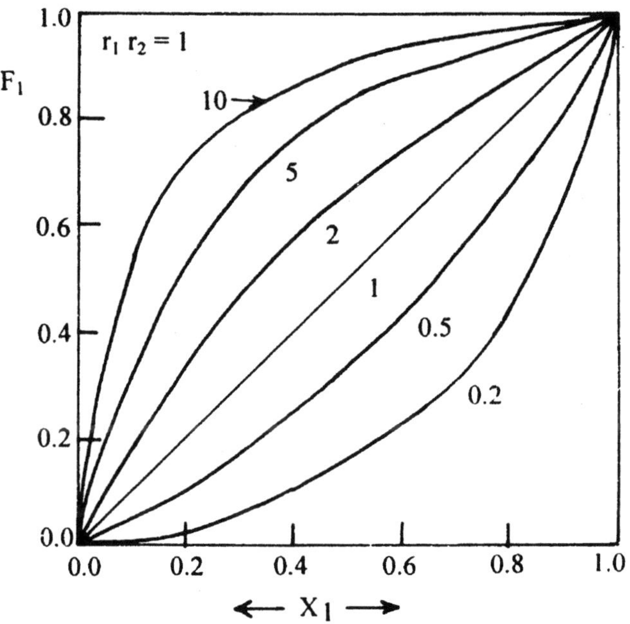

Fig. 4.3 : F_1—X_1 curves for ideal cases of copolymerisation : $r_1 r_2 = 1$.

If X_1 and X_2 are mole fractions of monomers M_1, M_2 in the feed, the copolymer equation (eq. 4.5) gives for the mole fraction F_1 of monomers M_1 in the copolymer—

$$F_1 = \frac{r_1 X_1^2 + X_1 X_2}{r_1 X_1^2 + 2X_1 X_2 + r_2 X_2^2} \qquad ...(4.6)$$

With the progress of polymerisation F_1 and X_1 both change and equation 4.6 may be used to calculate copolymer composition vs feed composition curves for different reactivity ratios. Fig. 4.3 gives such curves for random copolymerisations ($r_1 r_2 = 1$). Curves for nonideal cases of copolymerisation are given in Fig. 4.4. These curves exhibit the effect of increasing tendency of alternation. As alternation increases, more and more feeds yield a copolymer containing a good deal of each component. In those cases in which $r_1 < r, r_2 < 1$, the curves cross the line corresponding to $F_1 = X_1$.

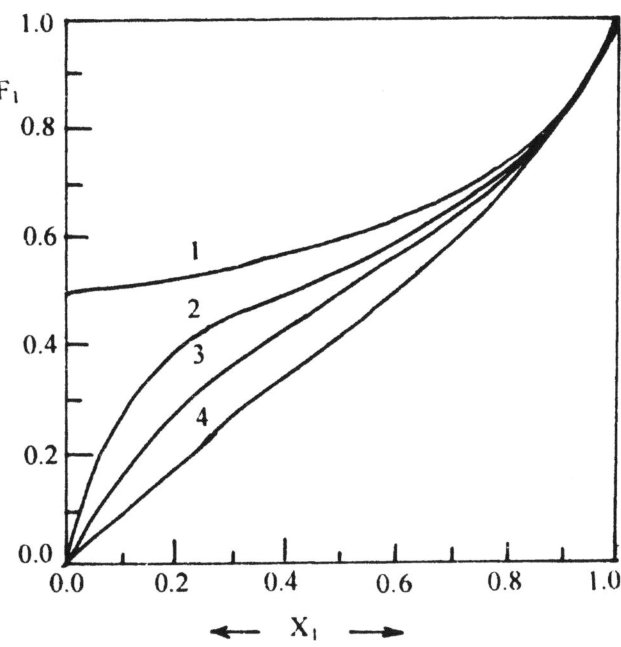

Fig. 4.4: F_1—X_1 curves for nonideal cases of copolymerisations with $r_1 = 0.5$ and r, as indicated.

At the point of intersection $F_1 = X_1$ and polymerisation takes place without change in composition of feed or of polymer. This is known as *azeotropic* copolymerisation. Azeotropic composition can be obtained from eq. 4.4 or eq. 4.5 as–

$$F_1^{az.} = X_1^{az.} = \frac{1 - r_2}{2 - r_1 - r_2} \qquad ...(4.7)$$

5

Ionic Polymerisations

5.1. Introduction

Ionic polymerisation is an **addition polymerisation** and is of commercial importance.

History of ionic polymerisation is very old and dates back to as early as 1789 A.D. when resinnification of terpentine by concentrated sulphuric acid was first reported.

In 1866 A.D. a polymeric product was formed from styrene and sulphuric acid. Another breakthrough was the production of synthetic rubber from butadiene by using metallic sodium or potassium by German scientists during 1911–22. In 1929, Ziegler reported polymerisation of vinyl monomers using butyllithium.

In 1936, Lewis acids were used as catalysts for polymerisation. Grignard reagents were used for polymerisation in 1945. Since 1956, there are innumerable paper on different types of ionic polymerisations.

Ionic polymerisation forms high molecular weight products and reactions can be easily carried out at room temperature or low temperature. Ionic polymerisations are chain reactions and are analogous to radical chain reactions. They also involve initiation and propagation steps.

Ionic polymerisation can be categorised into two classes, (i) Cationic polymerisation and (ii) Anionic polymerisation, depending upon the nature of ions used for the initiation of polymerisation. The cationic polymerisation refers to the process in which a positive ion is used for initiation. When a negative ion is used for initiation of polymerisation, then process is referred to as anionic polymerisation.

In this type of polymerisation propagation is not by radical but by either a carbonium ion (in cationic polymerisations) or by a carbanion (in anionic polymerisations). In anionic polymerisation, termination does not take place unless we add a transfer agent.

In the absence of transfer agent, polymer chains with active, ends which are known as **'living polymers'** can be synthesized. In case of cationic polymerisations, termination step is very slow. Ionic reactions are largely affected by solvents used.

5.2. Cationic Polymerisation

Cationic polymerisation involves carbonium ion active species and takes place with monomers which contain electron releasing substituents such as phenyl, vinyl, alkoxy and 1, 1 dialkyl.

Cationic polymerisations of vinyl derivatives proceeds by a polar or heterolytic cleavage of the double bond by the action of initiators which act as electron acceptors. Initiation is carried out by proton donation by acids or Co-catalysts

$$BF_3 + H_2O + \underset{\substack{|\\X}}{CH_2 = CH} \longrightarrow CH_3\text{—}\overset{\oplus}{\underset{\substack{|\\X}}{C}}\text{—}H(BF_3OH)\text{—}$$

The olefin molecule acts as an electron donor, i.e., as a "base" and after interacting with the electron acceptor, it becomes the carrier of positive charge.

$$\begin{matrix} R_1 \\ R_2 \end{matrix}\!\!>C = CH_2 + \overset{\oplus}{X} \longrightarrow \oplus \overset{\substack{R_1\\|}}{\underset{\substack{|\\R_2}}{C}}\text{—}CH_2X$$

This is followed by subsequent additions of the monomer or growth, leading to the formation of a macro cation. The polymerisation is carried out by catalysts of the type of approtonic Lewis acids, such as BF_3, $AlCl_3$, $AlBr_3$, $SnCl_4$, $TiCl_4$, H_2SO_4 and other strong Lewis acids. Propagation occurs by addition of monomer to carbonium ion.

The monomer molecule acts like electron donor, and reacts with the catalyst giving rise to polymer ions. The successive addition of the monomer to the polymer ion is the propagation reaction which may be represented as follows:

Propagation

$$XCH_2-\overset{\overset{\displaystyle R_1}{|}}{\underset{\underset{\displaystyle R_2}{|}}{C}}\oplus + H_2C=C\begin{matrix}\nearrow R_1\\ \searrow R_2\end{matrix} \dashrightarrow XCH_2\ \overset{\overset{\displaystyle R_1}{|}}{C}-CH_2-\overset{\overset{\displaystyle R_1}{|}}{\underset{\underset{\displaystyle R_2}{|}}{C}}\oplus$$

$$X\left(-CH_2-\overset{\overset{\displaystyle R_1}{|}}{\underset{\underset{\displaystyle R_2}{|}}{C}}-\right)_{n-2}CH_2-\overset{\overset{\displaystyle R_1}{|}}{\underset{\underset{\displaystyle R_2}{|}}{C}}\oplus + H_2C=\overset{\overset{\displaystyle R_1}{|}}{\underset{\underset{\displaystyle R_2}{|}}{C}} \dashrightarrow$$

$$X\left(-CH_2-\overset{\overset{\displaystyle R_1}{|}}{\underset{\underset{\displaystyle R_2}{|}}{C}}-\right)_{n}CH_2-\overset{\overset{\displaystyle R_1}{|}}{\underset{\underset{\displaystyle R_2}{|}}{C}}\oplus$$

Chain Termination

Termination can occur by proton transfer to negatively charged counter ion.

However, the chain termination is much more complicated in case of cationic polymerisation. There cannot be mutual termination in case of cationic polymerisation because the growing chains carry similar charges, and repel each other. Hence there can be three types of termination in case of cationic polymerisation.

(i) *Disproportionation Termination.* The termination may take place by the abstraction of hydrogen atom from the adjacent carbon atom, *i.e.*, termination through disproportionation.

It can be represented as under:

$$X\left(-\overset{|}{\underset{|}{C}}-\overset{|}{\underset{|}{C}}-\right)_n-\overset{|}{\underset{\underset{\displaystyle H}{|}}{C}}-\overset{|}{\underset{|}{C}}\oplus \rightarrow X\left(-\overset{|}{\underset{|}{C}}-\overset{|}{\underset{|}{C}}-\right)_n-\overset{|}{\underset{\underset{\displaystyle H}{|}}{C}}=\overset{|}{\underset{|}{C}}+\overset{+}{H}$$

(ii) *Termination by combination with base.* The termination may occur due to the recombination of the growing cation with the conjugate base, $\overset{\ominus}{B}$. It can be represented as under:

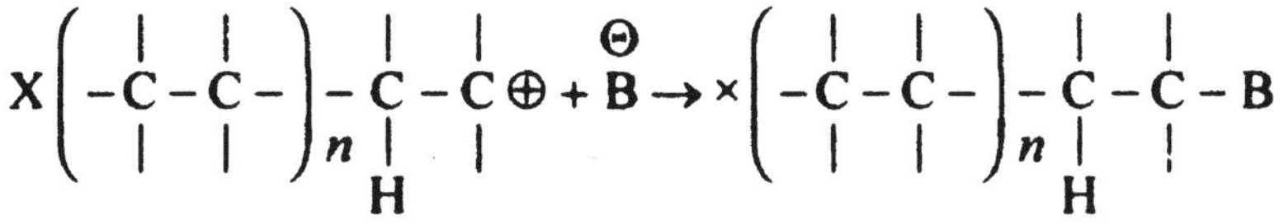

$$X\left(-\overset{|}{\underset{|}{C}}-\overset{|}{\underset{|}{C}}-\right)_n-\overset{|}{\underset{H}{C}}-\overset{|}{\underset{|}{C}}\oplus+\overset{\ominus}{B}\rightarrow X\left(-\overset{|}{\underset{|}{C}}-\overset{|}{\underset{|}{C}}-\right)_n-\overset{|}{\underset{H}{C}}-\overset{|}{\underset{|}{C}}-B$$

(iii) *Chain-transfer Termination.* The macro carbonium ion may transfer its positive charge by interaction with other molecules in the reaction medium.These molecules could be those of impurities or of monomer molecules themselves. The resulting polymer in such a cases have low molecular weight.

The process designated as termination in case of type (ii), is not very probable and some times, is even impossible. For example, $\overset{\ominus}{ClO_4}$ can not from a new chemical link with the carbon of the macro ion, nor can $\overset{\ominus}{BF_4}$ or $\overset{\ominus}{AlCl_4}$.

5.3. Reactivity of Vinyl Monomers

In cationic polymerisation monomer molecule acts as electron donor and higher activity is shown by vinyl derivatives which have an increased electron density at the double bond.

In most of the cases these are ethylene derivatives which have electropositive substituents. The order of reactivity of vinyl monomers is as under:

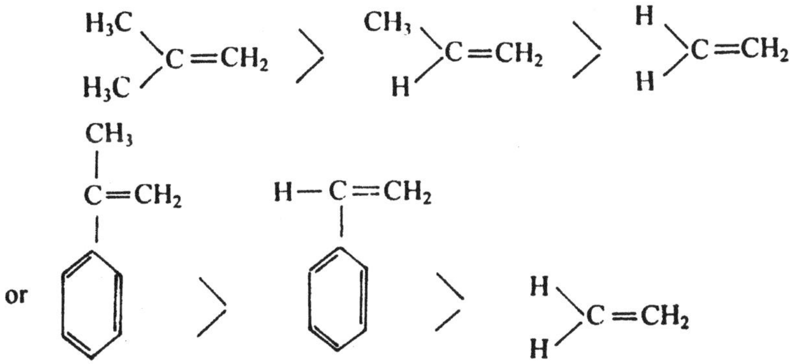

$$(H_3C)(H_3C)C{=}CH_2 > (CH_3)(H)C{=}CH_2 > (H)(H)C{=}CH_2$$

or

$$C_6H_5-\underset{CH_3}{C}{=}CH_2 > C_6H_5-\underset{H}{C}{=}CH_2 > (H)(H)C{=}CH_2$$

Typical monomers which polymerize through cationic mechanisms are isobutene, styrene, α-methylstyrene, vinyl ethers and vinyl carbazoles. At present, about 100 vinyl monomers are known that can be polymerised by the cationic initiators.

5.4. Catalyst

Both the monomers and the catalyst influence the mechanism of cationic polymerisation. Electrophilic substances which under certain conditions are capable of inducing cationic polymerisation can be classified as under:

(i) Protonic acids; HCl, H_2SO_4, Cl_3C COOH, $HClO_4$ etc.
(ii) Aprotonic acids; BF_3, AlC_{13}, $TiCl_4$, $SnBr_4$, $SbCl_3$, $SnCl_4$, $ZnCl_2$, $BiCl_3$ etc., with co-initiators such as H_2O, organic acids etc.
(iii) Carbonium salts; $Al(Et)_3$, $Al(Et_2)Cl$ or alongwith alkyl or aryl chlorides or mineral acid co-initiators.
(iv) Cationogenic substances; t-B_4ClO_4, I_2, Pb_3CCl, ionizing radiations.
(v) High energy radiation.

These catalysts and their reactivity in cationic polymerisation will now be taken up for discussion.

(i) Protonic Acids

Hydrochloric acid at 80°C. Does not react with propene but adds to isobutene. Hydrobromic acid is less reactive than hydrochloric acid. Anhydrous perchloric acid, whose anion can not add to an olefin reacts with cis-butene-2 to produce a complex having two to three molecules of olefin in every molecule of the acid. Sulphuric acid cannot induce the low-temperature polymerisation of even the most reactive monomer, *i.e.*, isobutene, although at higher temperatures it initiates polymerisation.

Hydroionic acid does not induce polymerisation even in polar solvents. In ethylene dichloride at 30°C, only the addition of HI to styrene can take place.

Styrene derivatives show a higher tendency to polymerize under the action of protonic acid in comparison to aliphatic olefins. α-methyl styrene polymerizes in ethyl chloride when catalysed by sulphuric acid at -78.5°C but isobutene does not polymerise.

The overall initiation by HCl follows the following reaction pathways.

$$HCl \dashrightarrow \overset{+}{H} + \overline{C}l$$

Initiation:

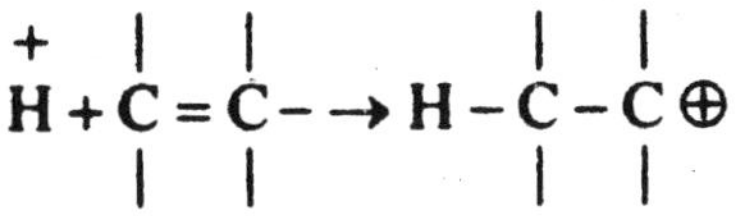

$$\overset{+}{H} + C = C \dashrightarrow H - C - C\oplus$$

Propagation:

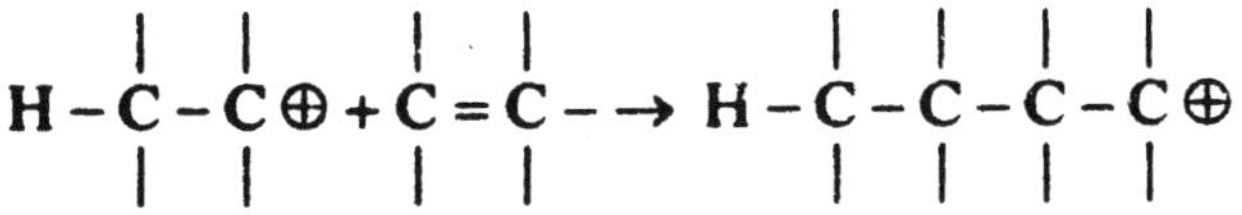

$$H - C - C\oplus + C = C \dashrightarrow H - C - C - C - C\oplus$$

Termination:

$$H - C - C - C - C\oplus + \overset{\ominus}{Cl} \dashrightarrow H - C - C - C - C - Cl$$

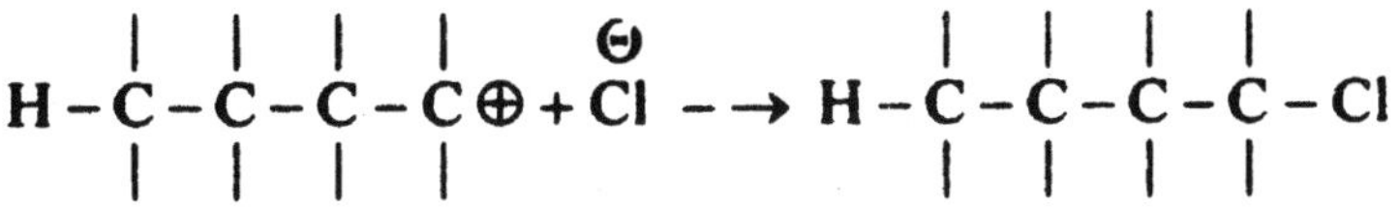

(ii) Aprotonic Acids

The aprotonic acid catalysts are made up of the halides of the metal of Group III and IV of the periodic table. Catalytic activity is also shown by halides of As, Sb, S, Se and many other metals.

The cationic polymerisation induced by aprotonic acids occurs only when a catalyst is present. For example, the cationic polymerisation of isobutylene initiated BF_3 complex at -100°C in liquid ethylene. It may be represented as follows:

Chain Initiation:

$$BF_3 + HOH \rightleftharpoons H[BF_3OH] \rightleftharpoons [BF_3OH]^{\ominus} \ldots H\oplus$$

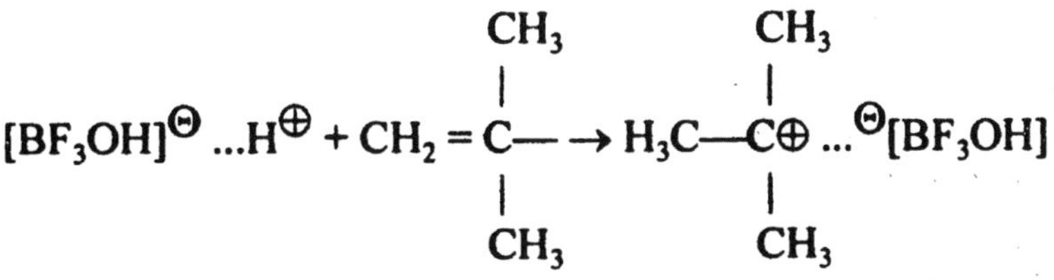

$$[BF_3OH]^{\ominus} \ldots H^{\oplus} + CH_2 = \underset{CH_3}{\overset{CH_3}{C}} \dashrightarrow H_3C - \underset{CH_3}{\overset{CH_3}{C}}\oplus \ldots {}^{\ominus}[BF_3OH]$$

Chain Propagation:

$$H_3C-\overset{\displaystyle CH_3}{\underset{\displaystyle CH_3}{\overset{|}{\underset{|}{C}}}}\oplus \ldots\ldots {}^{\ominus}[BF_3OH] + H_2C=\overset{\displaystyle CH_3}{\underset{\displaystyle CH_3}{\overset{|}{\underset{|}{C}}}}- \longrightarrow$$

$$H_3C-\overset{\displaystyle CH_3}{\underset{\displaystyle CH_3}{\overset{|}{\underset{|}{C}}}}-CH_2-\overset{\displaystyle CH_3}{\underset{\displaystyle CH_3}{\overset{|}{\underset{|}{C}}}} + {}^{\ominus}[BF_3OH] \xrightarrow{\text{etc.}}$$

Chain Transfer:

$$-CH_2-\overset{\displaystyle CH_3}{\underset{\displaystyle CH_3}{\overset{|\oplus\ominus}{\underset{|}{C}}}}\ldots\ldots[BF_3OH] + CH_2=\overset{\displaystyle CH_3}{\underset{\displaystyle CH_3}{\overset{|}{\underset{|}{C}}}}- \longrightarrow$$

$$\sim CH_2-\overset{\displaystyle CH_2}{\underset{\displaystyle CH_3}{\overset{\|}{\underset{|}{C}}}} + H_3C-\overset{\displaystyle CH_3}{\underset{\displaystyle CH_3}{\overset{|\ \ominus}{\underset{|}{C}}}} + [BF_2OH] \xrightarrow[\text{propagation}]{\text{chain}}$$

$$\sim CH_2-\overset{\displaystyle CH_3}{\underset{\displaystyle CH_3}{\overset{|}{\underset{|}{C}}}}\oplus \ldots\ldots \overset{\ominus}{[BF_3OH]} + CH_2=\overset{\displaystyle CH_3}{\underset{\displaystyle CH_3}{\overset{|}{\underset{|}{C}}}}- \longrightarrow$$

$$\sim CH=\overset{\displaystyle CH_3}{\underset{\displaystyle CH_3}{\overset{|}{\underset{|}{C}}}} + HC_3-\overset{\displaystyle CH_3}{\underset{\displaystyle CH_3}{\overset{|}{\underset{|}{C}}}}\oplus \ldots\ldots \overset{\ominus}{[BF_3OH]} \xrightarrow[\text{propagation}]{\text{chain}}$$

$$\sim CH_2-\overset{\overset{\displaystyle CH_3}{|}}{\underset{\underset{\displaystyle CH_3}{|}}{C}}\oplus \ldots \overset{\ominus}{[FB_3OH]} \longrightarrow \sim CH_2-\overset{\overset{\displaystyle CH_3}{\|}}{\underset{\underset{\displaystyle CH_3}{|}}{C}} + HB[F_3OH] \rightleftharpoons$$

$$H\oplus \ldots {}^{\oplus}[BF_3OH] \xrightarrow[\text{propagation}]{\text{chain}} \sim CH_2-\overset{\overset{\displaystyle CH_3}{|}}{\underset{\underset{\displaystyle CH_3}{|}}{C}}\oplus \ldots {}^{\ominus}[BF_3OH] + HOH$$

$$\longrightarrow \sim CH_2-\overset{\overset{\displaystyle CH_3}{|}}{\underset{\underset{\displaystyle CH_3}{|}}{C}}-OH + H^{\oplus} \ldots {}^{\ominus}[BF_3OH] \xrightarrow[\text{propagation}]{\text{chain}}$$

Chain Termination:

$$\sim H_2C-\overset{\overset{\displaystyle CH_3}{|}}{\underset{\underset{\displaystyle CH_3}{|}}{C}}{}^{\oplus} \ldots {}^{\oplus}[BF_3OH] + NH_3 \longrightarrow$$

$$\sim CH_2C-\overset{\overset{\displaystyle CH_3}{|}}{\underset{\underset{\displaystyle CH_3}{|}}{C}}-OH + H[BF_3NH_2]$$

It is quite reasonable to assume that all polar substances which, under the given conditions, and after combination with the aprotonic acid, can form an initiating ion, remain bound to the initiated end of the macromolecule.

It has been found in some cases that the rate of the reaction can be enhanced by substances which do not participate in the initiating process but influence in some way the dissociation of the active complex. Fragments of such molecules however, cannot be incorporated in the initiated end of the chain molecule.

In such a case the polymerisation stops before completion which indicates that side reactions cause a rapid consumption of the catalyst. The solvent ability and the dielectric constant of the solvent also play a major role in determining the course of such a polymerisation reaction.

(iii) Carbonium Salts

This class is made up of a number of catalyst-cocatalyst system. The first step is the ionization of the catalyst and the second step is the reaction of proton with the olefins.

For example in case of triethyl aluminium and alkyl or aryl halide the formation of the carbonium ion occurs as under:

$$H_3C-\overset{\overset{\large CH_3}{|}}{\underset{\underset{\large CH_3}{|}}{C}}-Cl + Al(C_2H_5)_3 \longrightarrow H_3C-\overset{\overset{\large CH_3}{|}}{\underset{\underset{\large CH_3}{|}}{C}}\oplus\,\overline{[Al(C'H_5)_3\,Cl]}$$

The carbonium ion generated in this way can add on to monomer molecules forming the polymer.

5.5. Mechanism of Cationic Polymerisation

Mechanism

Various mechanisms proposed for the cationic polymerisation are discussed one by one.

Whitmore Mechanism

Whitmore proposed a mechanism of proton initiated polymerisation in 1932. It can be represented as under:

I. Initiation:

$$\begin{matrix} R_1 \\ R_2 \end{matrix}\!\!>C=CH_2 + HB \longrightarrow \oplus\overset{\overset{\large R_1}{|}}{\underset{\underset{\large R_2}{|}}{C}}-CH_3 = \overset{\ominus}{B}$$

Growth:

$$H_3C-\underset{R_2}{\overset{R_1}{\underset{|}{\overset{|}{C}}}}\oplus + H_2C = C\begin{matrix} \diagup R_1 \\ \diagdown R_2 \end{matrix} \dashrightarrow H_2\underset{R_2}{\overset{R_1}{\underset{|}{\overset{|}{C}}}}-CH_2-\underset{R_2}{\overset{R_1}{\underset{|}{\overset{|}{C}}}}\oplus$$

II. Recovery of acid:

$$\sim\underset{R_2}{\overset{R_1}{\underset{|}{\overset{|}{C}}}}-CH_2-\underset{R_2}{\overset{R_1}{\underset{|}{\overset{|}{C}}}}\oplus + \overset{\ominus}{B} \;\Bigg|\; \begin{matrix} \rightarrow -\underset{R_2}{\overset{R_1}{\underset{|}{\overset{|}{C}}}}-CH_2-C\begin{matrix} \diagup R_1 \\ \diagdown R_2 \end{matrix} + HB \\ \rightarrow -\underset{R_2}{\overset{R_1}{\underset{|}{\overset{|}{C}}}}-CH=C\begin{matrix} \diagup R_1 \\ \diagdown R_2 \end{matrix} + HB \end{matrix}$$

III. Termination:

$$H_3C-\underset{R_2}{\overset{R_1}{\underset{|}{\overset{|}{C}}}}\oplus + \overset{\ominus}{B} \dashrightarrow H_3C-\underset{R_2}{\overset{R_1}{\underset{|}{\overset{|}{C}}}}-B$$

Reaction (I) is a bimolecular reaction involving a protonic acid on the olefin. The proton attacks the monomer and adds to a carbon atom that has the maximum electron density.

The carbonium ion so produced then adds more monomer molecules and grows until there is abstraction of a proton from the growing chain end of the macromolecule and its recombination with conjugated base $\overset{\ominus}{B}$. This process in case of the Whitmore mechanism, appears to be improbable because $\overset{\ominus}{ClO_4}$ is incapble of forming a new chemical link with carbon of the macrocarbonium ion, similarly neither $B\overset{\ominus}{F_4}$ nor $Al\overset{\ominus}{Cl_4}$ can form a new chemical link with carbon of the macrocarbonium ion.

Hunter and Yohe Mechanism

In 1930, Hunter and Yohe proposed the following mechanism for the cationic polymerisation. They assumed the formation of a complex between the Lewis acid and the monomer.

Initiation:

$$AlCl_3 + \begin{matrix} R_1 \\ R_2 \end{matrix}\!\!>C=CH_2 \rightarrow \begin{matrix} Cl \\ Cl \\ Cl \end{matrix}\!\!>\overset{\ominus}{Al} - CH_2 - \overset{\overset{\large R_1}{|}}{\underset{\underset{\large R_2}{|}}{C}}\oplus$$

Growth:

$$\overset{\ominus}{AlCl_3} - CH_2 - \overset{\overset{\large R_1}{|}}{\underset{\underset{\large R_2}{|}}{C}} \div + \begin{matrix} R_1 \\ R_2 \end{matrix}\!\!>C=CH_2 \rightarrow$$

$$\overset{\ominus}{AlCl_3} - CH_2 - \overset{\overset{\large R_1}{|}}{\underset{\underset{\large R_2}{|}}{C}} - CH_2 - \overset{\overset{\large R_1}{|}}{\underset{\underset{\large R_2}{|}}{C}}\oplus$$

The above scheme can not explain some experimental facts such as effect of cocatalysis, chain transfer termination etc.

To overcome these defects another modified mechanism was proposed by **Ciskowski,** in 1938, for a $SnCl_4$—catalysed polymerisation of styrene.

Complex Formation:

$$SnCl_4 + CH_2 = CHPh \rightarrow Cl_4\overset{\ominus}{Sn} - CH_2 - \overset{\oplus}{C}HPh$$

Growth:

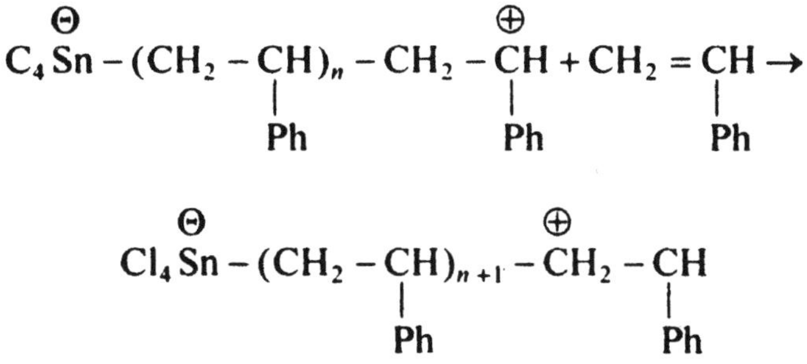

$$\overset{\ominus}{C_4}Sn-(CH_2-\underset{\underset{Ph}{|}}{CH})_n-CH_2-\overset{\oplus}{\underset{\underset{Ph}{|}}{CH}}+CH_2=\underset{\underset{Ph}{|}}{CH}\rightarrow$$

$$\overset{\ominus}{Cl_4}Sn-(CH_2-\underset{\underset{Ph}{|}}{CH})_{n+1}-CH_2-\overset{\oplus}{\underset{\underset{Ph}{|}}{CH}}$$

Termination:

$$\overset{\ominus}{Cl_4}Sn-(CH_2-\underset{\underset{Ph}{|}}{CH})_2-CH_2-\overset{\oplus}{\underset{\underset{Ph}{|}}{CH}}\rightarrow Cl_3Sn-(CH_2-\underset{\underset{Ph}{|}}{CH})_x-CH=\underset{\underset{Ph}{|}}{CH}+HCl$$

$$\rightarrow Cl_3Sn-(CH_2-\underset{\underset{Ph}{|}}{CH})_x-CH_2-\underset{\underset{Ph}{|}}{C}\ HCl$$

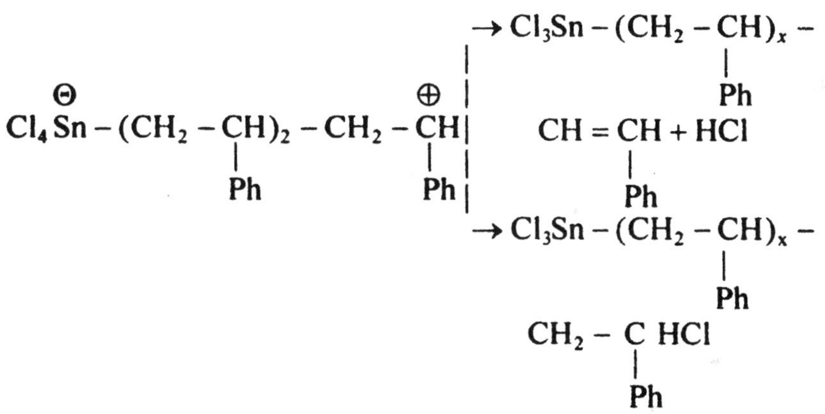

In this reaction scheme it is assumed that the splitting of a chloride anion occurs forming the negatively charged macrocation which is then transfered to the positively charged chain end, that is separated by a long hydrocarbon chain.

The above reaction scheme could not explain the marked effect of water on the reaction rate. Another drawback of the above reaction scheme is that if each monomer addition were followed by the shift of a hydride ion and isomerisation, the two charges should always remain in the vicinity and it should allow the elimination of HCl or the addition of chloride ion to proceed easily otherwise the reaction scheme will not be feasible.

Korshak and Lebedev Mechanism

This mechanism proposes a direct interaction of a Lewis acid with the double bond. It can be represented as under:

Initiation:

$$\underset{\substack{\vdots\ \delta- \\ Cl_4Al}}{\overset{\delta+}{Cl_2Al}} + H_2C{=}C\begin{matrix} R_1 \\ R_2 \end{matrix} \rightarrow \begin{matrix} \overset{\delta\oplus}{Cl_2Al} & \cdots\cdots & \overset{\delta\ominus}{CH_2} & \\ \vdots & & | & \\ \underset{\delta\ominus}{Cl_4Al} & \cdots\cdots & \underset{\delta\oplus}{C} & \begin{matrix} R_1 \\ R_2 \end{matrix} \end{matrix}$$

Growth:

$$\begin{matrix} & & R_1 & & \\ \delta\ominus & & |\ \delta+ & & \\ CH_2 & - & C & - & R_2 \\ \vdots\ \delta+ & & \vdots\ \delta\oplus & & \\ Cl_2Al & \cdots\cdots & Cl_4Al & & \end{matrix} + H_2C{=}C\begin{matrix} R_1 & R_2 \end{matrix} \rightarrow$$

$$\begin{matrix} & R_1\ R_2 & & \delta\odot & R_1 & \\ & \diagdown\diagup & & & |\ \delta\oplus & \\ H_3C - & C & --- & CH & - C & --- R_2 \\ & & & \vdots\ \delta\oplus & \vdots\ \delta\ominus & \\ & & & Cl_2Al & \cdots\cdots\ Cl_4Al & \end{matrix}$$

Chain Transfer:

$$\begin{matrix} & & R_1 & \\ \delta\ominus & & |\ \delta\oplus & \\ \sim\sim CH & --- & C & --- R_2 \\ \vdots\ \delta\oplus & & \vdots\ \delta\ominus & \\ Cl_2Al & \cdots\cdots & Cl_4Al & \end{matrix} + \overset{R_1\ \ R_2}{C}{=}CH_2 \rightarrow$$

$$\sim\sim CH{=}\overset{R_1\ \ R_2}{C} + \begin{matrix} R_1\ R_2 & & \delta\ominus \\ C & -- & CH_2 \\ \vdots\ \delta\ominus & & \vdots\ \delta+ \\ Cl_4Al & -- & Cl_2Al \end{matrix}$$

Catalyst recovery:

$$\begin{matrix} & & R_1 & \\ \delta\ominus & & |\ \delta\oplus & \\ \sim\sim CH & \cdots\cdots & C & --- R_2 \\ \vdots\ \delta\oplus & & \vdots\ \delta\ominus & \\ Cl_2Al & \cdots\cdots & AlCl_4 & \end{matrix} \longrightarrow \sim CH{=}\overset{R_1\ \ R_2}{C} + Al_2Cl$$

This mechanism does not require the presence of a cocatlyst and also fails to explain its function.

Chmelir Mechanism

In 1965 Chmelir proposed another mechanism for the polymerisation of isobutene catalysed by $AlBr_3$ in heptane solution. If the reaction system is dried it is found that upon addition of BF_3 or $TiCl_4$, no polymerisation takes place and only addition of water can bring about polymerisation.

However, in case of $AlBr_3$, polymerisation was obtained even without addition of water. The mechanism is illustrated below. Formation of ion pairs :

$$2AlBr_3 \rightleftharpoons \overset{\oplus}{Al}Br_2.\overset{\ominus}{Al}Br_4$$

(i) Initiation:

$$\overset{\oplus}{Al}Br_2.\overset{\ominus}{Al}Br_4 + \begin{matrix} R_1 \\ R_2 \end{matrix}\!\!>\!C=CH_2 \rightarrow$$

$$AlBr_2-CH_2-\underset{R_2}{\overset{R_1}{\underset{|}{\overset{|}{C}}}}\oplus\overset{\ominus}{Al}Br_4$$

(ii) Growth:

$$AlBr_2-(CH_2-\underset{R_2}{\overset{R_1}{\underset{|}{\overset{|}{C}}}}-)_n-CH_2-\underset{R_2}{\overset{R_1}{\underset{|}{\overset{|}{C}}}}\oplus\overset{\ominus}{Al}Br_4 + \begin{matrix} R_1 \\ R_2 \end{matrix}\!\!>\!C=CH_2$$

$$\rightarrow AlBr_2-(CH_2-\underset{R_2}{\overset{R_1}{\underset{|_{n+1}}{\overset{|}{C}}}}-)_{n+1}-CH_2-\underset{R_2}{\overset{R_1}{\underset{|}{\overset{|}{C}}}}\oplus\overset{\ominus}{Al}Br_4$$

(iii) Termination:

$$AlBr_2-(CH_2-\underset{R_2}{\overset{R_1}{\underset{|}{\overset{|}{C}}}}-)_n-CH_2-\underset{R_2}{\overset{R_1}{\underset{|}{\overset{|}{C}}}}\oplus \overset{\ominus}{AlBr_4}$$

$$\rightarrow AlBr_2-(CH_2-\underset{R_2}{\overset{R_1}{\underset{|}{\overset{|}{C}}}}-)-CH_2-C\begin{matrix}\diagup R_1\\ \diagdown R_2\end{matrix} + \overset{\ominus}{H}\ \overset{\ominus}{AlBr_4}$$

The complex protonic acid libcrated by reaction (ii) can initiate another reaction chain, so that no true termination is involved. The cahin-transfer with the monomer takes place as follows:

Chain Transfer

$$AlBr_2-(CH_2-\underset{R_2}{\overset{R_1}{\underset{|}{\overset{|}{C}}}}-)_n-CH_2-\underset{R_2}{\overset{R_1}{\underset{|}{\overset{|}{C}}}}\oplus \overset{\ominus}{AlBr_4}+H_2C=C\begin{matrix}\diagup R_1\\ \diagdown R_2\end{matrix}\rightarrow$$

$$\sim CH_2-C\begin{matrix}\diagup R_1\\ \diagdown R_2\end{matrix}\quad +CH_3-\underset{R_2}{\overset{R_1}{\underset{|}{\overset{|}{C}}}}\oplus \overset{\ominus}{AlBr_4}$$

5.6. Kinetics of Cationic Polymerisation

The mechanism of cationic polymerisation gets highly influenced by changes in the concentration of the reaction components as also in the composition of the solvent and in reaction conditions.

A further complication may some times arise due to the high reaction rate and difficult reaction conditions, *i.e.*, low temperatures which makes it impossible to measure the reaction rates accurately.

A simple kinetic scheme which involves the initiation propagation and termination steps may be represented as under:

Initiation:

$$AX \xrightarrow{k_i} \overset{+}{A}\overset{-}{X}$$

$$M + \overset{+}{A}\overset{-}{X} \xrightarrow{k_{ii}} \overset{+}{A}M_1\overset{-}{X} \qquad ...(5.1)$$

Propagation:

$$A\overset{+}{M}_n\overset{-}{X} + \overset{+}{M} \xrightarrow{k_p} A\overset{+}{M}_{n+1}\overset{-}{X} \qquad ...(5.2)$$

Termination:

$$A\overset{+}{M}_{n+1}\overset{-}{X} \xrightarrow{k_t} \text{Polymer} \qquad ...(5.3)$$

Transfer to monomer:

$$A\overset{+}{M}_{n+1}\overset{-}{X} + M \xrightarrow{k_{tr}} \text{Polymer} \qquad ...(5.4)$$

The rate of initiation;

$$V_i = k'_i [C][M] \qquad ...(5.5)$$

where [C] is the catalyst concentration. If we consider the ionisation of the catalyst as the slowest step then the rate of initiation is given by,

$$V_i = k_i [C] \qquad ...(5.6)$$

The rate of termination

$$V_t = k_i [M^+] \qquad ...(5.7)$$

Considering steady-state principle,

$$k'_i [C][M] = k_i [M^+]$$

or

$$[M^+] = (k'_i / k_t)[C][M^+] \qquad ...(5.8)$$

The rate of polymerisation

$$R_p = k_p[M^+][M] = (k_p k'_t / k_i)[C][M]^2 \qquad ...(5.9)$$

The kinetic chain length is given by

$$n = \frac{\text{rate of propagation}}{\text{rate of formartion of dead chains}}$$

$$= \frac{k_p[M^+][M]}{k_i[M^+] + k_{tr}[M^+][M]} \quad ...(5.10)$$

In case termination predominates over transfer, then

$$n = \frac{k_p[M^+][M]}{k_t[M^+]} = \frac{k_p}{k_t}[M] \quad ...(5.11)$$

In case transfer reaction predominates, then

$$n = \frac{k_p[M^+][M]}{k_{tr}[M^+][M]} = \frac{k_p}{k_{tr}} \quad ...(5.12)$$

The average degree of polymerisation, $\overline{DP}$, ($\overline{DP} = n$, for cationic polymerisation) is found to be independent of monomer and initiator concentrations.

In some examples of cationic polymerisation, the degree and the rate of polymerisation increase as the temperature decreases. This can be explained as follows.

The Arrhenius equation for k_i, k_p, k_t are given by,

$$k_i = A_i e^{E_i/RT}, k_p = A_p e^{-E_p/RT}$$

and $$k_i = A_i e^{-E_i/RT.}$$

As in case of radical polymerisation, the overall rate constant, k_R, for the polymerisation reaction is given by

$$k_R - \frac{A_r A_p}{A_t} = e^{E_t - E_i - E_p} \quad ...(5.13)$$

It is found that the propagation reaction has a rather low activation energy causing E_p to be much lower than E_i, E_t. Therefore, the overall activation energy terms in equation (5.13) would be positive for which k_R increases with decrease of the temperature.

5.7. Anionic polymerisation

Anionic polymerisation was known for many years before the nature of the polymerisation was predicted. The production of buna type synthetic rubbers in Germany and Russia by the polymerisation of butadiene with the sodium or potassium as the catalyst was known.

In anionic polymerisation π-electrons of the monomers are attached by negatively charged ion, an **anion,** so the anionic polymerisation is initiated by a negative ion or anion. An anion generates a carbanion on reaction with a monomer and carbonion is an active centre which propagates the reaction.

$$R-CH_2-\underset{X}{\underset{|}{\overset{\ominus}{C}H}} + CH_2 \div \underset{X}{\underset{|}{C}} \rightarrow R-CH_2-\underset{X}{\underset{|}{CH}}-CH_2-\underset{X}{\underset{|}{\overset{\ominus}{C}H}}$$

$$R-CH_2-\underset{X}{\underset{|}{CH}} + CH_2-\underset{X}{\underset{|}{\overset{\ominus}{C}H}} + CH_2 \div \underset{X}{\underset{|}{CH}} \rightarrow$$

$$R-CH_2-\underset{X}{\underset{|}{CH}}-CH_2-\underset{X}{\underset{|}{CH}}-CH_2-\underset{X}{\underset{|}{\overset{\ominus}{C}H}}$$

Or Simply

$$R-CH_2-\underset{X}{\underset{|}{\overset{\ominus}{C}H}} + nCH_2=\underset{X}{\underset{|}{\overset{\ominus}{C}H}} \rightarrow R[CH_2-\underset{X}{\underset{|}{CH}}]_n\, CH_2-\underset{X}{\underset{|}{\overset{\ominus}{C}H}}$$

Highly electropositive metals and their compounds are used as initiators *e.g.*,—Li, BuLi, n-BuMg Br etc., where Bu represents a butyl group (C_4H_9). Anionic reactions are very rapid and hence are usually carried at low temperatures (–100 to —70°C). Reactions are greatly influenced by solvating power of the solvents and usually ethers and amines are used for the purpose.

Polymerisation are very similar in nature except in their termination reactions. Terminations are very easy in case of cationic polymerisation whereas it is absent in case of anionic polymerisation. That is why anionic polymers are called lthe 'living polymers'.

We have only two elementary reaction paths in case of anionic polymerisation; (i) initiation and (ii) propagation.

(i) Initiation

Anionic polymerisation initiated by electron transfer

The synthesis of the exact molecular structure of natural rubber using a simple alkali metal focussed increase attention on the mechanism of anionic polymerisations.

It became apparent that the mechanism of anionic polymerisation induced by the alkali metals should also involve initiation by electron transfer from the metal to the monomer. For example, in the Lithium—initiated polymerisation of butadiene, we have,

$$\text{(I)}\quad L_i + CH_2 = CH - CH = CH_2 \dashrightarrow \dot{C}H_2 - CH = CH - \ddot{C}\bar{H}_2\overset{+}{L}_i$$

$$\text{(II)}\quad 2\dot{C}H_2 - CH = CH - \ddot{C}\bar{H}_2\overset{+}{L}_i \dashrightarrow \overset{+}{L}_i\bar{C}\ddot{H}_2 - CH = CH - CH_2 - CH_2 - CH = CH - \ddot{C}\bar{H}_2\overset{+}{L}_i$$

$$\text{(III) and/ or}\quad \dot{C}H_2 - CH = CH - \ddot{C}\bar{H}_2\overset{+}{L}_i + L_i \dashrightarrow \overset{+}{L}_i\ddot{C}\bar{H}_2 - CH = CH - \ddot{C}\bar{H}_2\overset{+}{L}_i$$

Reaction (I) involves the initial electron transfer from the metal to the monomer, which leads to the formation of a radical anion which could then participate in the reactions shown in (II) and (III) *i.e.*, by coupling of two radical ions or by transfer of another electron from the metal, respectively, both processes leading to a dianionic species.

Which of these reactions will be favoured depends on the prevalent conditions, *i.e.*, the concentration and the reactivity of the alkali metal.

Some alkali metal complexes are also capable of initiating anionic polymerisation. For example, sodium and naphthalene in tetrahydrofuran is a homogeneous solution initiates polymerisation as follows:

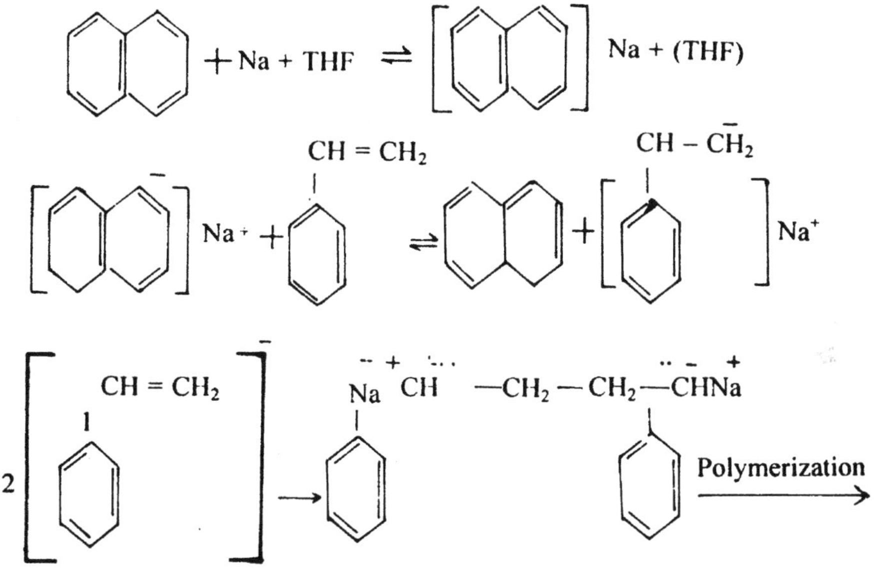

+ Na + THF ⇌ [ ] Na + (THF)

CH = CH₂ CH – C̄H₂

[ ]⁻ Na⁺ + ⇌ + [ ] Na⁺

2 [CH = CH₂]⁻ → Na CH —CH₂ — CH₂—CHNa $\xrightarrow{\text{Polymerization}}$

This mechanism is based on initiation by electron-transfer which leads to a styrene radical anion, which couples rapidly due to its high concentration, and forms a dimeric styrene dianion that is capable of further propagation by anionic attack on styrene monomer.

The unique features of this polymerisation are:

(i) the extreme rapidity of the initiation step. That is evidenced by the spontaneous colour change from greenish blue to the deep red of styrene anion.

(ii) the absence of any termination or, transfer, processes due to which all the chains continue growing until complete depletion of the monomer.

(iii) the resulting distribution of chain, lengths because of simultaneous growth of all the chains. Thus, the phenomenon of a 'living polymer', *i.e.*, the absence of any termination of transfer reaction was discovered in these homogeneous systems, because it was possible to study their stoichiometry.

Initiation by organometallic compounds

Some organo metallic compounds and Lewis bases can also act as initiators. In such cases initiation occurs by a direct attack of these compounds on the double bond of the monomer molecule.

However before the Lewis base can attack the monomer, it must ionise and only then a carbanion can be produced. The process of initiation is shown below:

$$BG \dashrightarrow B + \overset{+}{G}$$

$$B + H_2C = CHR \dashrightarrow BCH_2 - \underset{H}{\overset{R}{\overset{|}{\underset{|}{C}}}} \ominus$$

$$\overline{B}CH_2 - \underset{H}{\overset{R}{\overset{|}{\underset{|}{C}}}} \ominus + G^{\oplus} \dashrightarrow BCH_2 - \underset{H}{\overset{R}{\overset{|}{\underset{|}{C}}}}^{\ominus} \ldots G^{\oplus}$$

where BG is a Lewis acid and $G^{\oplus}$ is the ion that must remain near the carbonion produced in the initiation process.

Termination

The termination in anionic polymerisation is, generally not a spontaneous process, and unless some impurities are present or some strongly ionic substances are added deliberately, no termination takes place. Thus, if polymerisation is done under controlled conditions and impurities are avoided, the reaction proceeds till whole of the monomer is consumed. In this case, while there is no more monomer left to be polymerised, the carbonions as the chain ends remain potentially active. Which means that if a fresh quantity of monomer is added the polymerisation again goes on until all the freshly added monomer is consumed. It has been confirmed that polymerisation can be restarted in this way, even after weeks, by adding fresh monomer. The polymers are 'alive' and keep on growing, as long as we supply a fresh quantity of the monomer, just as many of the living organism grow as long as food is available. The 'living polymerisation technique' is useful for many applications. Block copolymers, for example, are prepared by using this technique.

5.8. Kinetic of Anionic Polymerisation

It is possible to illustrate the kinetics of anionic polymerisation by the polymerisation of styrene with potassium amide in liquid ammonia. The first step involves the bread-down of initiato:s into ions.

$$KNH_2 \dashrightarrow \overset{\oplus}{K} + \overset{\ominus}{NH_2}$$

Initiation:

In the initiation step, there occurs the addition of the amide ion, $\overset{-}{NH_2}$ to the monomer molecule yielding the active centre.

$$\overset{-}{NH_2} + M \overset{k_i}{\dashrightarrow} NH_2 - \overset{\ominus}{M} \qquad ...(5.14)$$

Propagation:

In this step there occurs the addition of the carbanion to the monomer molecule yielding the growing chain carrying a negative charge.

$$NH_2\overset{\ominus}{M_n} + M \overset{k_p}{\dashrightarrow} NH_2 - M_{n+1}{}^{\ominus} \qquad ...(5.15)$$

Termination:

In the termination there step occurs the transfer of a proton from a solvent molecule.

$$NH_2\overset{\ominus}{M_n} + NH_3 \overset{k_{tr}}{\dashrightarrow} NH_2M_nH + NH_2{}^{\ominus} \qquad ...(5.16)$$

The rate of initiation may be put as follows

$$V_i = -\frac{d[M_i]}{dt} = k_i[KNH_2][M] \qquad ...(5.17)$$

The rate of termination is given by

$$V_t = -\frac{d[M_t]}{d_t} = k_{tr}[\overline{M}][NH_3] \qquad ...(5.18)$$

At the steady-state, the rate of initiation is equal to the rate of termination.

Hence, $$k_{tr}\left[\overline{M}\right]\left[NH_3\right] = k_i\left[KNH_2\right]\left[M\right]$$

or, $$\left[\overline{M}\right] = \frac{k_i}{k_{tr}}\frac{\left[KNH_2\right]\left[M\right]}{\left[NH_3\right]} \qquad ...(5.19)$$

The rate of polymerisation,

$$R_p = -\frac{d\left[M\right]}{dt} = k_p\left[M\right]\left[\overline{M}\right]$$

$$= \frac{k_p k_i}{k_{tr}}\frac{\left[KNH_2\right]\left[M\right]^2}{\left[NH_3\right]} \qquad ...(5.20)$$

Hence, R_p is proportional to the square of monomer and first power of initiator concentration.

The degree of polymerisation, $\overline{X}_n$ is given by,

$$\overline{X}n = \frac{\text{rate of polymerisation}}{\text{rate of termination}}$$

From equation (5.18) and (5.20) we get,

$$\overline{X}_n = \frac{k_p\left[M\right]}{k_t\left[NH_3\right]}$$

Hence, the degree of polymerisation depends upon the solvent since the solvent ammonia takes part in the transfer mechanism.

The transfer reaction is highly competitive with the propagation step and the value of k_p/k_t is not very large and the overall reaction produces polymers with low molecular weight.

In case of the anionic polymerisation, since all the chains grow at the same rate, a monodisperse system is obtained in this case.

These days anionic polymers have attracted attention for the synthesis of monodisperse polymers and the specialized types of copolymers.

6

Polymerisation Coordination

6.1. Introduction

Polymerisation reaction, particularly of olefins and dienes catalysed by organometallic compounds fall under the category of coordination polymerisation, or **Ziegler-Natta polymerisation.**

K. Ziegler (Germany) and G.Natta (Italy) discovered this method for ionic polymerisation. Ziegler discovered a method for polymerising ethene (ethylene) which did not need the high temperature and pressure needed for free radical polymerisation. He made use of organometallic catalysts with very high activity. Guilio Natta used such catalysts for polymerisation of α-alkenes like propene and found that the catalysts were stereo directing. The catalysts are made by reacting titanium tetrachloride or trichloride with an alkyl aluminium compound. Ziegler-Natta catalysts bring about polymerisation under mild conditions and allow a considerable degree of control over the structure of the polymer.

6.2. Ziegler-Natta Polymerisation

The three phases that are present in the Ziegler-Natta polymerisation are (i) the monomer (ii) the solvent and (iii) the catalyst. Reactions take place at certain points on the surface of catalyst particles. The polymer molecule grows as the monomer units join the chain where earlier monomer is attached to the catalyst particle. The precise nature of the action of catalyst is not yet known. However, the first step in the polymerisation process proposed is the formation of a monomer-catalyst complex between the organometallic compound and the monomer.

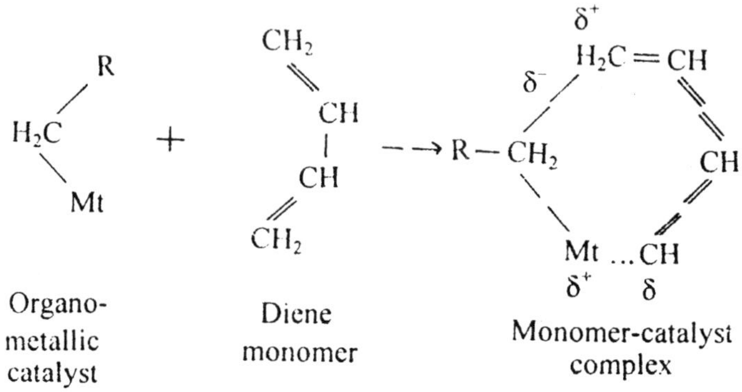

R
H_2C
Mt
+
CH_2
CH
CH
CH_2
⟶ R—CH_2
δ^{-}
δ^{+}
$H_2C=CH$
CH
Mt ...CH
δ^{+} δ

Organo-metallic catalyst

Diene monomer

Monomer-catalyst complex

where Mt represents transition metals such as Ti, Mo, Cr, V, Ni or Rh.

Ziegler and Natta discovered that in the presence of a combination of a transition-metal halide like $TiCl_4$ or $TiCl_3$ with an organometallic compound triethylaluminium (Ziegler-Natta catalyst), stereospecific polymerisation can be done. A coordination complex containing an alkyl group coordinated to Ti was reported to be formed. The olefin forms a π-complex with Ti and the olefin is held at the reaction site. Polymerisation amounts to the insertion of alkene molecules between the Ti---C bond. The polymer chain migrates back to the original position for the cycle to continue.

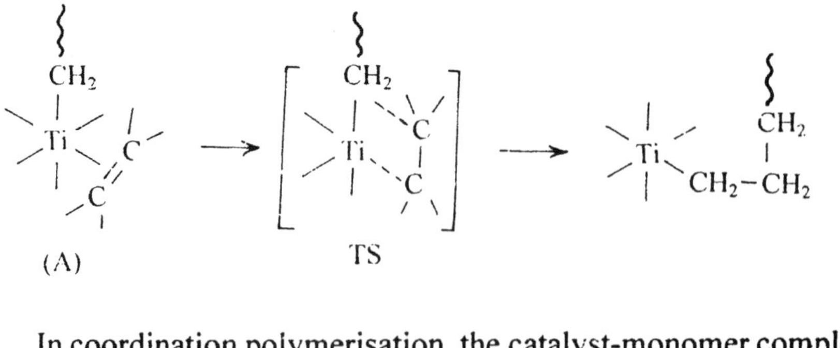

CH_2
Ti
C
C
(A)
⟶
CH_2
Ti
C
C
TS
⟶
Ti
CH_2
CH_2-CH_2

In coordination polymerisation, the catalyst-monomer complex forms a heterogeneous system in which the metal ion is in the solid phase and the carbanion of the alkyl group is in the solvent phase. The monomer is inserted in between the metal ion and the carbanion and the polymer chain formed is pushed out from the solid catalyst surface. Because of this coordination polymerisation is also known as insertion polymerisation.

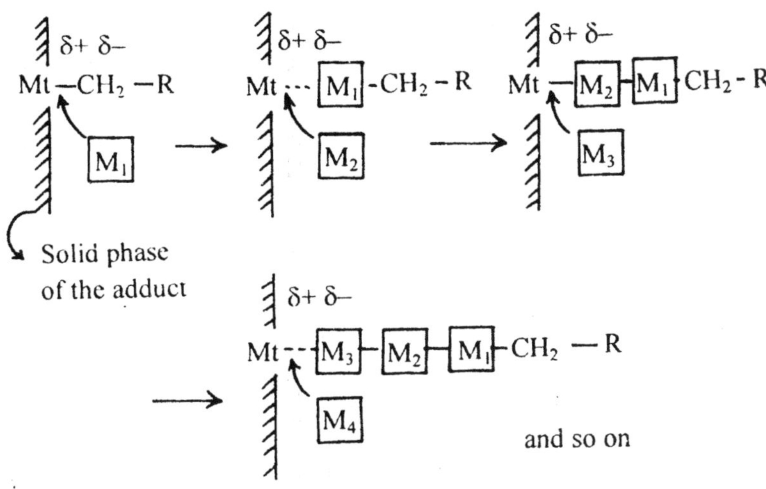

Here M_1, M_2, M_3 etc. are first, second, third etc. monomer units added to the growign chain.

It is proposed that polymerisation occurs at active sites involving transition metal atom having an octahedral configuration with one ligand valency. The monomer is having a π bond with titanium and undergoes insertion between the titanium and alkyl group.

□ = ligand valency

$$\mathrm{Cl{-}\overset{R}{Ti}(Cl)(Cl)(Cl){-}{-}\square} + CH_2 = CH - CH_3 \dashrightarrow \mathrm{Cl{-}\overset{R}{Ti}(Cl)(Cl)(Cl){-}{-}{-}\overset{CH_3}{CH}{=}CH_2}$$

$$\dashrightarrow \mathrm{Cl{-}\overset{\square}{Ti}(Cl)(Cl)(Cl){-}{-}CH_2{-}\overset{CH_3}{CH}{-}R} \xrightarrow{CH_2=CH-CH_3}$$

$$\dashrightarrow \mathrm{Cl{-}Ti(Cl)(Cl)(Cl)(CH_2{-}CH(CH_3){-}CH_2{-}CH(CH_3){-}R){-}{-}{-}\square}$$

Termination occurs on reaction with an active hydrogen compound. High density linear polyethylene is formed by this method in large quantities. This polymerisation is used for nonpolar unsaturated monomers like ethylene, propylene, 1-butene, 3-methyl-1-pentene, Styrene, butadiene, isoprene etc. Copolymerisation of ethylene with propylene, 2-butene or butadiene can also be done by this method.

Some polar monomers such as vinyl chloride, vinyl-acetate and acrylonitrile may be polymerised by this method using an active solvent such as tetrahydrofuran.

Depending on the polarity of the metal-carbon bond and that of the solvent medium, the metal counter ion is placed in a particular spatial arrangement with respect to anion. The specific spatial arrangement of the monomeric units inserted into growing chain imparts stereoregularity to the polymer formed. Because of this Ziegler-Natta Polymerisation can give stereoregular polymers. For monomer such as $CH_2 = CHX$ three different types of head to tail polymers are possible.

(1) Polymers in which all the X-substituted carbon atoms have same configuration. These polymers are called isotactic polymers.

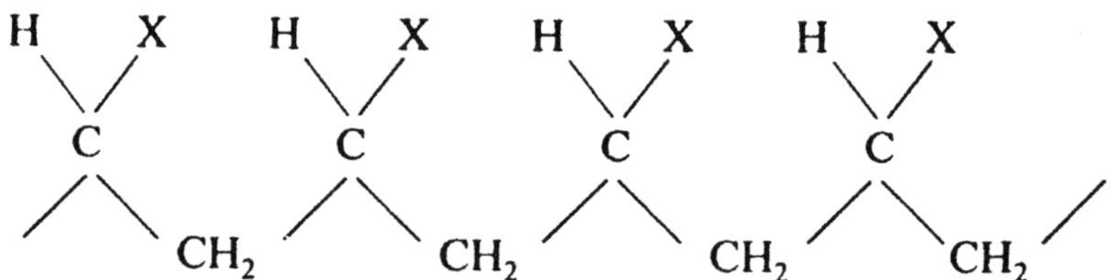

(2) Polymers in which the configuration of X substituted C-atom alternates are known as **Syndiotactic polymers**

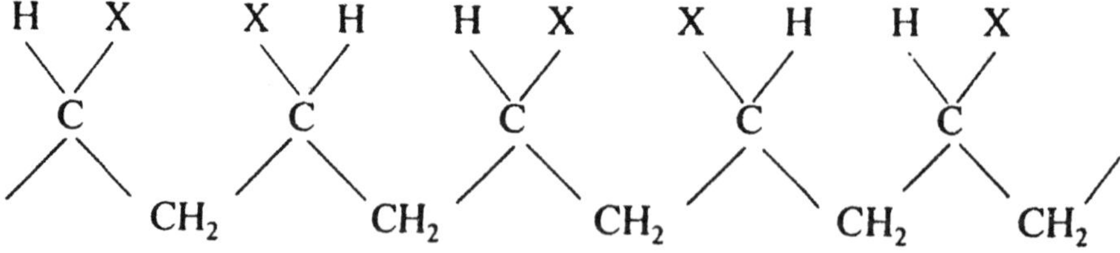

(3) Polymers in which the configuration of the X-substituted carbon atoms are randomly distributed are known as **atactic polymers.**

Free radical polymerisation is largely atactic while

polymerisation with Ziegler-Natta catalysts can result in **isotactic** or **syndiotactic polymers.**

Diene Polymerisation

The reaction of the diene with a free radical produces an allyl radical having unpaired electron delocalised over more than one carbon atom. The allyl free radical can undergo 1, 2, or 1, 4 addition.

$$R + CH_2 = CH - CH = CH_2 \rightarrow R - CH_2 - CH - CH_2$$

$$\rightarrow R - CH_2 - \underset{\underset{CH}{|}}{CH} - CH - \underset{\underset{CH}{|}}{CH} \sim\sim\sim \quad \text{1, 2, structure}$$

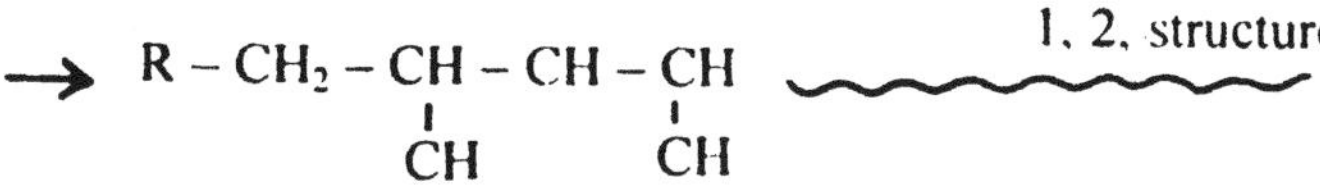

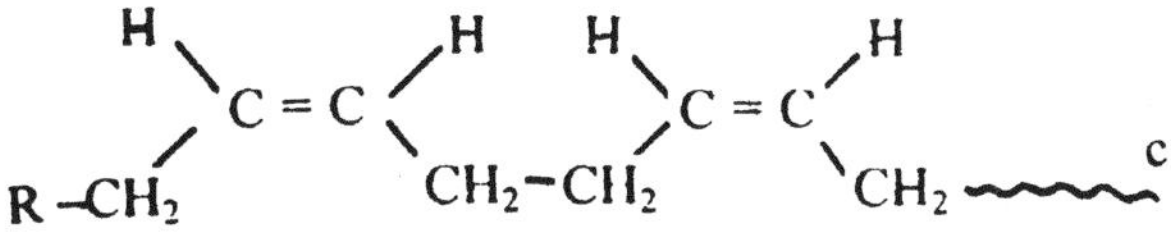

$$R-CH_2-CH{=}CH-CH_2-CH_2-CH{=}CH-CH_2 \sim\sim \quad \text{cis} - \text{1, 4 structure}$$

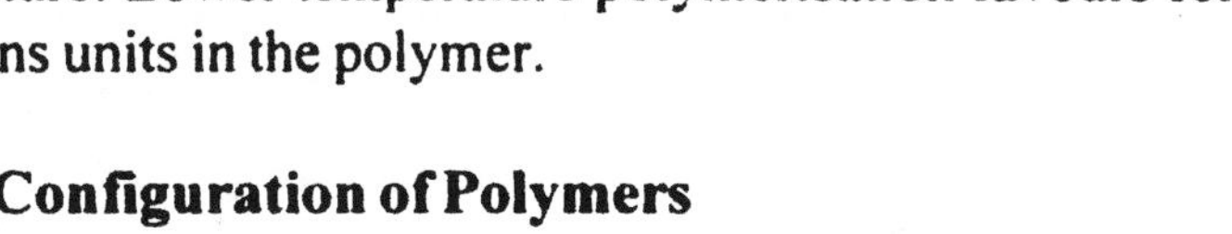

$$R-CH_2-CH{=}CH-CH_2-CH_2-CH{=}CH-CH_2 \sim\sim \quad \text{trans} - \text{1, 4 structure}$$

Free radical polymerisation yields a polymer with 15-20% 1, 2 structure. Lower temperature polymerisation favours formation of 1, 4-trans units in the polymer.

6.3. Configuration of Polymers

In stereo-regular polymers, each monomer segment is in regular configuration, it gives a structural regularity to the polymer molecule as a whole.

The structural regularity in such polymers gives rise to optical (D-L isomerism) and tacticity of polymers and geometrical cross-trans isomerism.

Optical Isomerism of (D-L-Isomerism) and Tacticity of Polymers. Optical isomerism has its origin in the way different substituents occupy positions on an asymmetric carbon atom in a polymer molecule. For example, polyethylene molecule has fully saturated carbon atoms as shown in the following chemical formula:

$$—CH_2—CH_2—CH_2—CH_2—CH_2—CH_2—$$

This formula can be written in a linear structural way, showing the manner in which various atoms are attached as follows:

$$\begin{array}{cccccccccccccc} & H & H & H & H & H & H & H & H & H & H & H & H & \\ & | & | & | & | & | & | & | & | & | & | & | & | & \\ — & C— & C— & C— & C— & C— & C— & C— & C— & C— & C— & C— & C & — \\ & | & | & | & | & | & | & | & | & | & | & | & | & \\ & H & H & H & H & H & H & H & H & H & H & H & H & \end{array}$$

If we assume that the carbon-carbon bonds are all on the plane of this paper and that the bonds between the carbon atom and its substituent atom are either below or above the plane of the paper. Then the structure of polyethylene in a planar zigzag configuration can be represented as shown in Fig. 6.1.

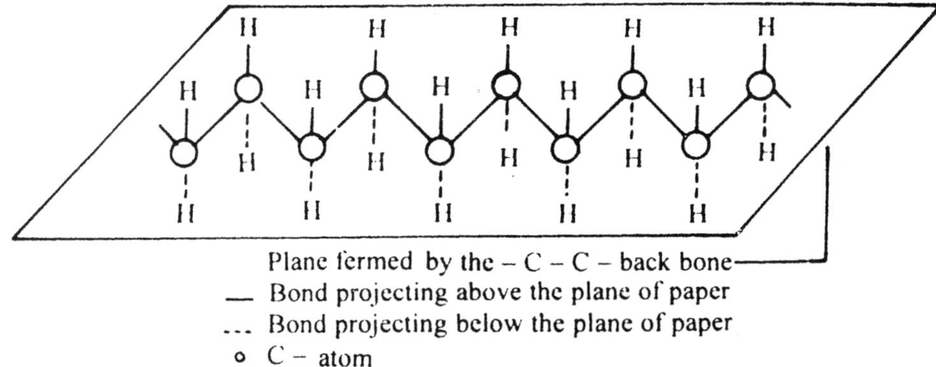

Fig. 6.1: Planar zigzag structure of polyethylene molecule.

It is seen from Fig. 6.1 that all carbon atoms are attached to each other through single bonds that are arranged in a zigzag manner. All carbon atoms lie in the plane of the paper but hydrogen atoms attached to the carbon atoms lie up or down, the plane of the paper. The hydrogen atoms above or below the plane of the paper are interchangeable without altering the over all polyethylene structure. This is possible because in polyethylene all hydrogen atoms attached to carbon atoms indistinguishable.

—H, R and two polymer chain segments of chain length *m* and *n* as shown below:

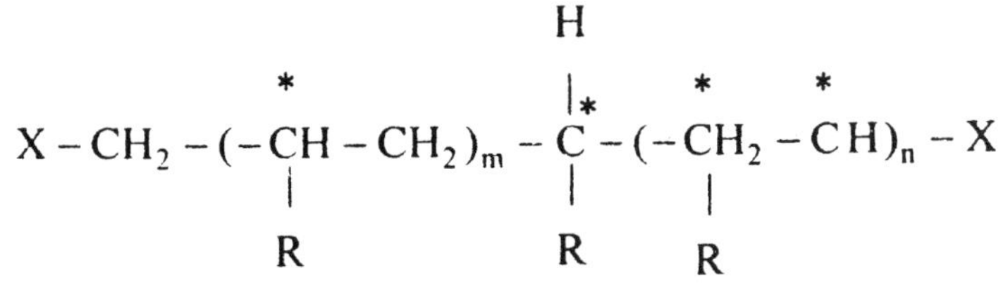

$$X-CH_2-(-\overset{*}{\underset{\displaystyle R}{C}}H-CH_2)_m-\overset{\displaystyle H}{\underset{\displaystyle R}{\overset{*}{C}}}-(-\overset{*}{C}H_2-\overset{*}{\underset{\displaystyle R}{C}}H)_n-X$$

where X denotes end groups.

However, if one of the hydrogen atoms in all ethylene units of this polymer is substituted by a substituent R like CH_3, Cl or CN. The polymer will then have the chemical formula —CH_2—CHR—CH_2—CHR—CH_2—CHR-— and its structural formula would be as under:

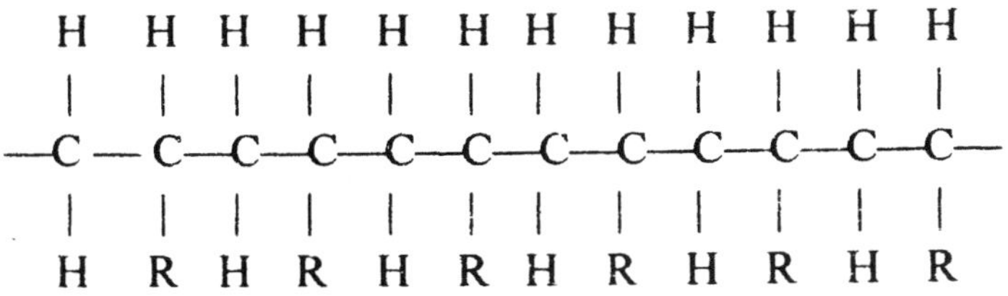

$$-\underset{H}{\overset{H}{C}}-\underset{R}{\overset{H}{C}}-\underset{H}{\overset{H}{C}}-\underset{R}{\overset{H}{C}}-\underset{H}{\overset{H}{C}}-\underset{R}{\overset{H}{C}}-\underset{H}{\overset{H}{C}}-\underset{R}{\overset{H}{C}}-\underset{H}{\overset{H}{C}}-\underset{R}{\overset{H}{C}}-\underset{H}{\overset{H}{C}}-\underset{R}{\overset{H}{C}}-$$

Bot these C* atoms provides a site for optical isomerism. Each such site can exhibit either *d*· or *l*-type isomerism which depends on whether the R group is located below the plane of carbon-carbon chain or above. The regularity or the order in which the successive asy:nmetric carbon sites, C*, exhibit their *d*- or *l*- form leads to three different types of isomeric structure in the polymer molecule. The structures shown in Fig. 6.2. In structure I, we have all the R groups located on one side of the plane of the carbon-carbon chain. In structure II, we have the R groups located alternatively above and below the plane of the carbon-carbon chain and in structure III, we have them located randomly. Polymers having structure I are known as **isotactic;** those having structure II are known as **syndiotactic;** and the third types are known as **actactic** or **hetrotactic.** These three types of polymers have the same chemical structure, but exhibit entirely different properties due to their differing configurations and the resulting geometrical structure. **Atactic polymers** are, generally low melting and easily soluble, while **isotactic** and **syndiotactic** polymers are high melting and less soluble.

(I)

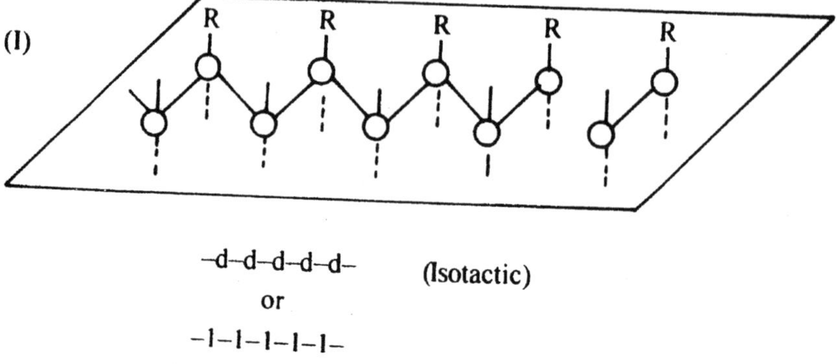

–d–d–d–d–d– (Isotactic)
or
–l–l–l–l–l–

(II)

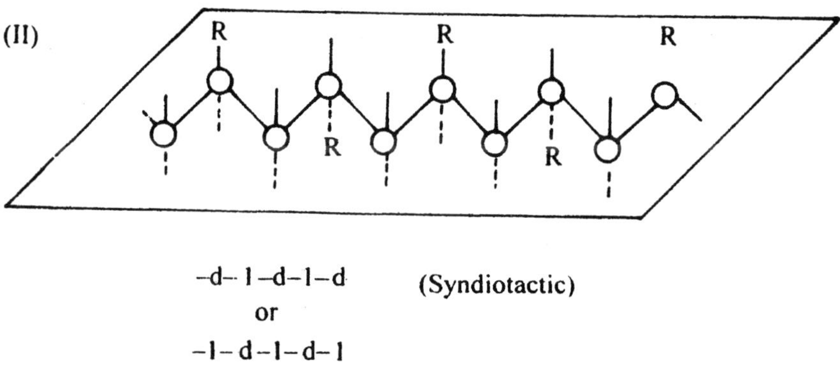

–d– l –d–l–d (Syndiotactic)
or
–l– d –l– d– l

(III)

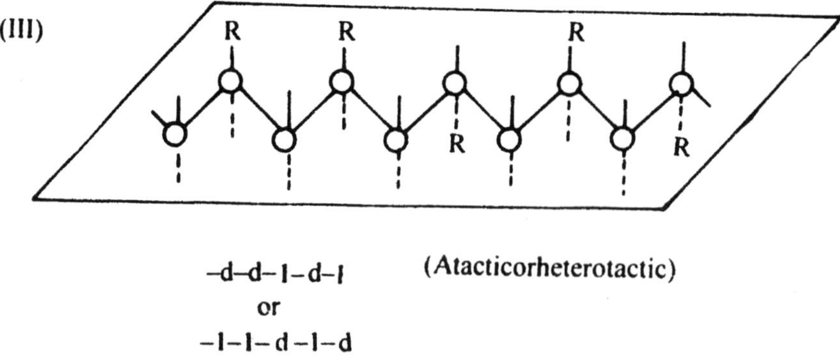

–d–d–l–d–l (Atacticorheterotactic)
or
–l–l– d –l–d

Fig. 6.2: Planar zigzag structure of polymer molecules showing: (I) isotactic, (II) syndiotactic, and (III) heterotactic configurations, (Hydrogen atoms are not shown for the purpose of calrity.)

Geometric Isomerism

Optical isomerism, involves single bond C—C atoms. The geometric isomerism involves double bond C = C atoms. Geometric isomerism arises from different configurations of the substituents on a carbon double bond and depends on the positions occupied by the substituent groups in space. For example, considering 1, 3-butadiene polymerisation. The butadiene monomer has two double bonds in its structure:

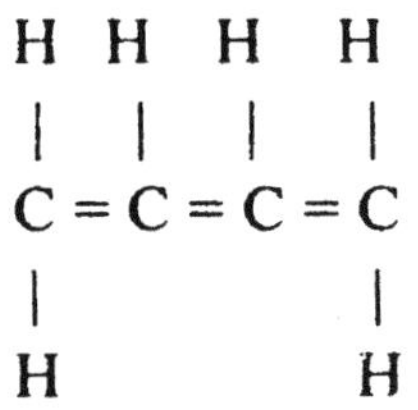

```
H   H   H   H
|   |   |   |
C = C = C = C
|           |
H           H
```

It can polymerise as under

$$nCH_2 = CH—CH = CH_2 \longrightarrow —(—CH_2—CH = CH—CH_2—)—n$$

The resultant polymer structure has a double bond in each repeat unit as shown below:

$$—(—CH_2—CH = CH—CH_2—)—$$

Each of these double bonds provides a site for a steric isomerism, depending on whether the CH_2 groups attached to the carbon atoms on either side of the doubld bond are close to or away from each other. The two possible configurations (Fig. 6.3) are: (a) where both the—CH_2 groups are on the same side of the double bond close to each other (called cis-configuration), and (b) where they are on the opposite sides and quite apart (called trans-configuration).

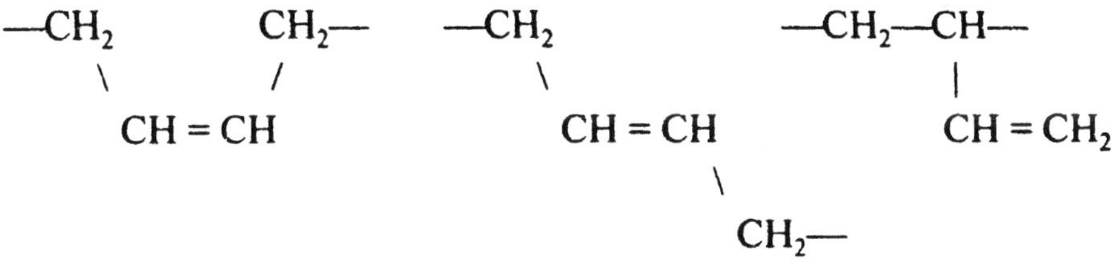

```
—CH2         CH2—     —CH2                    —CH2—CH—
    \        /            \                          |
     CH = CH               CH = CH                   CH = CH2
                                  \
                                   CH2—
```

cis-configuration   trans-configuration   1, 2 vinyl-configuration

Fig. 6.3: Cis-, trans-and 1, 2 vinyl configurations of butadiene units in polybutadiene molecule.

A third configuration known as 1, 2 vinyl configuration, is also possible (see Fig. 6.3). In the case of 1, 2 vinyl configuration, the three types of optical isomerism, *i.e.*, atactic, isotactic and syndiotactic configurations, are possible. It can also be seen from Fig, 6.3 that in cis-configuration, there is a bending back of the carbon-carbon chain, but in trans-configuration, there is a straightening out of the carbon-carbon chain. During the polymerisation of 1, 3-butadiene, if all repeat units take the cis-configuration, we obtain a 100% cis-polybutadiene and because of bending back of all the successive carbon-carbon chain segments, the molecule as a whole assumes a shape of a spring (Fig. 6.4) and shows good elongation. In case, all the repeat units take trans-configuration, the resultant polymer is a 100% trans-polybutadiene and because of the straightening out of all the successive carbon-carbon chain segments, the molecule acquires a straight and stiffened rod-like structure (Fig. 6.4) and shows lower elongation.

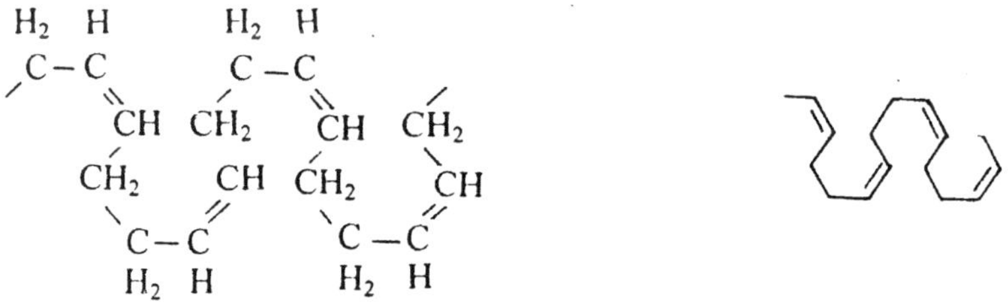

H₂ H H₂ H
C–C C–C
CH CH₂ CH CH₂
CH₂ CH CH₂ CH
C–C C–C
H₂ H H₂ H

(Cis–Polybutadienee)

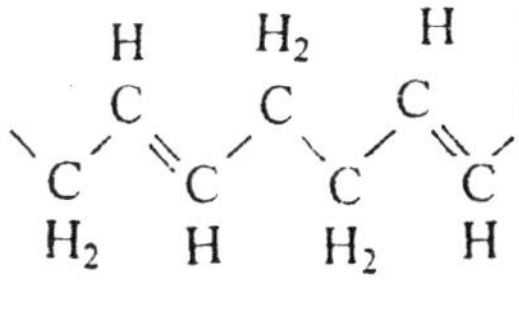

H H₂ H H₂
C C C C
C C C C
H₂ H H₂ H

(Trans – Polybutadienee)

H H H H
C C C C
C CH C CH C CH C CH
H₂ CH₂ H₂ CH₂ H₂ CH₂ H₂ CH₂

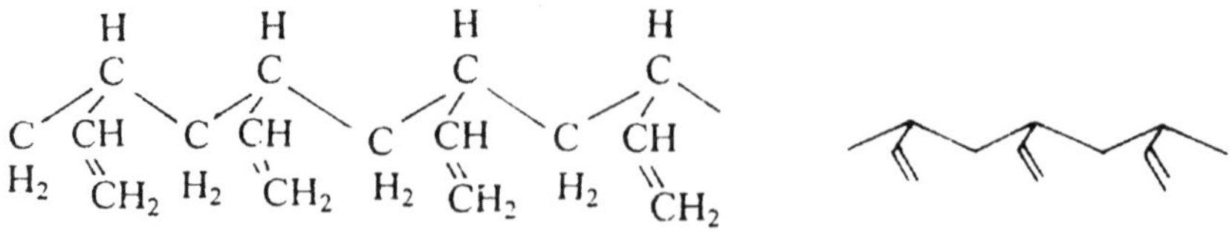

(1, 2 – Polybutadienee)

Fig. 6.4: Configurations of 100% cis-, 100% trans, and 100% 1,2-polybutadiene molecules.

In actual practice, however, polymerisation of butadiene to 100% cis or 100% trans is extremely difficult, and we generally obtain mixtures of cis and trans-configurations, randomly distributed throughout the chain length (presence of 1, 2-vinyl structure is also possible) Fig. 6.4. Depending on the ratio of cis to trans-chain segments present, the polymer can show either high or low elongation.

The cis- and trans-arrangements in case of natural rubber and gutta-parcha are shown below:

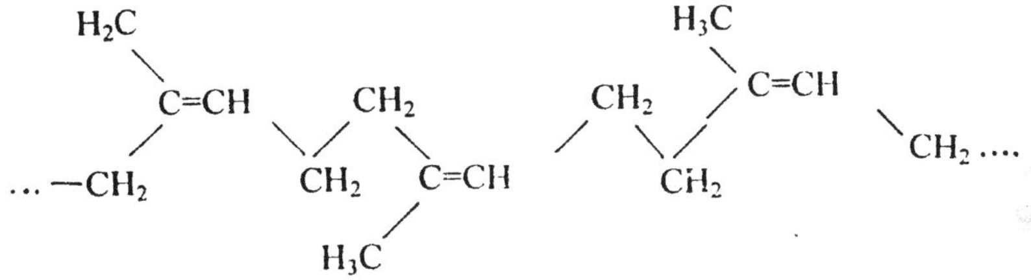

H_2C \ C=CH / ...—CH_2 \ / CH_2 — CH_2 \ C=CH / H_3C ... CH_2 / \ CH_2 / C=CH (H_3C) \ CH_2....

Cis – isomer (natural rubber)

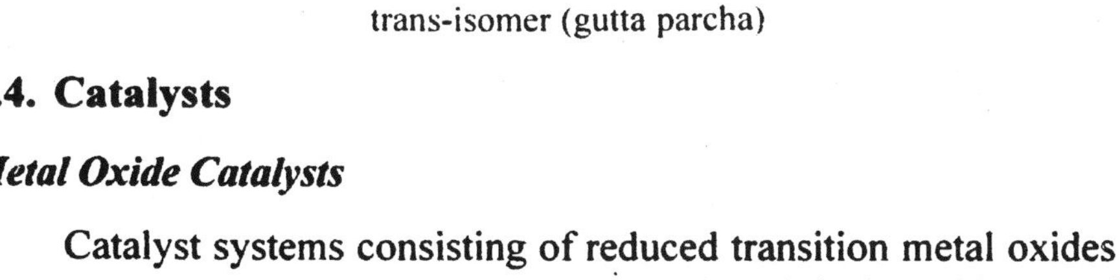

...– CH_2 / C=CH (H_3C) / \ CH_2 (H_3C) \ C=CH / CH_2 / \ CH_2 (H_3C) C=CH / CH_2 —...

trans-isomer (gutta parcha)

6.4. Catalysts

Metal Oxide Catalysts

Catalyst systems consisting of reduced transition metal oxides on supports such as alumina or silica developed during 50's are of considerable importance for the polymerisation of ethylene.

Common catalyst compositions include oxides of chromium or molybdenum, or cobalt and nickel metals, supported on silica, alumina, titania, zirconia, or activated carbon.

Ziegler-Natta Catalysts

These are a special type of coordination catalysts, made up two components. The two components are generally referred to as the

catalyst and the **co-catalyst.** The catalyst component consists of halides of IV-VIII group elements having transition valence and the co-catalysts are organo-metallic compounds like alkyls, aryls and hydrides of group I-IV metals. Although there are hundreds of such catalyst co-catalyst systems listed in Table 6.1. Systems based on the organo-aluminium compounds such as triethyl aluminium ($AlEt_3$) or diethyl aluminium chloride ($AlEt_2Cl$) in combination with titanium chlorides-both tri and tetra ($TiCl_3$ and $TiCl_4$)—are, the most commonly used. Aluminium alkyls act as the electron acceptor and the electron donor is titanium halides and the combination, therefore, readily forms coordination complexes. The complex formed is insoluble in the

Table 6.1 : Components of Typical Ziegler-Natta Catalysts

Organometallic Compound	*Transition-Metal Salt*
Triethyl aluminium	Titanium tetrachloride
Diethyl aluminium chloride	Vanadium trichloride
Diethyl aluminium chloride	Triacetyl acetone vanadium
Diethyl aluminium chloride	Triacetyl acetone chromium
Diethyl aluminium chloride	Cobalt chloride-pyridine complex
Butyl lithium	Titanium tetrachloride
Butyl magnesium iodide	Titanium trichloride
Ethyl aluminium dichloride	Dichlorodicyclopentadienyl titanium

solvent and is, hence heterogeneous in nature. Many structures have been proposed for these complexes. A few are shown below :

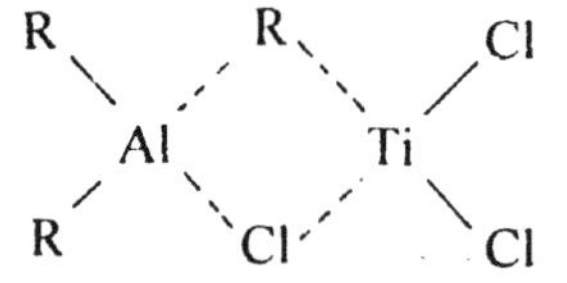

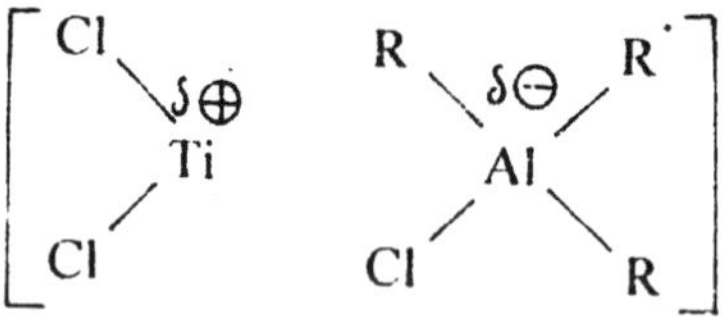

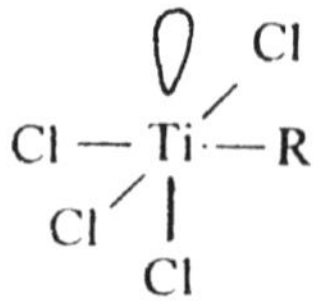

The active centres, from where the polymer chain growth propagates, are formed at the surface of the solid phase of the catalyst complex, and the monomer is complexed with the metal ion of the active centre before its insertion into the growing chain. The actual mechanism by which the propagation occurs and the factors governing the formation of stereo-regular polymers is state debatable. Among the several mechanisms proposed, the bimetallic mechanism of Natta and the monometallic mechanism of Cossee have received much attention. Cossee's mechanism, however, is widely accepted at present.

Bimetallic Mechanism

Natta's bimetallic mechanism stipulates that when the catalyst and co-catalyst components are mixed, the chemisorption of the aluminium alkyl (electropositive in nature) occurs on the titanium chloride solid surface which results in the formation of an electron-deficient bridge complex of the structure shown below:

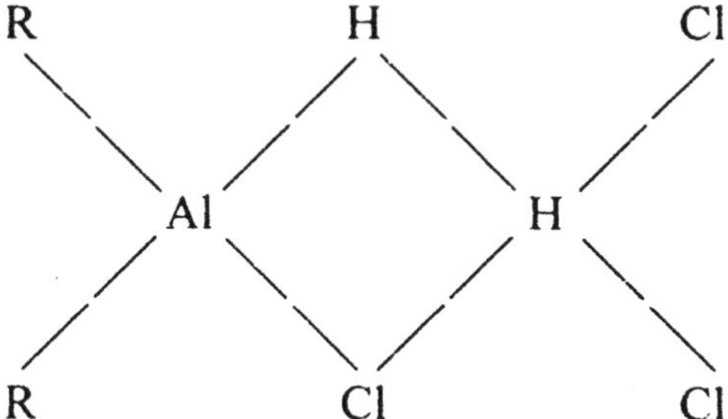

This complex then acts as the active centre. The monomer is then attracted towards the Ti—C bond (C from the alkyl group R) in active centre, when it formes a π complex with the Ti ion as under:

$$>Al\overset{R}{\underset{Cl}{\langle\rangle}}Ti< + CH_2 = CH \quad CH_3 \longrightarrow >Al\overset{R}{\underset{Cl}{\langle\rangle}}Ti< \uparrow CH_2 = CH\ CH_3$$

The bond between R and Ti opens up, producing an electron deficient Ti and a carbanion at R:

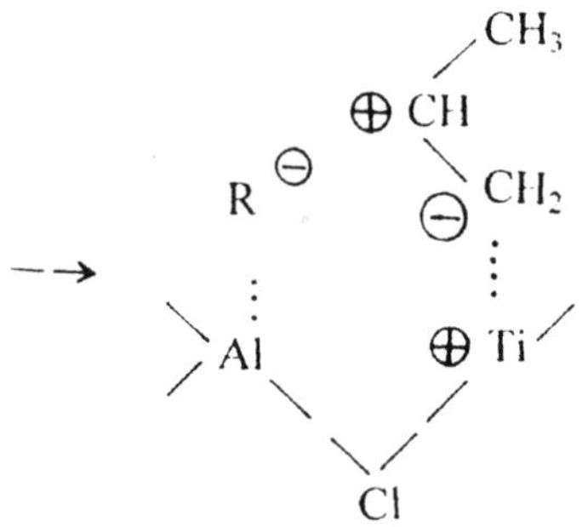

$$\longrightarrow \begin{array}{c} \text{R}^{\ominus} \quad \oplus\text{CH(CH}_3\text{)}-\text{CH}_2^{\ominus} \\ \vdots \qquad\qquad\qquad \vdots \\ >\text{Al} \qquad\qquad \oplus\text{Ti}< \\ \diagdown \quad \diagup \\ \text{Cl} \end{array}$$

The Ti ion attracts the π electron pair of the monomer and forms a σ bond, while the counter-ion attracts the electron-deficient centre of the monomer. Thus the monomer, is 'inserted' into a transition state ring structure shown below:

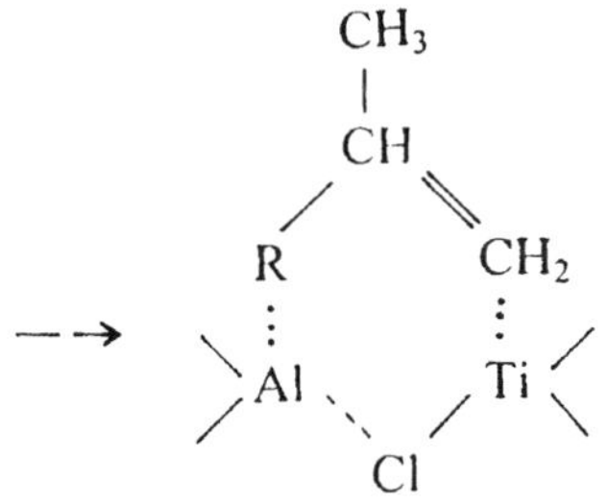

$$\longrightarrow \begin{array}{c} \text{CH}_3 \\ | \\ \text{CH} \\ \diagup \quad \diagdown\!\!\diagdown \\ \text{R} \qquad\quad \text{CH}_2 \\ \vdots \qquad\qquad \vdots \\ >\text{Al} \qquad\quad \text{Ti}< \\ \diagdown \; \diagup \\ \text{Cl} \end{array}$$

This transition state then gives rise to the chain growth at the metal carbon bond, regenerating the active centre:

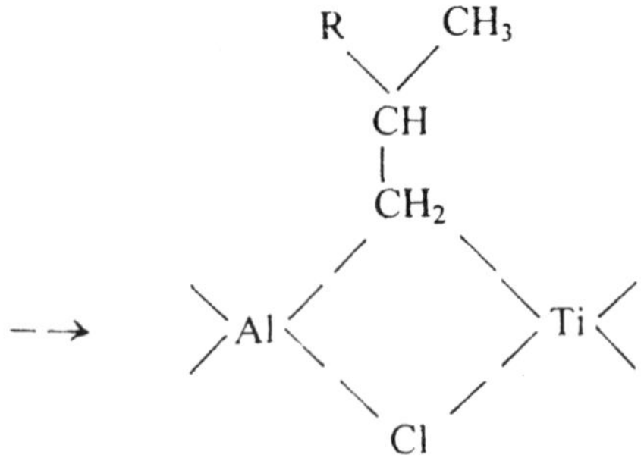

$$\longrightarrow \begin{array}{c} \text{R} \qquad \text{CH}_3 \\ \diagdown \; \diagup \\ \text{CH} \\ | \\ \text{CH}_2 \\ \diagup \qquad \diagdown \\ >\text{Al} \qquad\qquad \text{Ti}< \\ \diagdown \qquad \diagup \\ \text{Cl} \end{array}$$

Repeating the whole sequence, with the addition of a second monomer molecule, we will get the structure of the resultant chain growth as:

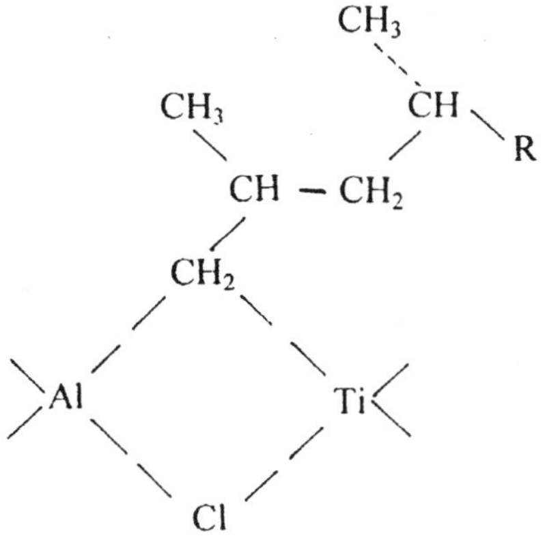

CH_3

CH — R

CH_3

CH — CH_2

CH_2

Al Ti

Cl

The monomer insertion is repeated in this way and the orientation of the substituent group of the monomer is always taken from the metal-ion end, resulting in a stereo-regulated polymer.

Monometallic Mechanism

The monometallic mechanism proposed by *Cossee* assumes that the active centre is at the Ti-R part of the catalyst, while the aluminium alkyl acts only as an alkylating agent for the $TiCl_3$. When the catalyst and co-catalyst are mixed, the aluminium is chemisorbed on the solid $TiCl_3$. The five-coordinated titanium ion on the surface of the catalyst is assumed to have a vacant *d*-orbital, shown at 6 in the structure shown below:

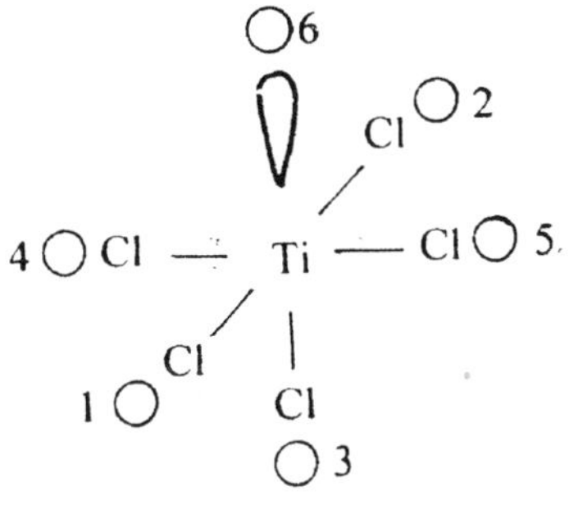

○6

Cl ○2

4 ○ Cl — Ti — Cl ○ 5

Cl

1 ○ Cl

○ 3

The Cl at 1 and 2 are considered to be attached to another titanium in the crystal lattice of $TiCl_3$, and, except the Cl at 5 which can be replaced by an alkyl group, all the other four Cl are non-exchangeable. Soon after the chemisorption of the aluminium alkyl on the $TiCl_3$ crystal, Ti^{3+} gets alkylated by an exchange mechanism, as under:

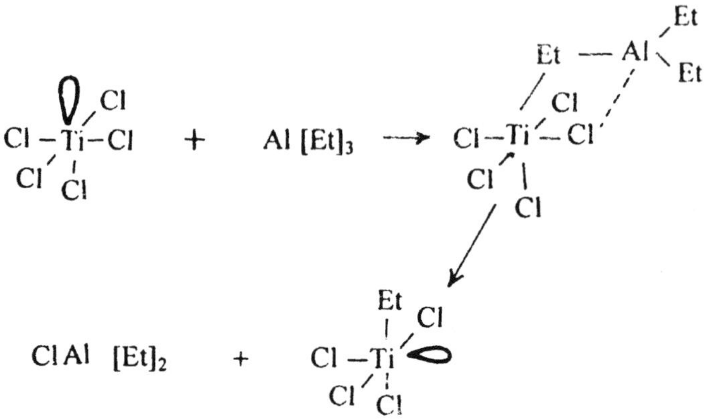

$$\text{Cl}_5\text{Ti}\,(\text{vacant orbital}) + \text{Al[Et]}_3 \longrightarrow [\text{Cl}_5\text{Ti}\cdots\text{Cl}\cdots\text{Al(Et)}_2\text{–Et}] \longrightarrow \text{ClAl [Et]}_2 + \text{EtTiCl}_4\,(\text{vacant orbital})$$

The active catalyst formed then has an octahedral structure, with the four Cl attached to the latice and the ethyl group attached to the Ti through a σ bond, and leaving a vacant orbital at the position where originally the fifth chlorine was attached. Once the active catalyst is formed, the monomer is attracted towards the vacant *d*-orbital which then forms a transition π complex with the Ti, as shown below:

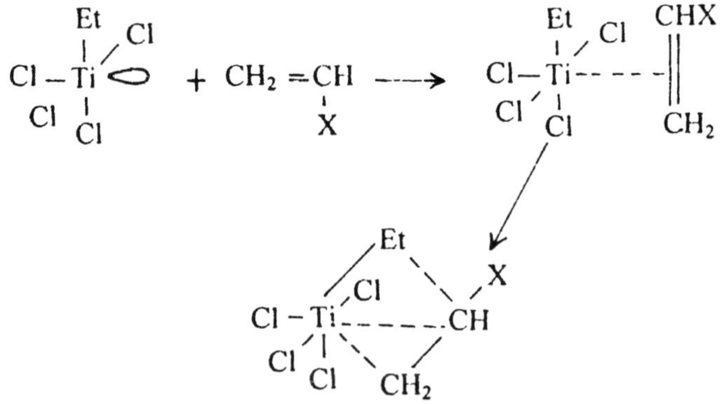

$$\text{EtTiCl}_4\,(\text{vacant orbital}) + CH_2{=}CH\text{–}X \longrightarrow \text{EtTiCl}_4\cdots(CHX{=}CH_2) \longrightarrow \text{Cl}_4\text{Ti(Et)}\cdots CH(X)\cdots CH_2$$

(Transition state)

The transition state quickly give rise to the growth of the polymer chain by the monomeric insertion at the Ti-Et bond. While the monomeric group is inserted and the chain grows, the vacant *d*-orbital can either be regenerated at the same position or at the position where the ethyl group was originally attached:

$$\text{Et}\cdots\text{CH(X)}\cdots\text{CH}_2\cdots\text{Ti} \;\xrightarrow{\boxed{\text{OR}}}\; \begin{cases} -\text{Ti}(\square)-\text{CH}_2-\text{CH(H)}-\text{Et} \\ (\square)-\text{Ti}-\text{CH}_2-\text{CH(X)}-\text{Et} \end{cases}$$

When the vacant *d*-orbital is regenerated at the same position all the time, the incoming monomeric units will be inserted with the same spatial arrangement, resulting in the formation of an **isotactic** polymer. With the insertion of each monomeric unit, if the *d*-orbital migrates from one position to the other alternatively, the monomers get inserted with alternating spatial arrangement, resulting in the formation of a **syndiotactic** polymer. Whether the vacant site is regenerated at the same position or migrates to the other site alternatively depends on the interactive forces that exist between (i) the substituent group of the incoming monomeric unit and the already inserted monomeric unit and (ii) the substituent group of the monomer and the chlorine atoms of the active catalyst.

Factors like the type of catalyst system, catalyst crystal structure, molar ratios of the catalyst components and temperature of the catalyst complex influence the shift or migration of the vacant *d*-orbital and, so the stereo-regularity of the polymer obtained.

The polymerisation is always characterised by the initiation, propagation and termination reactions as under:

$$\text{(i)}\quad \underset{\text{(Active centre)}}{M_t\ R} + \underset{\underset{X}{|}}{CH_2 = CH} \longrightarrow \underset{\underset{X}{|}}{M_tCH_2CHR}\ \text{[Initiation]}$$

(ii) $$M_t CH_2 \underset{\substack{|\\ X}}{CH}R + n\,CH_2 = \underset{\substack{|\\ X}}{CH} \longrightarrow M_t CH_2 CH\text{—}[\text{—}CH_2\underset{\substack{|\\ X}}{CH}\text{—}]_n\text{—}R$$

[Propagation]

(iii) $$M_t CH_2 \underset{\substack{|\\ X}}{CH}\text{—}[CH_2 \underset{\substack{|\\ X}}{CH}]_n R + R'H \longrightarrow M_t\text{—}R' + CH_3 \underset{\substack{|\\ X}}{CH}\text{—}[CH_2 \underset{\substack{|\\ X}}{CH}]_n R$$

[Termination by an active hydrogen compound]

(iv) $$Mt\,CH_2 \underset{\substack{|\\ X}}{CH}\text{—}[CH_2 \underset{\substack{|\\ X}}{CH}]_n R + CH_2 = \underset{\substack{|\\ X}}{CH} \longrightarrow Mt\,CH_2 \underset{\substack{|\\ X}}{CH_2} + CH_2 = \underset{\substack{|\\ X}}{C}\text{—}[CH_2 \underset{\substack{|\\ X}}{CH}]_n\ R$$

[Termination by Transfer with monomer]

(v) $$Mt\,CH_2 \underset{\substack{|\\ X}}{CH}\text{—}[CH_2 \underset{\substack{|\\ X}}{CH}]_n\ R \rightarrow Mt + CH_2 = \underset{\substack{|\\ X}}{C}\ [CH_2 \underset{\substack{|\\ X}}{CH}]_n\ R$$

[Termination by spontaneous internal transfer]

Here, Mt denotes transition metals such as Ti, Mo, Cr, V, Ni or Rh.

"Alfin" Polymerisation

The "Alfin" catalyst (Morton 1964; Reich 1966) is made up to a suspension, in an inert solvent like pentane, of a mixture of an alkylenyl sodium compound (such as allyl sodium), an alkoxide of a secondary alcohol (such as sodium isopropoxide), and an alkali halide (such as sodium chloride). The catalyst is highly specific for the polymerisation of dienes into the 1, 4-forms.

6.5. Mechanism of Coordination Polymerisation

The studies were carried out by Henrice-Olive (1981) on the system bis-cyclo-pentadienyl titanium (iv) dichloride (Cp_2TiCl_2) and an aluminium alkyl (generally trischloroethyl aluminium). In this case the active catalyst species that was identified can be represented as under:

$$\begin{matrix} & Et & & & \\ CP & | & Cl & & Et \\ & Ti & & Al & \\ Cp & & Cl & & Cl \end{matrix}$$

The transition metal acts as the active site and the chain growth step in polymerisation of ethylene can be represented as below:

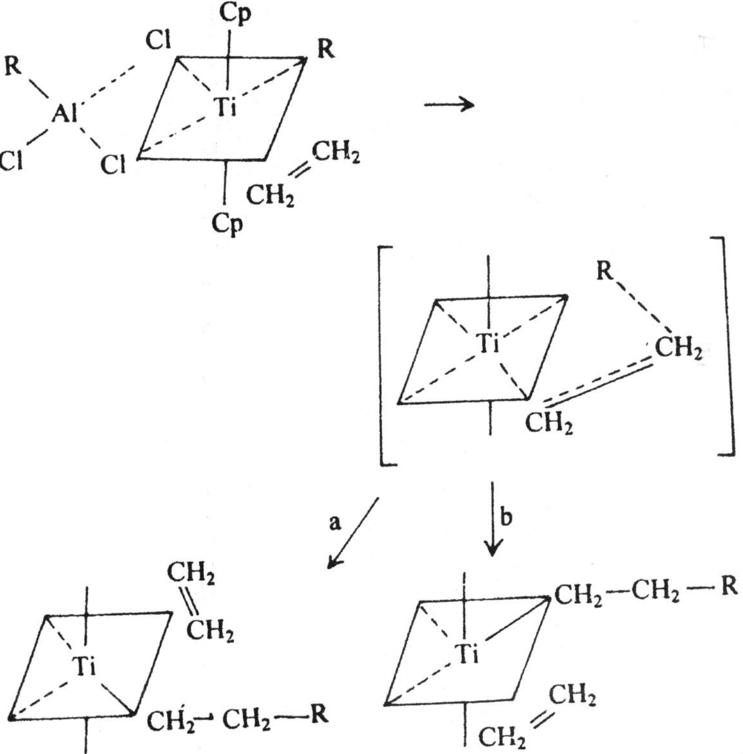

Fig.6.5: Growth step postulated for the polymerisation of ethylene by a soluble Ziegler-Natta catalyst.

The ethylene is coordinated to the free site at the transition metal and inserted between the metal and the alkyl group R. In this way the chain grows.

It was also observed that the present of other ligands affect the course as also the rate of the reaction.

A chain-transfer reaction exists in which a β-hydrogen is transfered from the growing chain to the transition metal, leaving the chain with a vinyal end group. This is a molecular-weight-determining step, though in the commercial production of Ziegler polyethylene it is unimportant, as the molecular weights are generally quite high.

Only heterogeneous Ziegler-Natta catalysts form isotactic polymers from non-polar monomers like α-olefins. The nature of the catalyst surface seems to be important in imparting stereoregularity in such cases. Propagation probably occurs at active sites on the crystal surface of the transition-metal compound, $TiCl_3$, for example. The sites are formed by the reaction of the group I-III soluble metalorganic compound, for example, Al $(C_2H_5)_3$ or al $(C_2H_5)_2$ Cl, with the crystal surface, yielding species of the type shown in Fig. 6.6. The chemical and crystal structures of the catalysts determined the orientation of the monomer as it is added to the growing chain. The driving force for isotactic propagation thus results from interaction between the monomer and the ligands of the transition metal at the active site (Boor 1967). Several other models have been proposed, but none appear to be as widely applicable.

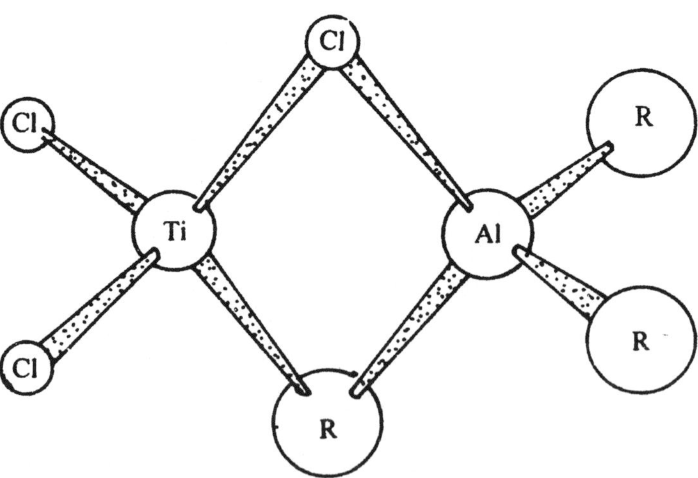

Fig. 6.6: Structure postulated for the metal halide-metal alkyl complex of a typical coordination catalyst.

7

Condensation or Step-Growth Polymerisation

7.1. Introduction

In the condensation polymerisation is also called step growth polymerisation. In it the polymer molecules are built up through many separate reactions of functional groups. Condensation polymerisation involves the condensation reactions between bi- or multi-functional monomer units.

The chemistry of condensation polymerisation deals generally with the chemistry of the functional groups. Condensation polymerisation takes place in steps, each step involves reaction between a monomer and the growing chain. At each step a small molecule like water is eliminated. Steady chain growth occurs continuously but reaction times are generally long. Most condensation polymers are copolymers involving two types of monomer units. Chain building can occur only if each reactive molecule possesses two active functional groups. Monomers various degrees of functional groups give polymers with various degrees of branching and cross-linking.

Though it is possible to use many known reactions with organic functional groups in step polymerisation however condensation, addition, ring-opening, amination and ester-interchange reactions are most commonly used. Whether the elemental composition of the polymer formed and that the monomers are the same or not depends on whether the reaction is accompanied by the elimination of small molecules or not.

7.2. Polycondensation

Polycondensation can be brought about by monomers having two or more reactive functional groups (like hydroxyl, carboxyl and amino) and condensing with each other. Basic reactions taking place are the same as between various functional groups in low molecular weight organic compounds. For example, the reaction between a hydroxyl group and a carboxylic group, yields an ester and a water molecule:

$$—OH + HO-\underset{\underset{O}{\parallel}}{C}- \dashrightarrow -O-\underset{\underset{O}{\parallel}}{C}- + H_2O$$

In case a monohydric alcohol *e.g.*, methanol is reacted with a monocarboxylic acid, *e.g.*, acetic acid, the following reaction occurs:

$$CH_3OH + HOOC-CH_3 \dashrightarrow CH_3COOCH_3 + H_2O$$

However, it may be noted that in this methyl acetate formation, the reactive functional groups (*viz.*, hydroxyl and carboxyl) are used up to give an unreactive functional group (*viz.*, ester). There is no more reactive functional group left with the product and, therefore, it cannot react further with any other reactant molecules so as to form a different product.

Again if we consider the reaction between ethylene glycol (EG), and acetic acid, we find that, one molecule of EG reacts with one molecule of acetic acid as follows:

$$\underset{\text{(EG)}}{HOCH_2CH_2OH} + HOOCCH_3 \dashrightarrow HOCH_2CH_2OOCCH_3 + H_2O$$

The product in this case, still contains another hydroxyl group that can react with another molecule of acetic acid to yield the ethylene glycol di-acetate as under:

$$CH_3COOCH_2CH_2OH-HOOCCH_3 \dashrightarrow CH_3COOCH_2CH_2OOCCH_3 + H_2O$$

This product formed contains no more reactive functional group and, therefore, cannot react further to give any other product. The reaction between methanol and acetic acid, provides an instance of one molecule of each of the reactants reacting to give a single molecule of the product and one water molecule and the product contains one ester group in its molecule. In the case of reaction between EG and acetic acid on the other hand, three reactant molecules are involved (but only two reactants) to give again a single molecule of the product. However in this case, two water molecules are lost and the product contains two ester groups in its molecule.

Now, if we consider the reaction between two molecules of adipic acid and two molecules of EG we find that one molecule of EG reacts with one molecule of adipic acid to form a monoester products as under:

$$HOCH_2CH_2OH + HOOC-(-CH_2-)-_4COOH \dashrightarrow$$
$$HOCH_2CH_2OOC-(-CH_2-)-_4COOH + H_2O$$

In addition to a non-reactive ester group, the product also contains two reactive groups (*i.e.*, one hydroxyl and one carboxyl) and each can react with one more molecule of EG or adipic acid. Both EG and adipic acid and still available because we started with two molecules of each of the reactants and so far have only consumed one of each and therefore, are still left with one molecule each of EG and adipic acid.

Say, there are *n* molecules of adipic acid and *n* molecules of EG. We will get the overall reaction as

$$n\ HOCH_2CH_2CH + n\ HOOC-(-CH_2-)_4-COOH$$
$$\dashrightarrow HO-[-CH_2CH_2OOC-(-CH_2-)_4-COO-]-n$$
$$+ (2n-1)H_2O$$

Following generalisations can be made regarding polycondensation or any step polymerisation reaction:

(i) That monomers should have two reactive functional groups for polymerisation to occur.

(ii) That polymerisation proceeds by step-wise reaction between reactive functional groups.

(iii) That only one type of reaction (*i.e.*, condensation reaction) between two functional groups is involved in polymer formation.

(iv) That the polymer produced still contains both the reactive functional groups at its chain ends (as end groups) and, hence, is 'active' and not dead', as in chain polymerisation.

In the above examples two different types of reactive functional groups, viz., – OH and COOH, were present on two different monomers. In some cases, both the reactive functional groups could be present on the same monomer, and in such a case, a self-polycondensation reaction takes place. For example, the polycondensation of amino enanthoic acid to yields polyenanthoamide:

$$n\,H_2N\,(\!-CH_2-)-{}_6COOH -$$

$$H-[-HN-(\!-CH_2-\!)\!-{}_6CO\!-]\!-nOH + (n-1)H_2O$$

In addition to amino acids, hydroxy acids and alkylene chlorohydrines can also undergoing self-polycondensation.

When a pair of bifunctional monomers (like dicarboxylic acid/dialcohol, dicarboxylic and/diamine or dialcohol/dihalide) undergoes polycondensation, this is generally described as an AA-BB type polycondensation and the reaction can be shown as follows:

$$n\,A-A+n\,B-B-\!\!\rightarrow A-(\,AB-)_{2n-1}\!-B + \text{byproduct}$$

When a single bifunctional monomer undergoes self-condensation, it is described as an A-B type polycondensation. The reaction, then, can be shown as follows:

$$n\,A-B-\!\!\rightarrow B-(-\,AB-)_{n-1}-A + \text{byproduct}$$

If in both the AA-BB and A-B types of polycondensation reactions intramolecular condensation takes place in preference to intermolecular condensation, one will end up with cyclic products of low molecular weight instead of, or in addition to, straight chain polymer molecules. During the preparation of polyethylene terephthalate (PETP), for example, cyclic products have also been isolated in addition to the required polymer, as shown:

$$n\ CH_3OC(=O)-C_6H_4-C(=O)OCH_3 \quad + \quad n\ HOCH_2CH_2OH$$

(Dimethyl terephthalate) (Ethylene glycol)

Intra molecular condensation

Inter molecular condensation

$$O\text{-}[C(=O)\text{-}C_6H_4\text{-}C(=O)\text{-}O\text{-}CH_2CH_2\text{-}O]_m\text{-}$$
$$CH\text{—}CH_2\text{-}O\text{-}C(=O)\text{-}C_6H_4\text{-}C(=O)\text{—}$$

where m = 2, 3 or 4

(Unwanted Cyclic Product)

$$H\text{-}[OCH_2CH_2\text{-}O\text{-}C(=O)\text{-}C_6H_4\text{-}C(=O)]_n\text{-}OCH_2CH_2OH$$

(Required Petp Polymer

If in the AA-BB type of polycondensation reaction, one of the monomers is trifunctional, the polymer obtained is a three dimensional network polymer. For example, glycerol reacts with a dibasic acid to form a complex structure of the intermediate molecule, even at the early stage of polymerisation. For every glycerol molecule reacted, the functionality in the resultant intermediate product increases by one. This means that the number of possible sites where the polymer build-up can occur increases rapidly and stupendously and several branching sites are produced on the growing polymer molecule. Many such branched growths containing reactive functional end groups on further consideration yield an enormously large, three-dimensional molecule of almost infinite molecular weight. When the three-dimensional network is just about to be produced the whole polymer becomes an isoluble and infusible mass and is known as 'gel'. The precise stage of the reaction at which **gelation** takes place is found to depend on the ratio of the bifunctional monomer to the trifunctional one. For instance, when an equimolar ratio of glycerol and dibasic acid is used, the **gel point** is reached when the extent of reaction reaches a value of 0.765.

We now take up some generic polycondensation reaction leading to different types of polymers.

(i) Dihydric alcohols result in the formation of polyethers

$$n\ HO - R - OH \longrightarrow HO\ \{RO\}_n\ H + (n-1)H_2O$$

(ii) Dihydric alcohols and dicarboxylic acid polyesters

$$n\ HO - R= OH + nHOOC-R'-COOH$$

$$\longrightarrow HO - \{R-OOC- R'-COO\} - nH+(N-1)H_2O$$

(iii) Dicarboxylic acids and diamines forming polyamides (nylons).

$$n\ HOOC - COOH+ nH2-R'-NH2$$

$$\longrightarrow HO\ \{OC-R-CONH-R-NH\} - nH+ (2n-1)\ H_2O$$

(iv) Dihydric alcohols and methylester of dicarboxylic acids forming polyesters

$$n\ HO - R - OH+n\ CH_3\ OOC-R'-COOCH_3$$

$$\longrightarrow HO\ \{R-OOC-R'-COO\} - nCH_3 + (2N-1)\ CH_3OH$$

(v) Amino acids undergoing self-condensation forming Polyamides

$$n\ H_2N-R-COOH \rightarrow H - \{NH-R-CO\}_n\ OH+ (n-1)\ H_2O$$

(vi) ω–hydroxy carboxylic acids wielding Polyesters

$$n\ HO-R-COOH \rightarrow HO - \{R\text{-}COO\}_n - H+ \{n\text{-}1\}\ H_2O$$

Examples for these type of polymers have been taken up in next article (Article 7.2)

7.2. Functionality and Polycondensation Reaction

Functionality of the monomer is the average number of reactive functional groups present per monomer molecule. Monofunctional monomers give only low-molecular-weight products. Bifunctional monomers give linear polymers and polyfunctional monomers having more than two functional groups per molecule yield branched or crosslinked (three-dimensional) polymers. The properties of the linear and the three-dimensional polymers differ widely.

Condensation polymers can be obtained from monomers which have two or more functional groups. The functional groups present may be –OH, COOH, –Cl, $–NH_2$ etc. To produce polymer molecules through condensation polymerization, it is essential that the monomer molecules should possess two or more functional groups. The monomer molecules react by condensation reaction with the elimination of water of other simple molecules such as NH_3, HCl or NaCl. Since the molecules are lost during the process the molecular formula of the polymer is not an integral multiple of the molecular formula of the monomer units. Some examples of condensation polymerisation are illustrated below:

Polyester (Polyethylene Terephthalate, PET)

"Terylene" which is a copolymer of terephthalic acid and ethylene glycol is a polyester:

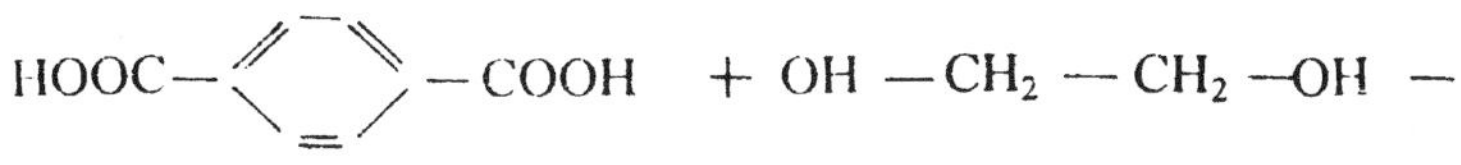

$$HOOC-C_6H_4-COOH \quad + OH-CH_2-CH_2-OH \; -$$

–polyester

Bifunctional monomers on reacting produce a bifunctional dimer, on further reaction each product formed is bifunctional. Condensation of bifunctional units generally leads to the formation of a linear polymer chains. Poly (ethene terephthalate) can be represented as under:

$$HO-[OC-C_6H_4-COOCH_2\,CH_2O-]_n-H$$

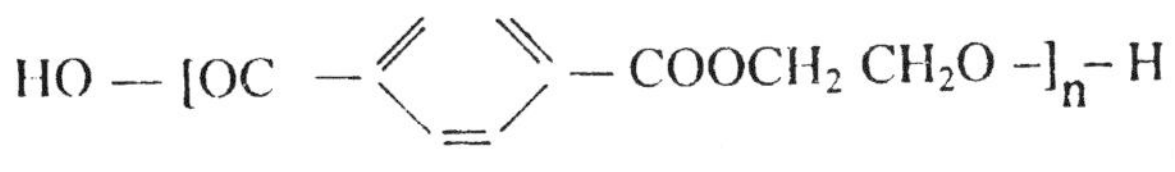

The brackets enclose the repeating unit.

Such reactions in general can be represented as under:

$$x\ HO–R–OH + x\ HOOC–R'\ COOH \rightarrow$$

$$HO\ [–R–OCO–R'.\ COO]_x\ H + (2x–1)H_2O$$

Polyethylene terephthalate

(R = phenyl group, and R' = methylene group)

In the following pages we will take up the discussion of some important such reactions.

Phenol-Formaldehyde Resins

These were the first synthetic polymers to be manufactured commercially in 1909 by the American chemist Bakeland who gave them the name *Bakelite*. These were produced by reacting phenol with formaldehyde. The chemistry of these reactions is quite complex and is still not fully understood.

The reaction between phenol and formaldehyde is catalysed either by acids or bases. The nature of the polymer formed has been found to depend on the type of catalyst used and the ratio of formaldehyde to phenol.

(a) The *first step* in the polymerisation is the formation of methylol derivatives (phenol alcohols) by a direct condensation according to the following reaction:

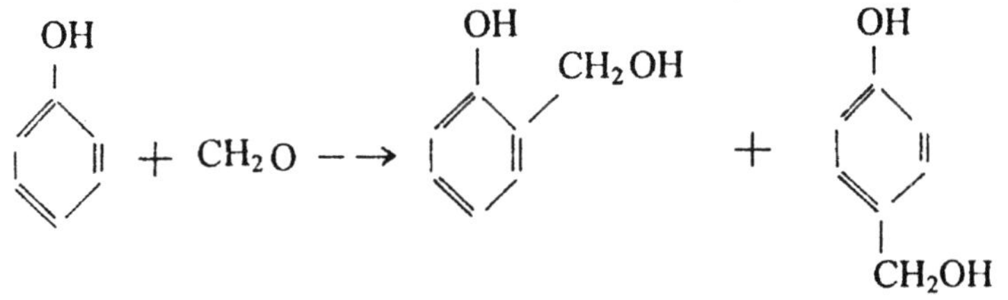

OH (phenol) + CH_2O ⟶ OH, CH_2OH (ortho) + OH, CH_2OH (para)

o-hydroxybenzyl alcohol p-hydroxybenzyl alcohol

The entrance of the methylol group is always either *ortho* or *para* to the hydroxyl group; one, two or three formaldehyde molecules may add to a phenol molecule depending upon the ratio of formaldehyde to phenol that is used. The various possibilities are as under:

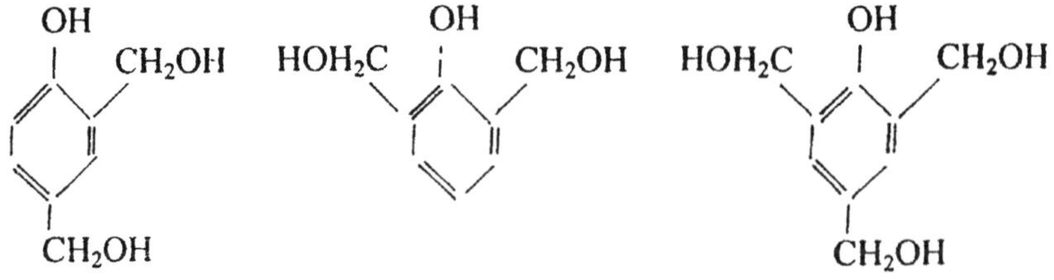

OH, CH_2OH, CH_2OH HOH_2C, OH, CH_2OH HOH_2C, OH, CH_2OH, CH_2OH

(b) The second step involves the condensation of the methylol group of a phenol molecule with the *ortho* or *para* position of another phenol molecule in the presence of an acid or base.

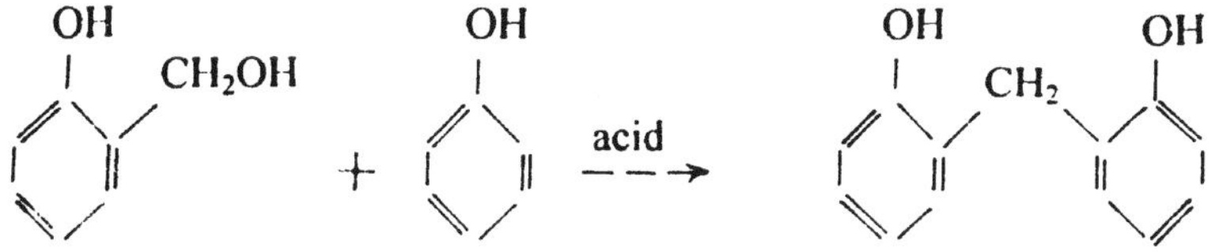

OH CH_2OH + OH $\xrightarrow{\text{acid}}$ OH CH_2 OH

Further, depending upon the phenol/formaldehyde ratio and the number of steps in the process, the following two types of phenol-formal dehyde resins have been synthesised:

(i) If the ratio of formaldehyde to phenol is less than 1:1 and if an acid catalyst is used. the methylol derivatives undergo condensation with phenol of form dihydroxydiphenyl-methane with the splitting off of water. Further condensation produces linear low molecular-weight polymers known as *novolaks* according to the following equation:

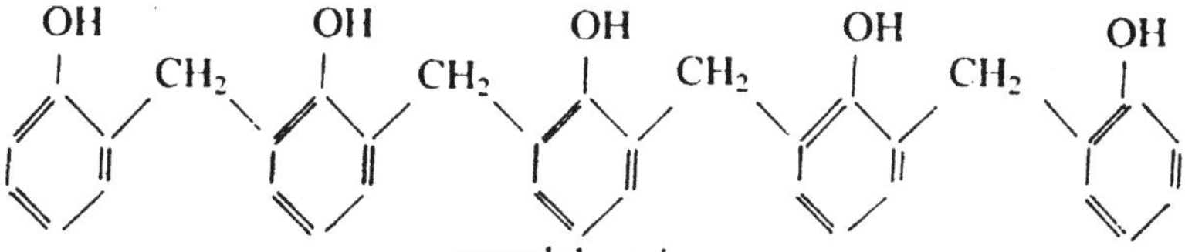

OH CH_2 OH CH_2 OH CH_2 OH CH_2 OH

novolak resin

In novolaks, phenol molecules are linked by methylene bridges in random *ortho* and *para* positions. In novolaks, there are not more than ten phenolic groups, corresponding to molecular weights up to about 1000.

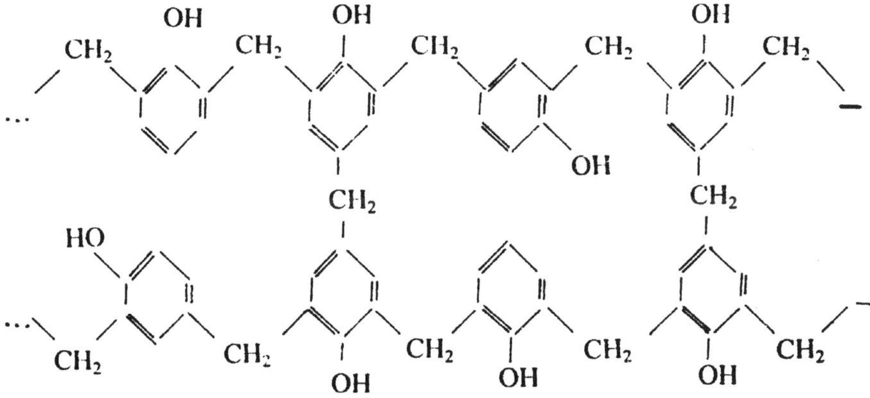

... CH_2 OH CH_2 OH CH_2 CH_2 OH CH_2 —

OH CH_2 CH_2

HO

... CH_2 CH_2 OH CH_2 OH CH_2 OH CH_2 —

cross-linked phenolic resin

The novolaks are permanently fusible and soluble. However, they can be converted to a cross-linked resin by heating with an alkaline solution of formaldehyde to provide additional methylol group for cross-linking.

(ii) With an alkaline catalyst (*e.g.*, Na_2CO_3) and a molar ratio of formaldehyde to phenol greater than 1:1 the phenol alcohols condense to form low molecular-weight polymers known as *resols*. Here the phenol molecules are linked either by methylene linkages or by ether linkages. Some typical equations are given as follows:

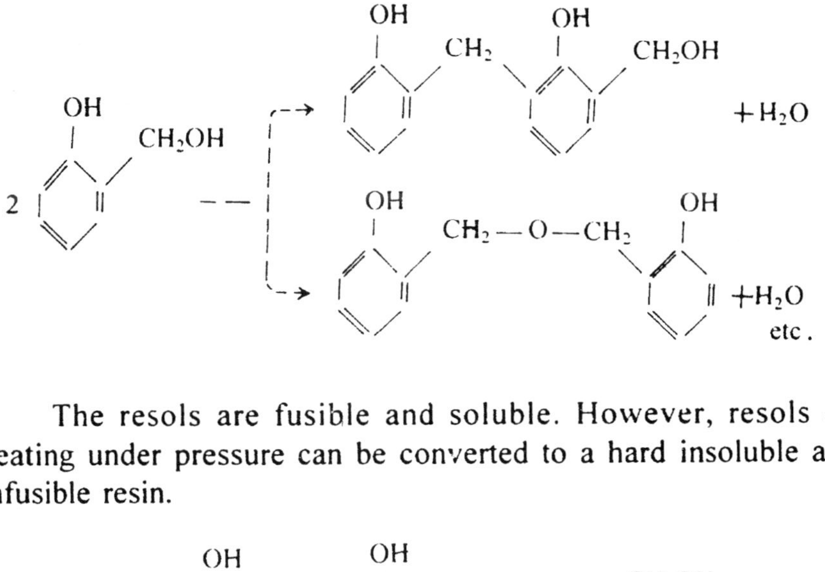

OH
CH_2OH
2 ——

OH CH_2 OH CH_2OH $+H_2O$

OH $CH_2—O—CH_2$ OH $+H_2O$
etc.

The resols are fusible and soluble. However, resols on heating under pressure can be converted to a hard insoluble and infusible resin.

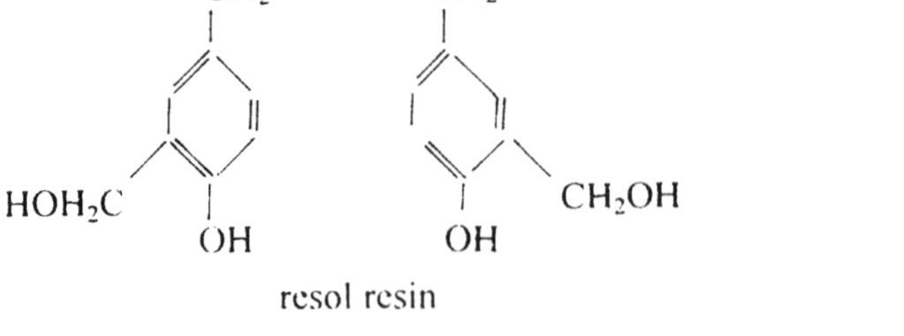

OH CH_2 OH CH_2 CH_2OH
OH
CH_2 CH_2
HOH_2C CH_2OH
OH OH

resol resin

Epoxy Resins

These are polymers of epoxy compounds with dihydric phenols. Theoretically, the epoxy resins are polyethers.

The most common epoxy resin is made by condensing epichlorohydrin with bisphenol A, according to the equation.

$$\underset{\text{epichlorohydrin}}{H_2\overset{\overset{O}{\diagup\ \diagdown}}{C\text{——}CH}\ CH_2Cl} + \underset{\text{bisphenol A}}{HO-\langle C_6H_4\rangle-\underset{CH_3}{\overset{CH}{\overset{|}{\underset{|}{C}}}}-\langle C_6H_4\rangle-OH}$$

$$\xrightarrow{NaOH} H_2\overset{\overset{O}{\diagup\ \diagdown}}{C\text{——}CH}.CH_2-\left[-O-\langle C_6H_4\rangle-\underset{CH_3}{\overset{CH_3}{\overset{|}{\underset{|}{C}}}}-\langle C_6H_4\rangle-O-\right.$$

$$\left.-CH_2.\overset{\overset{OH}{|}}{CH}.CH_2-\right]_n-O-\langle C_6H_4\rangle-\underset{CH_3}{\overset{CH_3}{\overset{|}{\underset{|}{C}}}}-\langle C_6H_4\rangle-$$

$$\text{——}OCH\quad \overset{\overset{O}{\diagup\ \diagdown}}{CH-CH_2}$$

A substantial excess of the epichlorohydrin is normally used to ensure that the chains are terminated by epoxide groups which will help in cross-linking the polymer during thermosetting of the resins.

The resins produced as above are thermoplastic and soluble. The resins may be converted to a cross-linked thermosetting plastic by heating under pressure with a suitable *curing agent* like polyamine, polysulphide, polyamide, acid anhydride, etc.

The main use of the epoxy resins is in surface coatings such as floor coatings, tank linings, etc. They are also used in the manufacture of reinforced plastics, in the "potting" of electrical components are as an adhesive for bonding many different substances.

Polysiloxane

Polysiloxanes can be obtained by the self-condensation of dialkyl silicic acid with alkyl chloro silanes. The dialkyl silicic acids can be prepared by the reaction of dichloro silane with water, dialkyl silicic acid is a very unstable compound.

Various reactions involved can be represented as under:

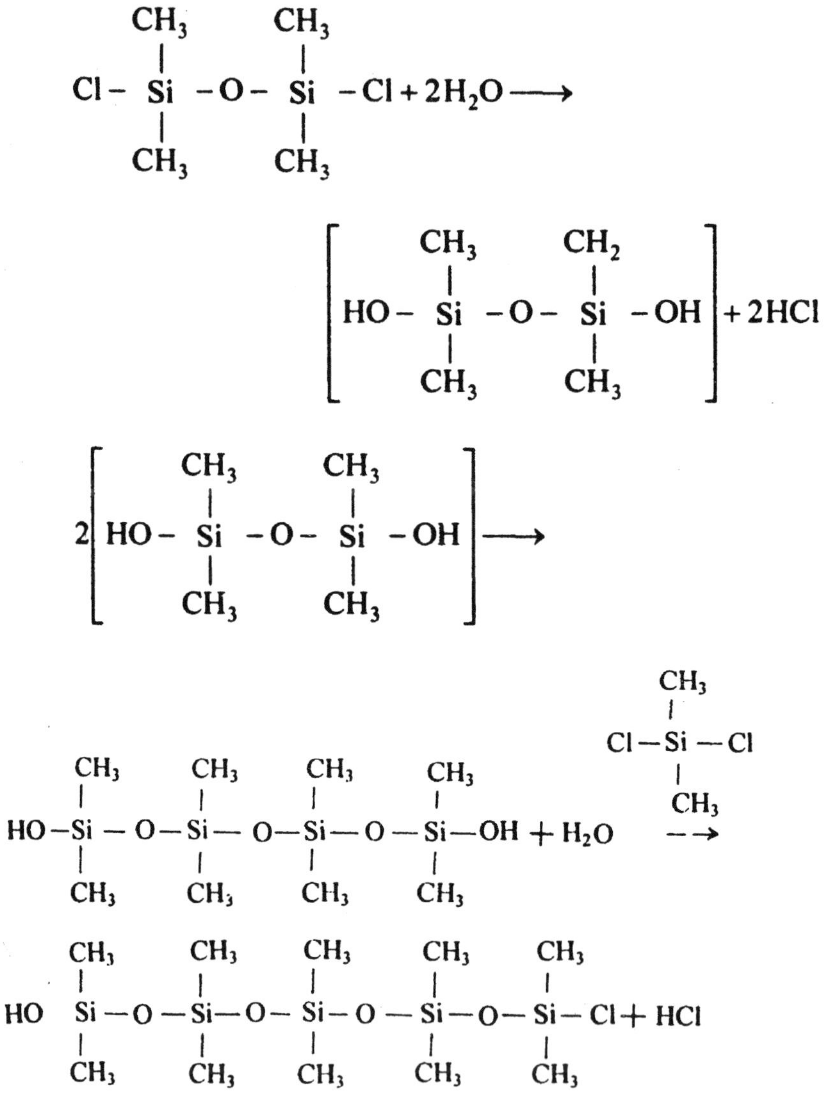

$$\begin{array}{ccccccc} & CH_3 & & CH_3 & & & \\ & | & & | & & & \\ Cl- & Si & -O- & Si & -Cl + 2H_2O \longrightarrow & & \\ & | & & | & & & \\ & CH_3 & & CH_3 & & & \end{array}$$

$$\left[\begin{array}{ccccc} & CH_3 & & CH_2 & \\ & | & & | & \\ HO- & Si & -O- & Si & -OH \\ & | & & | & \\ & CH_3 & & CH_3 & \end{array}\right] + 2HCl$$

$$2\left[\begin{array}{ccccc} & CH_3 & & CH_3 & \\ & | & & | & \\ HO- & Si & -O- & Si & -OH \\ & | & & | & \\ & CH_3 & & CH_3 & \end{array}\right] \longrightarrow$$

$$\begin{array}{cccccccccc} & CH_3 & & CH_3 & & CH_3 & & CH_3 & & \\ & | & & | & & | & & | & & \\ HO- & Si & -O- & Si & -O- & Si & -O- & Si & -OH + H_2O & \xrightarrow{\;Cl-Si(CH_3)_2-Cl\;} \\ & | & & | & & | & & | & & \\ & CH_3 & & CH_3 & & CH_3 & & CH_3 & & \end{array}$$

$$\begin{array}{cccccccccccc} & CH_3 & & CH_3 & & CH_3 & & CH_3 & & CH_3 & & \\ & | & & | & & | & & | & & | & & \\ HO & Si & -O- & Si & -O- & Si & -O- & Si & -O- & Si & -Cl + HCl \\ & | & & | & & | & & | & & | & & \\ & CH_3 & & CH_3 & & CH_3 & & CH_3 & & CH_3 & & \end{array}$$

Polysiloxane

Amino resins

Amino resins can be obtained by condensing melamine or urea with formaldehyde. The melamine resins have better properties as compared to urea resins but are costlier.

Preparation of urea resin

Following steps are involved in the preparation of urea resins.

Step I: Urea or melamine which is a trimer of cyanamide, is reacted with formaldehyde to get methylol compounds as under:

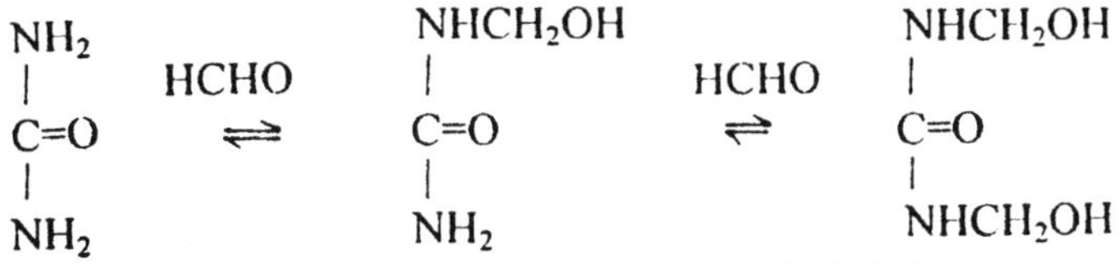

$$\begin{matrix} NH_2 \\ | \\ C=O \\ | \\ NH_2 \end{matrix} \underset{}{\overset{HCHO}{\rightleftharpoons}} \begin{matrix} NHCH_2OH \\ | \\ C=O \\ | \\ NH_2 \end{matrix} \underset{}{\overset{HCHO}{\rightleftharpoons}} \begin{matrix} NHCH_2OH \\ | \\ C=O \\ | \\ NHCH_2OH \end{matrix}$$

Methylolurea

Step II: Condensation of methylolurea to yield the resins:

In the second phase of the reaction, the condensation occurs in presence of an acid. The condensation occurs between the methylol groups and the amido hydrogen to yield the methylene links as shown below:

$$-CH_2-OH + \overset{+}{H} \rightleftharpoons CH_2 \quad \overset{+}{O}H_2 \rightleftharpoons -\overset{+}{C}H_2 + H_2O$$

$$-\overset{+}{C}H_2 + H_2N- \dashrightarrow -CH_2 + \overset{+}{N}H_2-$$

$$\dashrightarrow -CH_2-NH- + \overset{+}{H}$$

Since a typical resin solution involves species which contain methylol and amido groups, various methylene compounds can be obtained by using this type of condensation reaction.

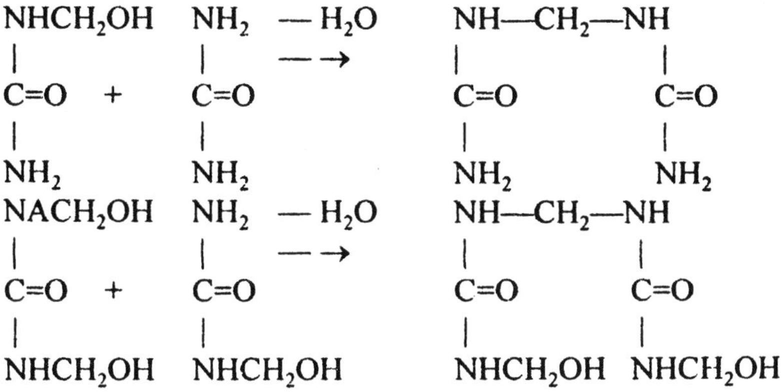

$$\begin{matrix} NHCH_2OH \\ | \\ C=O \\ | \\ NH_2 \end{matrix} + \begin{matrix} NH_2 \\ | \\ C=O \\ | \\ NH_2 \end{matrix} \xrightarrow{-H_2O} \begin{matrix} NH & -CH_2- & NH \\ | & & | \\ C=O & & C=O \\ | & & | \\ NH_2 & & NH_2 \end{matrix}$$

$$\begin{matrix} NACH_2OH \\ | \\ C=O \\ | \\ NHCH_2OH \end{matrix} + \begin{matrix} NH_2 \\ | \\ C=O \\ | \\ NHCH_2OH \end{matrix} \xrightarrow{-H_2O} \begin{matrix} NH & -CH_2- & NH \\ | & & | \\ C=O & & C=O \\ | & & | \\ NHCH_2OH & & NHCH_2OH \end{matrix}$$

The above condensation products contain methylol and amido groups. Hence the condensation can proceed to give polymeric methylene compounds.

The general structure of the methylene containing polymers can be represented as under:

$$HOCH_2—[—NH—CO—NH—CH_2—]_n—NH—CO—NH—CH_2OH$$

Structure of melamine resin

The general structure of melamine resins prepared by condensing melamine and formaldehyde is shown below:

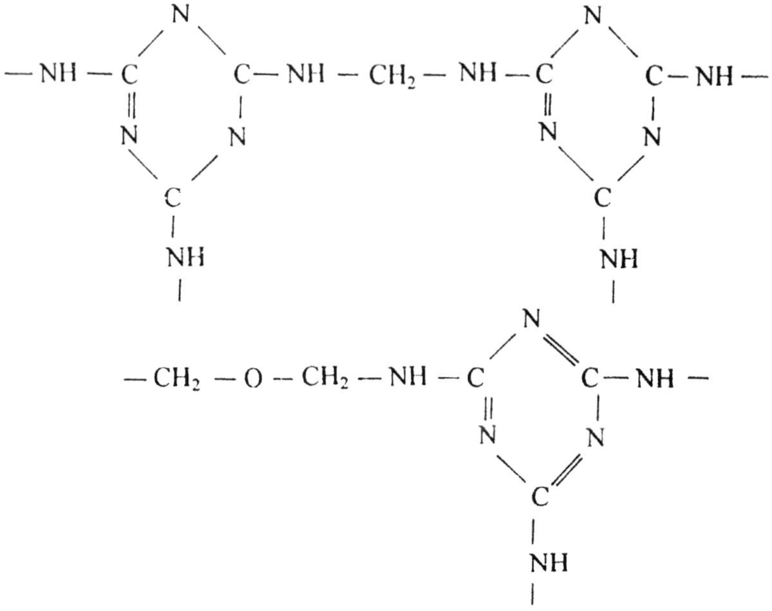

```
         N                                          N
       /   \                                      /   \
—NH — C     C— NH — CH₂ — NH — C          C— NH —
      ||    |                   ||         |
      N     N                   N          N
       \   /                     \        /
         C                          C
         |                          |
        NH                         NH
         |                          |
                              N
                            /   \\
 — CH₂ — O — CH₂ — NH — C         C— NH —
                          ||       |
                          N        N
                            \     //
                               C
                               |
                              NH
                               |
```

Melamine resin

Ring formation

When certain bifunctional monomers are used the reaction can proceed intermolecularly yielding cyclic polymers along with the linear polymers. For example, hydroxy acids can form either linear polymer or a lactone or both as shown below:

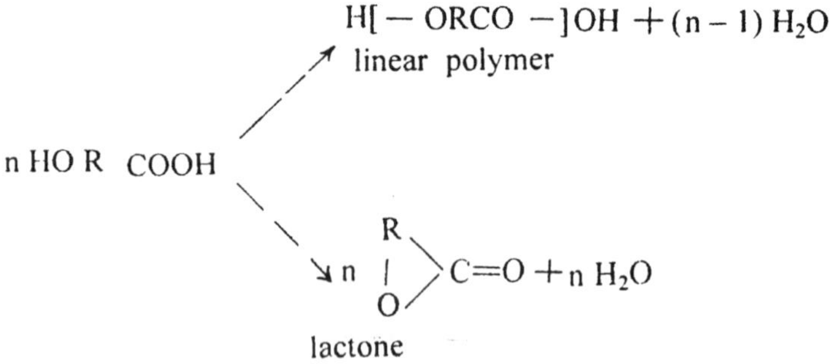

$$n\,HO\,R\,COOH \rightarrow H[-ORCO-]OH + (n-1)\,H_2O \quad \text{(linear polymer)}$$

$$n\,HO\,R\,COOH \rightarrow n\;\begin{matrix} R \\ | \\ O \end{matrix}\!\!>\!C{=}O + n\,H_2O \quad \text{(lactone)}$$

Example: α-Hydroxy acids, such as lactic acid can under go condensation polymerization to yield the dimeric cyclic ester, a lactide and a linear polymer.

```
                                     O
                                   /   \
                         CH3 — CH       O
                        ↗       |       |
        CH3            /        O       CH — CH3
         |            /          \     /
HO — C — COOH                      CO
         |            \          lactide
         H             \
                        ↘
                    H[OCH(CH3) CO]n - OH
```

Similarly amino acids can undergo condensation polymerization yielding a cyclic product or linear polymers.

The extent of the formation of the ring is found to depend strongly upon the size of the ring that would be produced. If the ring size is less than five atoms or more than seven, the product obtained would almost be a linear open-chain polymer. If the ring contains six or seven atoms in either or both, ring and chain polymers are obtained. Larger rings get produced only under special conditions, *i.e.*, by carrying out the reaction at high dilution or by heating the polymer in vacuum in the presence of catalysts.

7.3. Types of Condensation Polymerization

Several types of condensation polymers together with the inter-unit linkages are listed in Table 7.1. The polymers illustrated in Table 7.1 are polyesters, polynides, polythers etc. depending upon the inter-unit linkages or upon the reaction involved.

The condensation polymers can be broadly classified into two categories: (i) linear polymers, and (ii) non-linear polymers. Linear polymers are obtained from monomers that are bifunctional in nature. If the functionality of the monomer is greater than two, the polymerizing molecule can grow in more than one dimension yielding a non-linear or three-dimensional or cross-linked polymer.

7.4. Kinetics of Condensation Polymerization

The kinetics of the condensation polymerization is quite complicated because molecules of a variety of lengths are present in the reaction mixture. Taking the examples of the reaction between equivalent amounts of a dibasic acid and a glycol, the reaction occurs as under:

$$HOOC(CH_2)_xCOOH + HO(CH_2)_yOH \longrightarrow H...[OOC(CH_2)_xCOO(CH_3)_y]....OH + H_2O$$

Table 7.1: Some Condensation Polymers

Type of product formed	Inter-unit linkage	Examples	
(i) Polyamides	$-\overset{\overset{\displaystyle O}{\|}}{C}-NH-$	$NH_2(CH_2)_5\,COOH$	$\longrightarrow H-[NH(CH_2)_5\,CO]_x-OH$
(ii) Polyanhydride	$-\overset{\overset{\displaystyle O}{\|}}{C}-O-\overset{\overset{\displaystyle O}{\|}}{C}-$	$HOOC(CH_2)_8\,COOH$	$\longrightarrow HO-[CO(CH_2)_8\,COOH]-H$
(iii) Polyster	$-\overset{\overset{\displaystyle O}{\|}}{C}-O-$	$HO(CH_2)_9)COOH$	$\longrightarrow HO-[(CH_2)_9\,COOH]_x-H$
		$HO(CH_2)_2\,O(CH_2)_2OH + HOOC\,(CH_2)_4\,COOH \longrightarrow$	
		$HO-[(CH_2)_2\,O\,(CH_3)_2OCO(CH_2)_4\,COOH]_x-H$	
(iv) Polysulphide	$-S-$	$ClCH_2CH_2Cl + Na_2S$	$\longrightarrow Cl-[CH_2CH_2S]_x-Na$
	$-\overset{\overset{\displaystyle S}{\|}}{S}-\overset{\overset{\displaystyle S}{\|}}{S}-$	$ClCH_2CH_2Cl + Na_2S_4$	$\longrightarrow Cl-[CH_2CH_2\underset{\underset{\displaystyle S}{\|}}{S}-\underset{\underset{\displaystyle S}{\|}}{S}]_x-Na$

(V)	Silk fibroin	$-\overset{\overset{O}{\parallel}}{C}-NH-$	$NH_2CH_2COOH + NH_2CH\ R\ COOH \longrightarrow -[NHCH_3CONHCH\ R\ CO]_x-$
(VI)	Phenol –aldehyde	$-CHR-C_6H_4(OH)-$	$C_6H_5OH + CH_2O \longrightarrow$ (cross-linked network of OH-substituted benzene rings joined by $-CH_2-$ bridges, with HOH_2C- end groups)

This is an example of a polyesterification reaction. Kinetically polyesterification should be identical to simple esterification reactions if the increase in molecular weight accompannying the formation of polyester does not change the reactivity of end-groups, and any increase in the viscosity associated with the polymerization does not cause the process to become diffusion controlled. The kinetics of the condensation reaction can be divided into two categories, (i) catalysed and (ii) uncatalysed. Esterification reactions are catalysed by hydrogen ions. In the absence of a strong acid catalyst, a second molecule of the carboxylic acid can undergo esterification reaction and acts as the self catalysts. The reaction can be represented as under:

$$\sim\sim COOH + OH \sim\sim + \text{catalyst} \rightarrow \sim\sim COO + \sim + H_2O + \text{catalyst.}$$

Thus the rate of the reaction is proportional to the product of hydroxyl concentration and the square of carboxyl concentration.

$$-d[COOh]/dt = -d[OH]/dt = k\,[COOH]_2[OH] \qquad ....(7.1)$$

where k is the rate constant of velocity constant for the reaction.

If the concentration of the carboxyl groups and hydroxyl groups are equal, *i.e.*, if equal number of functional groups are used, then

$$[COOH] = [CH] = C$$

Thus equation (7.1) becomes,

$$-dc/dt = kc^3 \qquad ...(7.2)$$

Integrating equating (7.2) within the limits of $t = O$, $C = C_0$ we get

$$\int_{C=0}^{C} -\frac{dC}{C^3} = \int_{o}^{t} kdt \qquad ...(7.3)$$

or

$$\frac{1}{C^2} - \frac{1}{C_0^2} = 2kt$$

Generally the results are expressed in terms of the extent of reaction, p. The extent of reaction is defined as the fraction of the functional group that has reacted in time t. If $(1-p)$ is the reaction of the functional group that has not been reacted at time t then we have,

$$C = (1-p)\,C_0 \qquad ...(7.4)$$

where C_0 is the initial concentration of functional group. By making substitution, equation 7.3 becomes

$$2kt\,C_0^2 = \frac{1}{(1-p)^2} - \qquad ...(7.5)$$

Hence, a plot of $\frac{1}{(1-p)^2}$ versus reaction time 't' should be a straight line if k does not change with change of the size of the reacting species. From the slope, the value of *k*, the reaction constant, can be found. This relationship has been proved by the esterification reaction of glycol with adipic acid as shown in Fig. 7.1.

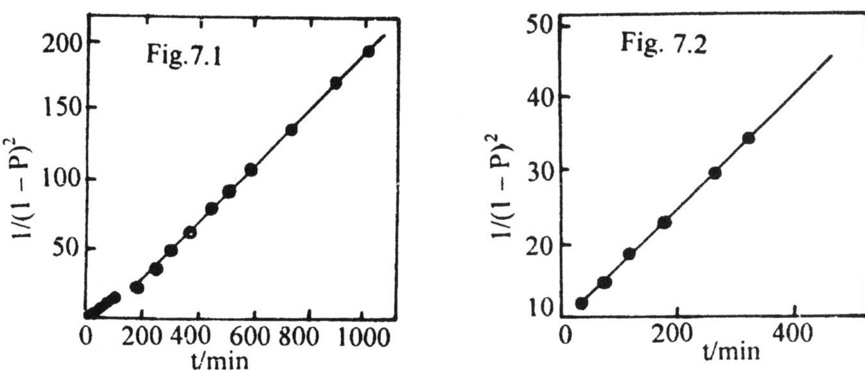

Fig. 7.1: Plot of $1/(t-p)^2$ as a function of time for the polymerization of ethylene glycol with adipic acid at 439 K.

Fig. 7.2: Plot of $1(1-p)$ as a function of time for the polymerization of decamethylene glycol with adipic acid as a function of time using *p*-toluene sulphonic acid as a catalyst at 399 K.

If a strong acid is used as a catalyst, the polyesterification reaction takes up a pseudo second order kinetics and therefore,

$$-dC/dt = k[\text{catalyst}]\,[\text{COOH}]\,[\text{OH}] \qquad ....(7.6)$$

Since the catalyst concentration is constant, the equation can be written as under:

$$-dC/dt = k'[\text{COOH}]\,[\text{OH}]$$

where $k' = k\,[\text{catalyst}]$

or $-dC/dt = k'\,C2$ ...(7.7)

where $[\text{COOH}] = [\text{OH}] = C$

Integrating the equation (7.7)

we get,

$$\frac{1}{C} - \frac{1}{C_0} = k't \qquad \text{....(7.8)}$$

on substituting concentration term for the extent of reaction, p, equation 7.8 becomes,

$$C_0 k't = \frac{1}{1-p} - 1 \qquad \text{...(7.9)}$$

The above equation (eq. 7.9) has been varified experimentally by plotting $1/(1-p)$ versus 't' for the reaction of decamethylene glycol with adipic acid using paratoluene acid as the catalyst as shown in Fig. 7.2.

7.5. Reactivity and Molecular Size

The similarity between monofunction and polyfunctional step reactions is illustrated in the chemical equations. Other similarities also exist, these are observed in the influence of temperature and catalysts on the rates of the reactions. Thus we can substitute the simple concept of reaction between functional groups for that of the myriad separate reactions of each molecular species with all the others. Hence the kinetics of stepwise polymerization is essentially identical with that of simple condensation; without it, analysis of the kinetics would be almost difficult.

Making the simplification we find that the rate of the reaction of a group is independent of the size of the molecule to which it is attached. This assumption is amply justified by experimental evidence. The rate constants of condensation reactions in a homologous series reach asymptotic values independent of chain length quite rapidly and show no tendency to drop off with increases in molecular size. The diluting effect of the large chain, of course should be taken into consideration by adjusting to constant molar composition of reactive groups. Moreover, the rate constants for monofunctional and bifunctional reagents are identical for sufficiently long chains separating the reactive groups of the bifunctional compound.

7.6. Carothers Equation

A relationship between the number average degree of polymerization, $\overline{X}_n$, and the extent of reaction, p was proposed by carothers for the first time. To understand this relationship considering the polymerization of an ω-hydroxy carboxylic acid we find that as the reaction proceeds, the number of molecular decreases as the average length of the individual molecule increases.

Let N_0 — Number of molecules initially present

N — Number of molecules remaining at time 't'

Then the total number of —COOH or —OH groups reacted = N_0—N

Thus extent of reaction:

$$p = \frac{\text{Number of groups which have reacted}}{\text{Number of groups initially present}}$$

Therefore,

$$p = \frac{N_0 - N}{N_0}$$

or

$$\frac{N_0}{N} = \frac{1}{1-p} \qquad ....(7.10)$$

The total number of molecules gets reduced and their average length becomes longer as the polymerization progresses.

Since the number average length of polymerization, $\overline{X}_n$ is given by,

$$\overline{X}_n = \frac{\text{Number of molecules present initially}}{\text{Number of molecules reacting after time t,}}$$

or

$$\overline{X}_n = \frac{N_0}{N} \qquad ...(7.11)$$

Combining equations (7.10) and (7.11), we get

$$\overline{X}_n = \frac{1}{1-p} \qquad ...(7.12)$$

This equation is known as the *Carothers equation* which gives $\overline{X}_n$ in terms of extent of reaction which increases with time of reaction.

The scale of DP in Fig. 7.1 shows that uncatalyzed esterification needs quite long times to reach high degrees of polymerization. Greater success can be achieved by adding to the system a small amount of catalyst, whose concentration is constant throughout the reaction. In such a case the concentration of the catalyst may be included in the rate constant:

$$-\frac{d[\text{COOH}]}{dt} = k'[\text{COOH}][\text{OH}]$$

$$-\frac{dc}{dt} = k'c^2 \qquad \text{....(7.13)}$$

$$c_0 k' c^2 = \frac{1}{1-p} + \text{const.}$$

Hence $\bar{x}_n$ increase linearly with reaction time (Fig. 7.2). Similar reactions have been shown to be linear up to at least $\bar{x}_n$ = 90, corresponding to molecular weights of 10,000.

Most of condensation reactions are rapid and go to almost 100% completion. At different % age of reaction *i.e.*, for different values of $p_1/\bar{x}_n$ can be calculated and $\bar{x}_n \longrightarrow \alpha$ as p $\longrightarrow$ 1:

p :	0.95	0.96	0.98	0.99	0.995	0.999	1.00
$\bar{x}_n$:	20	25	50	100	200	100	α

Statistics of Linear Step-Reaction Polymerization

Rate equations similar to Eq. 7.13 may be written for each molecular species in the reaction mixture. From such equations we can derive the distribution of molecular weights in a step-reaction polymer. The same relation can, more readily be derived from statistical considerations.

The statistical analysis of stepwise polymerization gives results which are equivalent to and in some cases more easily obtained than those of kinetic considerations. The method assumes independence of reaction rate and molecular size. The extent of reaction p has been defined as the probability that a functional group has reacted at time t. This definition is equivalent to the previous definition of p as the fraction of the functional groups that has reacted. Thus finding a functional group unreacted is $(1-p)$.

It is now essential to find the probability that a given molecular selected at random is an x-mer, that is, contains x repeating units. Such as molecule contains $(x-1)$ reacted functional groups (*e.g.*, carboxyls) and, on the end, one unreacted group of this type. The probability of finding a single reacted carboxyl group in the molecule is p, and that of finding $(x-1)$ of them in the same molecule is p^{x-1}. The presence of one unreacted group has a probability of $(1-p)$. Hence the probability of finding the complete molecule is $p^{x-1}(1-p)$. This is equal to the fraction of all the molecules that are x-mers. For N molecules, the total number of x-mers is given by

$$N_x = Np^{x-1}(1-p) \quad ...(7.14)$$

If the total number of units present is N_0, $N = N_0(1-p)$, and

$$N_x = N_0(1-p)2p^{x-1} \quad ...(7.15)$$

This is the **number-distribution function** for a linear stepwise polymerization at extent of reaction p.

The **weight fraction** w_x of x-mers is given by

$$w_x = \frac{xN_x}{N_0}$$
$$= x(1-p)^2p^{x-1} \quad ...(7.16)$$

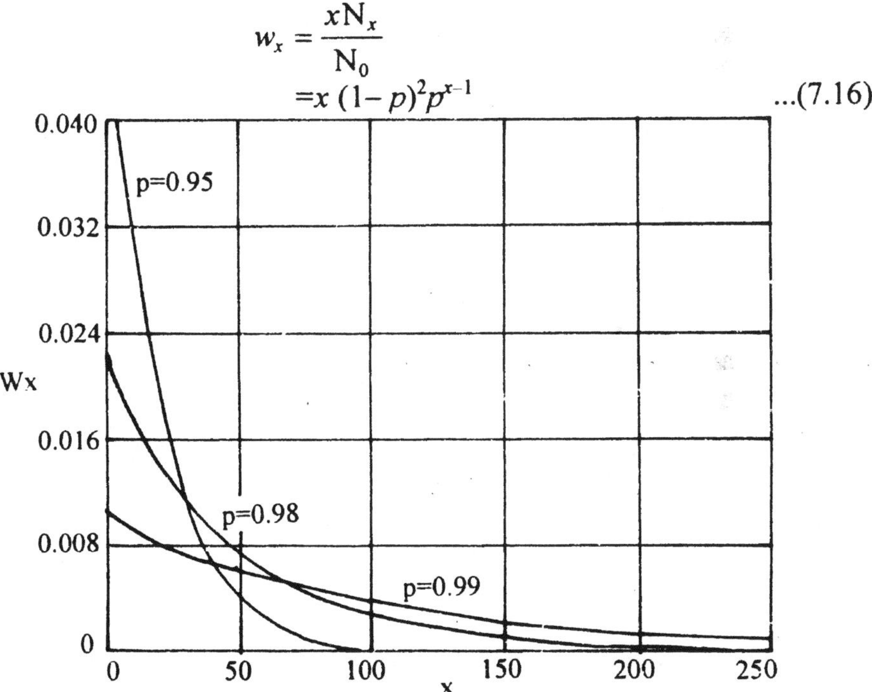

Fig. 7.3: Number or mole-fraction distribution of chain molecules in a linear step-reaction polymer for several extends of reaction p.

This is the **weight-distribution function** for a linear stepwise polymerization at extent of reaction p. Equations 7.15 and 7.16 are illustrated in Figs. 7.3 and 7.4, respectively. On a number basis, monomers are more plentiful than any other molecular species at all stages of the reaction. On a weight basis however, the proportion of low-molecular-weight material is very small and decreases as the average molecular weight increases. The maximum in the weight-distribution curve is present near the number-average molecular weight.

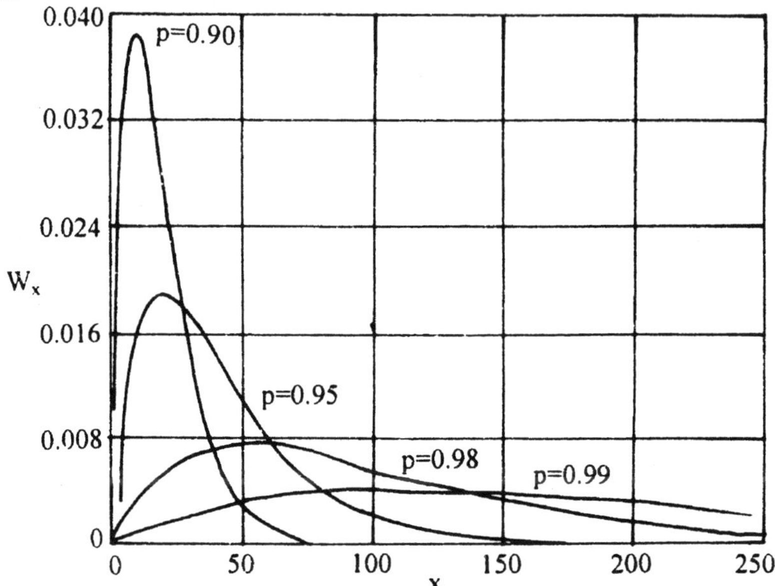

Fig. 7.4: Weight-fraction distribution of chain molecules in a linear step-reaction polymer for several extents of reaction *p*.

Various average degrees of polymerization can be obtained from the weight-or number-distribution functions according to the definitions of the corresponding molecular-weight averages. The number-average degree of polymerization $\bar{x}_n$ calculated like this is given in Eq. 7.12. The weight-average degree of polymerization is

$$\bar{x}_n = \frac{1+p}{1-p} \qquad \text{...(7.17)}$$

The breadth of the molecular-weight distribution curve $\bar{x}_w / \bar{x}_n = 1 + p$; at large extents of reaction $\bar{x}_w / \bar{x}_n \rightarrow 2.0$. This has been confirmed experimentally for nylon.

Conditions for obtaining High Polymers

A very high degree of reaction of functional groups must be achieved for obtaining polymers with high molecular weights. As these condensation reactions are reversible the other products of reaction must be removed during the reaction so that polymerisation continues at a fast rate. If an excess of one of the monomers is used then the other monomer can be completely consumed. But a large excess of one monomer gives a low degree of polymerisation. If 1% excess of one monomer is used the maximum degree of polymerisation attainable is 199. Monomers used should have a very high degree of purity to enable accurate proportioning of monomers. The requirements for obtaining high molecular weight linear polymers by condensation polymerisation are quite demanding and only a limited number of reactions are carried out commercially *e.g.*, polyesters, polyamides, polyurethanes etc. To attain large molecular weight the monomers must be largely consumed *i.e.*, more than 99.5%.

Molecular-Weight Control

Although molecular-weight control in a stepwise polymerization can be achieved initially by stopping the reaction (*e.g.*, by cooling) at the desired point, the stability of the product to subsequent heating may not be adequate with respect to changes in molecular weight.

This can be avoided easily by adjusting the composition of the reaction mixture slightly away from stoichiometric equivalence by adding either a slight excess of one bifunctional reactant or (as in molecular-weight stabilization of nylons by acetic acid) a small amount of a monofunctional reagent. Finally the functional group deficient in amount is completely consumed, and all chain and consist of the group present excess. When only bifunctional reactants are present, the two types of groups being designated A and B and initially present in numbers $N_A < N_B$ such that the ratio $r = N_A/N_B$, the total number of monomers present is $\frac{1}{2}(N_A + N_B) = \frac{1}{2} N_A(1+1/r)$. At extent of reaction p (defined for A groups, for B groups, extent of reaction $= rp$), the total number of chain ends is equal to $N_A(1-p) + N_B(1-rp) = N_A[1-rp/r]$ which is twice the number of molecules present,

$$\bar{x}_n = \frac{N_A(1+1/r)/2}{N_A[1-p+(1-rp/r]/2} = \frac{1+r}{1+r-2rp} \qquad \text{...(7.18)}$$

As $p \to 1$,

$$\bar{x}_n = \frac{1+r}{1-r} \qquad \text{...(7.19)}$$

Thus if 1 mol. % of stabilizing groups is added,

$$\bar{x}_n = \frac{1+(100/101)}{-r(100/101)} = 201$$

This illustrates the precision required in maintaining stoichiometric balance so as to obtain high degrees of polymerization. Loss of one ingredient, side reactions, or the presence of monofunctional impurities can severely limit the degree of polymerization likely to be achieved.

The analysis can be used to the ease of an added monofunctional reagent, retaining Eqs. 7.18 and 7.19 unchanged if r is appropriately defined in terms of the numbers of functional group present.

Instead of infinity as in case of equal initial concentrations of monomers is the initial concentrations of monomers if the initial concentration of monomers taken is exactly equal but the reaction mixture contains some monofunctional compound. Presence of monofunctional compound would decrease the degree of polymerization. If the fraction of monofunctional compound is r_1, then —

$$\bar{X}_n = \frac{1+r_1}{1+r_1-p}$$

$$= \frac{1+r_1}{r_1} (\text{as } p \to 1) \qquad ...(7.20)$$

For example, consider a mixture of glycol (1 mole), sebacyl chloride (1 mole) and acetyl chloride (0.01 mole, monofunctional). After completion of reaction ($p = 1$), the degree of polymerization will be given by,

$$\bar{X}_n = \frac{1+0.01}{0.01}$$

$$= 101.$$

The importance of exact stoichiometric concentration of monomers must be emphasized. As any error in taking equai mole ratio may yield unexpectedly low molecular weight product so also, any impurity of monofunctional compound also limits the molecular weight of the product.

7.7. Molecular Weight Distribution in Linear Polycondensations Reactions

From our discussion in previous article we have found that the most fundamental point for condensation polymerization is the equal reactivity of all the functional groups. At every stage of polymerization all unreacted functional groups have equal chances to participate in the reaction irrespective of the size of the molecule to which it is attached. At the end of the reaction, the probability that a given functional group has reacted is equal to the fraction of all the functional groups that have reacted.

Considering the polycondensation of a ω-hydroxy acid. Calculation about the probability of a polymer chain composed of x basic repeat units H[—ORCO—]$_x$OH at time, t, during the reaction can be very easily done by considering the number ester linkages.

$$\underset{1}{\text{H—ORCO}}\text{—}\underset{2}{\text{ORCO}}\text{—......—}\underset{x-1}{\text{ORCO}}\text{—}\underset{x}{\text{OH}}$$

In a polymer containing x repeat units, there would be $(x-1)$ ester linkages. The probability that the carboxyl group of the first unit is esterified $= p$.

The probability of the formation of the second ester linkage next to the first $= p^2$.

The probability of the formation of $(x-1)$ ester linkages next to $(x-2) = p^{x-1}$. Hence the probability that all the ester linkages have been formed is p^{x-1}. The polyster should also prossess the unreacted ester linkages at the end. The probability that a group has not reacted is $(1-p)$.

Hence, the probability, that the molecule is composed of x units is represented by,

$$P_x = p^{(x-1)}(1-p) \qquad ...(7.21)$$

If the total number of molecules present at time t is N, then the total number of x-mers is given by,

$$N_x = N\, p^{(x\ 1)}(1-p) \qquad ...(7.22)$$

Let, N_0 be the total number of units,

then, $$N = N_0(1-p) \qquad ...(7.23)$$

Hence, $$N_x = N_0(1-p)^2\, p^{(n-1)} \qquad ...(7.24)$$

Equation (7.24) represents the expression for the number of molecules of length x in terms of the initial number of molecules N_0 and the extent of reaction, p.

The weight fraction, Wx, for a particular length of the polymer is given by

$$W_x = \frac{\text{mass of molecule of length } x}{\text{total mass of all the molecules}}$$

$$= \frac{N_x \times (x - M_0)}{N_0 \times M_0} = \frac{xN_x}{N_0} \qquad \text{...(7.25)}$$

where M_0 is the molar mass of the repeat unit or monomer molecule.

If we ignore the end-groups and the difference in mass between repeat units and monomer molecules, then combining equations (7.24) and (7.25), we get,

$$W_x = xp^{(x-1)}(1-p)^2 \qquad \text{...(7.26)}$$

Equation (7.25) represents the weight distribution fraction for a linear condensation polymerization reaction.

The above equation is illustrated in Fig. 7.5

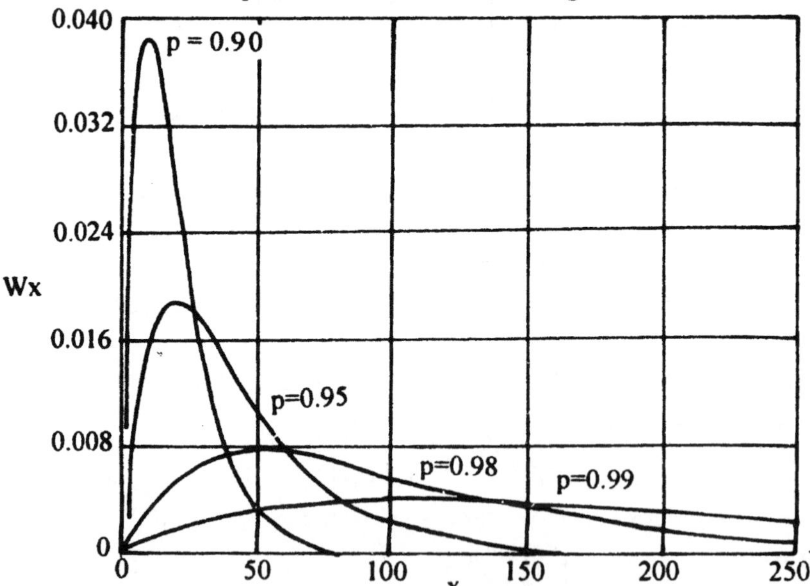

Fig. 7.5: Weight-fraction distribution of chain molecules in a linear stepreaction polymer for several extents of reaction p.

It can be easily conceived that, on a number basis, monomers are more plentiful than any other molecular species at all stage of the reaction. Considering the weight, the proportion of low-molecular weight material is very small and decreases as the average molecular weight increases. The maximum in the weight distribution curve occurs near the number average molecular weight.

7.8. The Number Average Molecule Weight, $\overline{M}_n$

The number average molecular weight is defined as the sum of the product of the molecular weight of each fraction multiplied by its mole of fraction.

$$M_n = \Sigma X_i M_i \qquad ...(7.27)$$

Where X_i is the mole fraction of the polymer molecules having length i.

Since
$$X_i = \frac{N_i}{N} \qquad ...(7.28)$$

Where N_i is the length of the ith molecule N is the total number of molecules

Thus,
$$\overline{M}_n = \Sigma N_i M_i / \Sigma N_i \qquad ...(7.29)$$

and
$$W_i = N_i M_i / \Sigma N_i M_i \qquad ...(7.30)$$

Where W_i is the mass fraction of ith molecules. This can be written as

$$\Sigma(W_i / M_i) = \Sigma N_i / \Sigma N_i M_i \qquad ...(7.31)$$

Combining equation (7.29) and (7.31), we get,

$$\overline{M}_n = 1 / \Sigma(W_i / M_i) \qquad ...(7.32)$$

The number average degree of polymerization can be written as,

$$\bar{x}_n = \overline{M}_n / M_0 \qquad ...(7.33)$$

From equation (7.27),

$$\overline{M}_n = \Sigma N_i M_i / \Sigma N_i \qquad ...(7.34)$$

If a polymer sample contains N molecules then $\Sigma N_i = N$.

Substituting equation (7.22) in equation (7.34), we get,

$$\overline{M}_n = M_0(1 - p)\Sigma x p^{x-1} \qquad ...(7.35)$$

Where M_0 is the molar mass of the polymer repeat unit.
Using the mathematical reaction

$$\sum_{1}^{x} xp^{x-1} = \frac{1}{(1-p)^2} \text{ for } p < \qquad ...(7.36)$$

Since $$\overline{M}_n = \frac{M_0}{1-p}$$

and since $\overline{X}_n = \overline{M}_n / M_0$,

So, $$\overline{X}_n = \frac{1}{1-p} \qquad ...(7.37)$$

In this way **Carothers equation** can be obtained on statistical consideration.

7.9. The Weight Average Molecular Weight $\overline{M}_w$

The weight average molecular weight is the sum of the products of the molecular weight of each fraction multiplied by weight fraction W_i.

$$\overline{M}_w = \Sigma W_i M_i \qquad ...(7.38)$$

Combining equation (7.38) and (7.30), we get,

$$\overline{M}_w = \Sigma N_i M_i^2 / \Sigma N_i M_i \qquad ...(7.39)$$

Since the weight average degree of polymerization can be expressed as,

$$\overline{X}_w = \overline{M}_w / M_0 \qquad ....(7.40)$$

$\overline{M}_w$ may be equal to or greater than $\overline{M}_n$. The ratio $\overline{M}_w/\overline{M}_n$ is used as a measure of breadth of the molecular weight distribution. For a homopolydisperse system, $\overline{M}_w/\overline{M}_n = 1$. For a polydisperse system it changes from 1.5 – 2 to 20 – 50.

For example, if there are 100 polymer molecules of molecular mass 10^4, and 200 molecular of molecular mass 10^5.

Then,

$$\overline{M}_n = \frac{100 + 10^3 + 200\times10^4 + 200\times10^5}{100 + 200 + 200}$$

$$= 10^5(1 + 20 + 200)$$

$$= 4.4\times10^4 = 4400$$

$$\overline{M}_w = \frac{100 + 10^6 + 200\times10^8 + 200\times10^{10}}{100\times10^3 + 200\times10^4 + 200\times10^5}$$

$$= \frac{10^8(1 + 200 + 20,000)}{10^5(1 + 20 + 200)}$$

$$= \frac{10^3 \times 20201}{221}$$

$$\approx 91000$$

It becomes clear from the above calculation that the number-average molecular mass and the weight-average molecular mass are not same for a polydisperse system. For a monodisperse system, the two are same *i.e.*, $M_w = M_n$.

We know that

$$M_w = \Sigma\, W_i M_i$$

Using equation (7.26),

$$Mw = M_0\,(1-p^2)\,\Sigma\, x^2 p^{x-1} \qquad ...(7.41)$$

Since $$\Sigma x^2 p^{(x-1)} = \frac{1+p}{(1-p)^3}$$

Making this substituting in equation (7.41), we get,

$$\overline{M}_w = M_0 \frac{(1+p)}{(1-p)} \qquad ...(7.42)$$

The weight average degree of polymerization,

$$\overline{X}_w = \frac{(1+p)}{(1-p)}$$

or $$\frac{\overline{M}_w}{\overline{M}_n} = (1+p) \qquad ...(7.43)$$

when p tends to unity

$$\overline{M}_w / \overline{M}_n = 2 \qquad ...(7.44)$$

A polymer with a heterogeneity index of 2 has an ideal distribution of molar mass.

7.10. Non-linear Condensation Polymerisation

When one or both the monomers have more than two functional groups, the polymer formed will not be linear. It will be made up of a three dimensional network. If we react a triol with a dibasic acid within a few reaction steps branched intermediates are formed. A very rapid increase in molecular weight can occur resulting in a very high molecular weight polymer that is infusible and insoluble. Such a polymer would be difficult to handle. These types of polymers are known as **thermosetting polymers.** Upto a certain stage in the polymerisation, the product remains in a liquid state known as **gel point.** Many times polymerisation is stopped at this stage and completion of polymerisation is done only during the fabrication of the finished product *e.g.*, phenol formaldehyde resins.

Condensation polymerisation is commercially important for fibres like polyesters, polyamides and polyurethanes, resins such as epoxides, phenolics, urea-formaldehyde, melamine formaldehyde, alkyl resins etc.

Assumption that all functional groups have the same reactivity may not be valid in many cases, but is needed for simplifying the kinetics. Introducing a functionality factor f (>2), as compared to the bifunctional monomers, the extent of reaction p can written as—

$$p = \frac{C^\circ - C}{C^\circ f/2}$$

$$\overline{X}_n = \frac{C^\circ}{C} = \frac{n^\circ}{n}$$

$$= \frac{2}{2 - fp}$$

or

$$\overline{X}_n = \frac{1}{1 - f_{2-} \cdot p} \qquad \text{...(7.45)}$$

and when $f = 2$, eq. (7.45) is identical with eq. (7.12). Consider a reaction between terephthalic acid (1 mole) with glycol (1 mole). At $p = 0.95$, $\overline{X}_n = 20$. Now consider the following composition—

Terephthalic acid	(1 mole)	: $C^\circ = 1$: $n^\circ = 2$
Glycol	(0.9 mole)	: $C^\circ = 0.9$; $n^\circ = 1.8$
Glycerol	(0.91 mole)	: $C^\circ = 0.9$; $n^\circ = 0.3$

Hence per 2 moles of —COOH groups, —OH groups from glycol and glycerol are 2.1 moles, hence $f = 2.1$. Therefore, at $p = 0.5$, $X_n = 400$. Thus, degree of polymerization has increased considerably at low extent of reaction. If the reaction proceeds slightly more *i.e.*, at $p = 0.952$, $\overline{X}_n = \alpha$, *i.e.*, $p = 0.952$ corresponds to **gel point.** Thus, **gel point** is *calculated by equating the denominator of eq. 7.45 to zero, i.e.,* $p = 2/f$.

Non-linear Condensation Polymers (Three Dimensional Polymers)

In the previous paragraphs we have studied that if the functionality of the monomer is greater than two, the polymerizing molecule can grow in more than one dimension forming a nonlinear or three dimensional or cross-linked polymers. Examples of such polymers are met with in the reactions of glycerol with a dibasic acid.

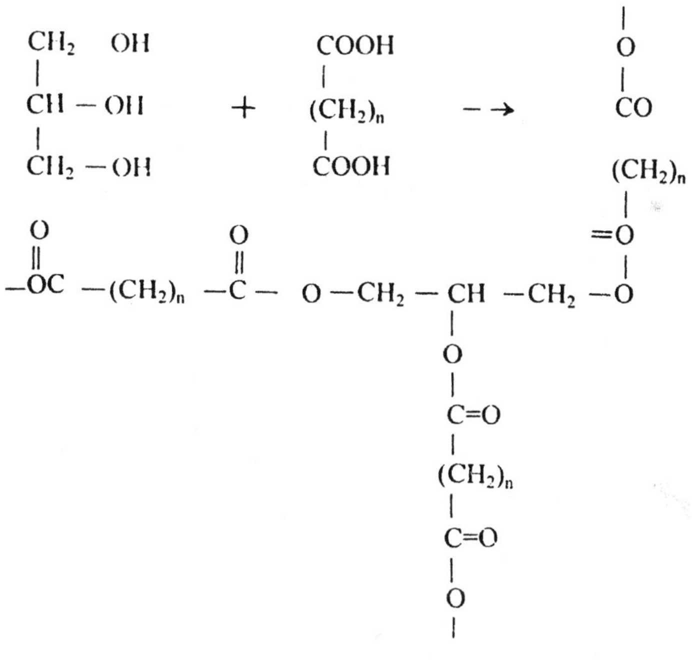

$$\begin{array}{l} CH_2\ \ OH \\ | \\ CH-OH \\ | \\ CH_2-OH \end{array} + \begin{array}{c} COOH \\ | \\ (CH_2)_n \\ | \\ COOH \end{array} \longrightarrow \begin{array}{c} | \\ O \\ | \\ CO \\ (CH_2)_n \\ | \\ =O \\ | \end{array}$$

$$-\overset{\overset{\displaystyle O}{\|}}{O}C-(CH_2)_n-\overset{\overset{\displaystyle O}{\|}}{C}-O-CH_2-\underset{\underset{\underset{\underset{\underset{\underset{\underset{\underset{\underset{\displaystyle |}{O}}{|}}{C=O}}{|}}{(CH_2)_n}}{|}}{C=O}}{|}}{O}}{\overset{}{CH}}\ -CH_2-O$$

The molecular weight distribution of nonlinear polymers are complicated and the polymerization of multifunctional monomers is complicated by the formation of a **gel.** The sudden transformation of the reaction mixture from a viscous liquid to an elastic material of infinite viscosity may be attributed to the formation of infinite network structure. Immediately after gelation, a small fraction of the polymer becomes insoluble in all solvents which do not react chemically in such a way as to degrade the polymer. The other fraction is soluble in solvents. Therefore, there will be two distinct type of materials, the soluble portion known as the **'soil'** and the insoluble portion known as the **'gel'**. As the condensation polymerization proceeds beyond the gel point, the percentage of gel increases at the expense of the sol. The three dimensional polymers that are produced at the gel point have the low number average molecular weight.

7.11. Interchange Reactions

In interchange reactions, which may take place at elevated temperature, as in polymer melts, the molecular weights of the reacting molecules can change, because of the fact that two molecules of average length may reaction to give one longer and one shorter than average. However, the total number of molecules and hence $\bar{x}_n$ remain unchanged.

It can be shown that if free interchange occur, the final molecular-weight distribution is always the most probable distribution. The change can be demonstrated by mixing two polymers of different $\bar{x}_n$ and following a weight-average property (such as melt viscosity) as a function of time at elevated temperature.

7.12. Multichain Polymer

Another type of step-reaction polymer that is of interest is produced by polymerizing a bifunctional monomer of type A-B with a small quantity of monomer of functionality f of the type R-Af. If the reaction is carried nearly to completion, the polymer formed is made up, of f chains growing out from a central unit R. Network structures cannot occur, since no units of the type B-B are present. The case f = 1 corresponds to the most probable distribution. The polymer produced when f = 2 is also linear, those with higher functionalities are branched.

Since the length of each of the branches is statistically determined independent of all the others, the probability of having a molecule with several branches all longer or shorter than average becomes smaller as the number of branches increases. The molecular-weight distribution thus becomes narrower with increasing functionality. The distribution breadth is given by

$$\frac{\bar{x}_w}{\bar{x}_n} = 1 + \frac{1}{f} \qquad \text{...(7.46)}$$

For $f = 2$, $\bar{x}_w / \bar{x}_n = 1.5$.

7.13. Gelation

The **gel point** is defined as that stage in polymerisation upto which the product remains in a liquid state.

Let f_{av} be the average functionality of all the monomer units. Suppose initially the system contains N_0 molecules. The total number of functional group present is $N_0 f_{av}$. If there are N molecules present at time *t*, then the number of functional groups that have reacted is $2(N_0 - N)$ since two different groups must react together to form one linkage.

Let p be the extent of reaction, *i.e.*, the probability that any functional group has reacted, then p is given by the relation

$$p = \frac{\text{Number of groups that have reacted at time } t}{\text{Number of groups present initially}}$$

i.e.,

$$p = \frac{2(N_0 - N)}{N_0 f_{av}} \qquad \text{...(7.47)}$$

But $\bar{X}_n = N_0 / N$

Hence, $\bar{X}_n = 2/(2 - pf_{av})$ ...(7.48)

If it is assumed that at the gel point X_n becomes infinity then the critical extent of reaction for gelation *i.e.* p_G would be,

$$p_G = 2/f_{av} \qquad \text{....(7.49)}$$

Thus any increase in the average functionality produces an excellent effect upon the degree of polymerization.

Calculation of the Gel Point

To calculate the point in the reaction at which gelation occurs a *branching coefficient* α is defined as the probability that a given functional group on a *branch unit* (*i.e.*, a unit of functionality greater than 2) is connected to another branch unit.

The value of α at which gelation is possible can be calculated from statistical considerations to be just $1/(f - 1)$; hence the critical value of α of gelation is

$$\alpha_0 = \frac{1}{f-1} \qquad \text{...(7.50)}$$

Here f is the functionality of the branch units; if more than one type of branch unit is present, an average f over all types of branch units may be used in Eq. 7.50.

The relation between α and the extent of reaction can be shown to be

$$\alpha = \frac{p_A p_B \rho}{p_A p_B (1-\rho)} \qquad \text{...(7.51)}$$

where the extents of reaction for A and B groups are p_A and p_B, and the ratio of A groups on branch units to all A groups in the mixture is ρ. We can eliminate either p_A or p_B from this expression by defining $r = N_A/N_B$, hence $p_B = rp_A$

Then

$$\alpha = \frac{rp_A^2 \rho}{1 - rp_A^2(1-\rho)} = \frac{p_B^2 \rho}{r - \rho_B^2(1-\rho)} \qquad \text{...(7.52)}$$

Simple relations can be obtained for several cases. When equal numbers of A and B groups are present.

$r = 1$ and $p_A = p_B = p$:

$$\alpha = \frac{p^2 \rho}{1 - p^2(1-\rho)} \qquad \text{...(7.53)}$$

When there are no A – A **units,** $\rho = 1$ and

$$\alpha = rp_A^2 = \frac{p_B^2}{r} \qquad \text{...(7.54)}$$

If both conditions apply, that is, $r = \rho = 1$,

$$\alpha = p^2 \qquad \text{...(7.55)}$$

Finally, with only branch units present the probability that a functional group on a branch units leads to another branch unit is just the probability that it has reacted:

$$\text{(branch units only)} \qquad \alpha = p \qquad \text{...(7.56)}$$

The equations hold good for all branch unit functionalities with the definitions given r and ρ.

Experimental Observations of the Gel Point

The gel point can be observed precisely at the time when the polymerizing mixture suddenly loses fluidity, for example, **when bubbles no longer rise in it.** If the extent of reaction has been followed as a function of time, say, by removing aliquots of the solution and titrating for the number of functional groups present, the value of p at the gel point can be determined.

In many cases investigated by Flory the observed values α_0 were always slightly higher than those theoretically required. This discrepancy can be attributed to the reaction of some functional groups to form intramolecular links, that do not contribute to, network structures. The reactions must therefore be done slightly further to reach the critical point.

7.14. Polyfunctional Step-Reaction Polymerization

Three-dimensional step-reaction polymers are obtained from the polymerization of reactants having more than two functional groups per molecule. The structures of such polymers are more complex as compared to those of linear step-reaction polymers. Three dimensional polymerization is complicated by the occurrence of gelation, or the formation of infinitely large polymer network in the reaction mixture. The gelation marks the division of the mixture into two parts: the gel, that is insoluble in all nondegrading solvents, and the sol, that remains soluble and can be extracted from the gel. As the polymerization proceeds beyond the gel point, the amount of gel increases at the expense of sol, and the mixture rapidly transforms from a viscous liquid to an elastic material of infinite viscosity. An important feature of the onset of gelation is the low number-average molecular weight of the mixture at the gel point, where its weight-average molecular weight becomes infinite.

In the statistical consideration of three-dimensional step polymerization, it is assumed that all functional groups are equally reactive, independent of molecular weight or viscosity, and that all the reactions between functional groups on different molecules. Errors arising from these assumptions affect only the numerical agreement between theory and experiment, and not the overall considerations of the theory.

7.15. Ring Opening Polymerisation

Molecules containing strained rings can undergo ring opening and polymerisation. Oxiranes, oxetanes, furans, lactarn and lactones can be cleaves and polymerised by acids or bases

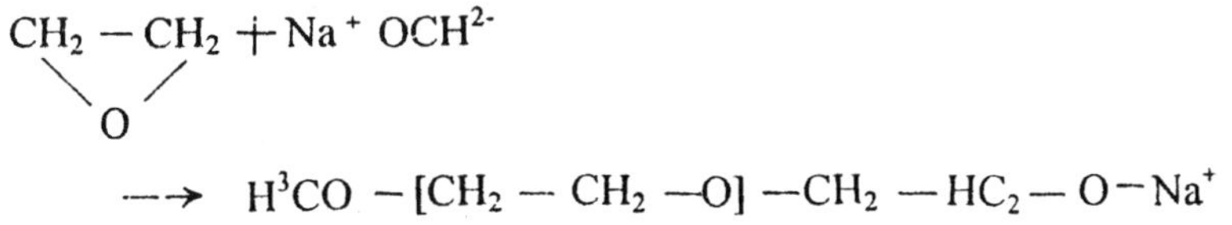

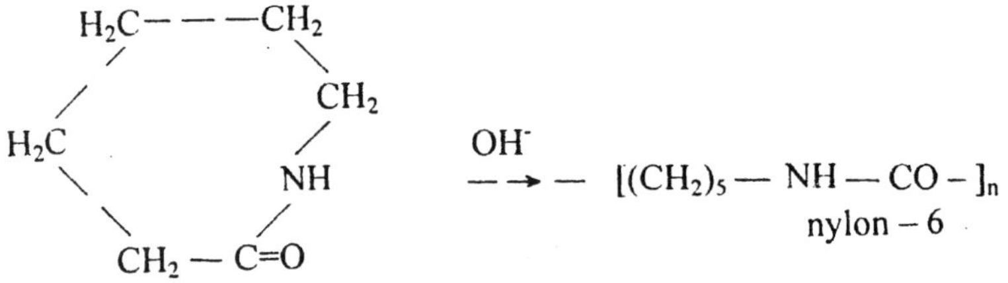

```
CH₂ — CH₂ + Na⁺ OCH²⁻
  \    /
    O

     ——→ H³CO — [CH₂ — CH₂ —O] —CH₂ —HC₂— O—Na⁺

      H₂C———CH₂
      /       \
     /        CH₂
  H₂C          /          OH⁻
     \        NH        ——→—  [(CH₂)₅— NH — CO — ]ₙ
      \      /                       nylon – 6
      CH₂ — C=O
      Caprolatom
```

8

Structure and Characterization of Polymers

8.1. Introduction

The characterization of polymers as very difficult because of various inherent weaknesses observed in case of various polymers. The difficulty in characterization can be well grasped from the fact that unlike low molecular weight compounds, like benzene (M=78), methacrylate (M=100.12), vinyl chloride (M=62.5) etc. which have a fixed value of molecular weight, the polymers which are macromolecules, do not have a certain definite and permanent molecular weight. For example, the molecular weights of polystyrene varies between 50,000 to 10 million.

Because of this difference in molecular weight the same macromolecular compound greatly differs in properties. For example, polystyrene, with a molecular weight below 10,000 is a bright material without any mechanical strength, that can be easily rubbed into fine powder, and at a temperature of 50°C it becomes an oily liquid. Whereas, moulded polystyrene of a molecular weight of 250,000 is a hard and strong glassy plastic, and polystyrene with a molecular weight over 1 million, gets precipitated out of its solutions; as a fibrous material and at temperatures overs 100°C it turns into a tough rubbery material. Generally all properties of macromolecular compounds change with the molecular weight, some only in a small region, upto a certain value of the molecular weight that is generally around 100,000.

However, other properties, like the viscosity, change continuously over the entire region of molecular weights obtainable.

Thus for characterization of a polymeric material its molecular weight must form part of its description. Thus by the designation of polystyrene of molecular weight 200,000, the number of possible polystyrenes is limited to a very smaller molecular weight range. Unfortunately, the characterization is still not complete, because macromolecules do not have a uniform length. Thus, a molecular weight of 200,000 can result from a number of different situations. All molecules could have molecular weights near 200,000 (for example from 190,000 to 200,000) in which case we speak of a narrow distribution range of molecular weight. In some other case, the molecular weights may lie between 10^4 and 5 million or more, and then one has a broad distribution. The manner in which the macromoleculars of a product are distributed over the different molecular weights can be seen from the molecular weight distribution curve, which has its maximum at the molecular weight most fraudulently present. However, in some cases two or more maximise may occur in distribution curves. Such a situations has been observed with natural products and also when we mix two polymers with different molecular weights.

However, before proceeding to the determination of molecular weights of polymers we will take up an elementary discussion of the use of *x*-ray diffraction, spectroscopic techniques and electromicroscopic techniques etc. in determining the structure of polymers.

First of all we will take up the chemical analysis of polymers.

8.2. Chemical Analysis of Polymers

The chemical analysis of polymers is very similar to the chemical analysis of low molecular weight organic compounds, if we make suitable modification to ensure solubility or the availability of sites for reaction (*e.g.*, insoluble specimens should be ground to expose a large surface area). The general methods used for functional group and elemental analysis are applicable. Chemical reactions of polymers also provide means of chemical analysis, as also their reactions of degradation.

The mass spectrometry and gas chromatography are the two powerful techniques that are used for this purpose. These are discussed briefly in the following section:

Mass Spectrometry

For mass spectrometry to polymer systems, the polymer is allowed to react to form low-molecular-weight fragments, which are condensed at liquid-air temperature. They are then volatilized, ionized and separated according to mass and charge by the action of electric and magnetic fields in typical mass spectrometer analysis. From the abundance of the various ionic species found, the structures of the low-molecular-weight species can be inferred.

Gas Chromatography

Gas chromatography is a method of separation wherein gaseous or vaporized components are distributed between a moving gas phase and a fixed liquid phase or solid adsorbent. By a continuous succession of adsorption or elution steps, occurring at a specific rate for each component, separation can be achieved. The components can be detected by one of various methods as they emerge successively from the chromatographic column. From the detector signal, proportional to the instantaneous concentration of the dilute component in the gas stream, information about the number, nature and amounts of the components present is obtained.

Disorder in the Crystal Structure

However, defects and distortions are present in crystals of polymers and they play an important role in the crystal structures of polymers, X-ray diffraction data provides a lot of information about the qualitative aspects of disorder in the crystal structure, because disordering results in the broadening of the diffraction maxima and a systematic study of the broadening gives detailed information about the types of disturbances in the crystals.

Quantitative measurement of the disorder is not possible because broadening of the diffraction maxima is produced both by small crystalline size and by distortions within larger crystals.

8.3. Polymers and X-ray Diffraction

The x-ray diffraction method is a powerful tool to investigate orderly arrangement of atoms or molecules through the interaction of electromagnetic radiation to give interference effects with structures comparable in size to the wavelength of the radiation. If the structures are arranged in an orderly array or lattice, the interferences can be sharpened and the condition can be found when the radiation is scattered or diffracted. From this it is possible to get information about the geometry of the scattering structures.

X-ray crystallographic techniques when extended to polymeric solids some interesting features of the internal structure of these substances. It was found that good majority of polymers diffract X-rays like any crystalline substance but many behave like amorphous materials giving very broad and diffuse X-ray diffraction patterns. This is seen in Fig. 8.1.

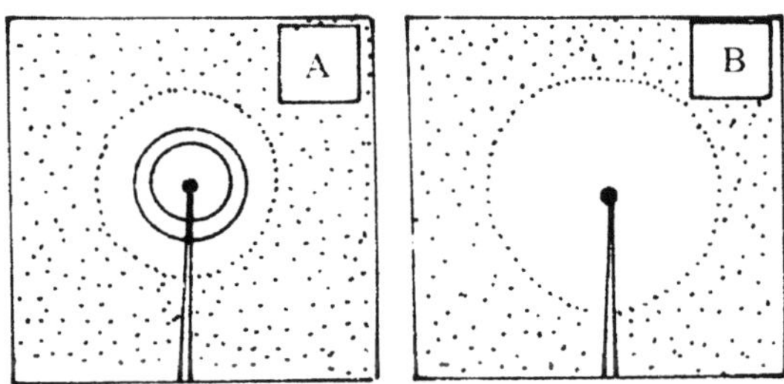

Fig. 8.1: X-ray diffraction patterns of (A) highly crystaliine polymer and (B) amorphous polymer.

Unlike simple inorganic compounds (*e.g.*, NaCl or KCl), polymers do not have a perfectly ordered crystal lattice formation and are not completely crystalline. In fact, they contain both crystalline and amorphous regions. Hence, the X-ray diffractions from them are found to be a mixture of sharp as well as diffused patterns.

The wavelengths of *x*-rays are comparable to interatomic distances in crystals; the information obtained from scattering at wide angles describes the spatial arrangements of the atoms. Low-angle *x*-ray scattering is useful in detecting large periodicities, which may arise from lamellar crystallites or from voids.

Experimental Method

Monochromatic x-rays are used and the diffracted x-rays are detected by their action on photographic flues or plates. They can also be detected by means of a radiation counter and electronic equipment feeding data to a computer. Since *x*-rays of a given wavelength are diffracted only for a certain specific orientations of the sample so if the sample is a single crystal it should be placed in all possible orientations during experiment. For this the sample is rotated or oscillated about one of its axes. This can also be achieved by using a sample made as a powder of a very small crystal, in such a powder the minute particles are randomly oriented and all possible orientations are included.

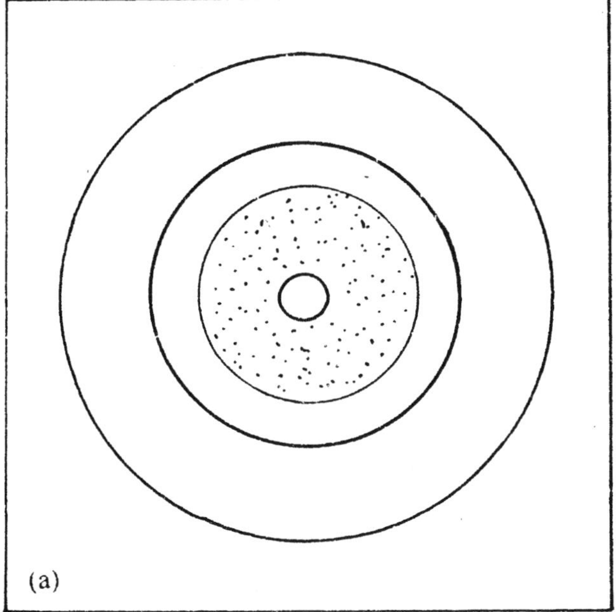

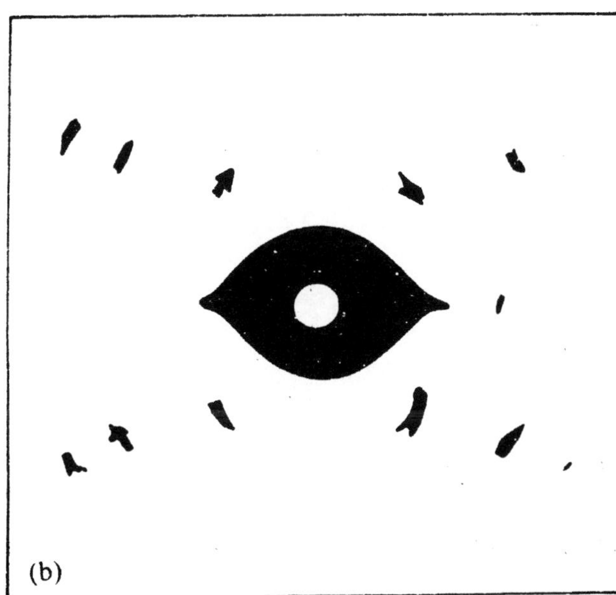

Fig. 8.2: X-ray diffraction patterns for unoriented (a) and a oriented (b) polyoxymethylene (courtesy of E.S. Clark).

Application to Polymers

Since polymer single crystals prepared these days are too small for x-ray diffraction experiments, the crystal structure of a polymer is generally determined from x-ray patterns of a fibre drawn from the polymer. Due to the alignment of the crystalline regions with the long axes of the molecules parallel to the fibre axis, the pattern is essentially identical to a rotation pattern from a single crystal. In such a pattern diffraction maxima occur in rows perpendicular to the fibre axis, known as *layer lines* (Fig. 8.2 *b*).

Chain Conformations

Since in molecular structure of a polymer a *repeat unit* exist and the periodicity of its crystal is characterized by a *repeat distance* so repeat distance is directly determined by measuring the distance between the layer lines. The greater the repeat distance, the closer together are the layer lines. Determination of the repeat unit is quite difficult. It proceeds through derivation, from the positions of the diffracted *x*-ray beams, of the dimensions of the *unit cell.* The positions of the atoms in the unit cell are derived from the relative intensities of the diffracted beams. However, the diffraction patterns of polymers, do not provide enough information to permit such analyses to be carried to completion. More structural information is used from other sources, like normal bond lengths and angles and atomic arrangements along the chain suggested by the chemical structure of the sample. For example, the repeat distance of 25.5 nm in the crystals of polyethylene is readily identified with a single repeat unit in the planar zigzag conformation.

Many polymer conformations can be described in terms of atoms regularly spaced along helices. Methods for analyzing the diffraction patterns from helical structures provide a highly versatile technique for the determination of the repeat unit in such polymers.

Chain Packing

The packing of the chains can be described most completely in terms of the unit cell and its contents. The volume of the unit cell, and thus the volume occupied in the crystal by a single repeat unit, can be obtained from the repeat distance and the positions of the diffraction spots on the layer lines. This volume can measure the density of the crystals, that is useful in determining the degree of crystallinity.

The atomic arrangement within the unit cell is more difficult to determine than the cell dimensions. Trial structures deduced from these dimensions and a knowledge of the chain conformation, are tested in terms of calculated and observed diffraction intensities, making use of well-established crystallographic techniques.

8.4 Microscopy

The application of election microscopy technique to polymer analysis involves sufficient extension beyond the ordinary techniques. Some salient points are discussed here.

Light Microscope

Reflected light microscopy can be used for examining the texture of solid opaque polymers. Materials which can be prepared as thin films are generally examined by transmitted light. Two common techniques used are (i) polarized-light microscopy, and (ii) phase contrast microscopy.

(i) *Polarized Light Microscopy*

In this we take advantage of the ability of crystalline materials to rotate the plane of polarized light. Thus the structure of spherulites is studied with the sample between crossed polarizers and the crystalline melting point is taken as the temperature of disappearance of the last traces of crystalinity when using a hot stage polarizing microscope.

(ii) *Phase Contrast Microscopy*

A second useful technique is phase-contrast microscopy which allows observations of structural features involving differences in refractive index rather than absorption of light as in the conventional case. Interference microscopy can be used to measure thicknesses as low as a few angstrom units and has proved valuable in the study of polymer single crystals.

Electron Microscopy and Electron Diffraction

Electron microscopy can be used for resolution of smaller objects the practical limit of resolution being a few angstrom units. Electron microscopy has been used in the study of the morphology of crystalline polymers. The usual techniques of replication, heavy-metal shadowing, and solvent etching are widely used. The direct observation of thin specimens, like polymer single crystals, is also possible and permits the observation of the electron-diffraction pattern of the some specimen area, which is invaluable for determining crystallographic directions and relating them to morphology. The damage of the specimen by the electron beam causes problems. Polymer single crystals get severely damaged in a few second to a minute. The alleviate from it the sample is maintained well below room temperature with a cold stage, by the use of accelerating voltages several times higher than the usual 50,000 – 100,000 V, or by the use of an image intensifier providing a cathode-ray tube image with far less beam current than that generally needed.

Scanning Electron Microscopy

In this type of microscopy a fine beam of electrons is scanned across the surface of an opaque specimen to which a light conducting film has been applied by evaporation. Secondary electrons, backscattered elections, or (in the *electron microprobe*) x-ray photons emitted when the beam strikes the specimen are collected to provide a signal that is used to modulate the intensity of the electron beam in a television tube, scanning in synchronism with the microscope beam. Since the latter maintains its small size over large distances relative to the specimen so the resulting images have great depth of field and a remarkable three-dimensional appearance. Resolution is limited at present to the order of 1000 nm.

8.5. Spectroscopic Methods

Infrared Spectroscopy

Emission or absorption spectra are produced when molecules undergo transitions between quantum states that correspond to two different internal energies. They energy difference ΔE between the states is related to the frequency of the radiation emitted or absorbed by the equation $\Delta E = h\nu$. Infrared frequencies in the wavelength range 1 – 50 μm are associated with molecular vibration and vibration-rotation spectra.

1 μm = 10,000 cm^{-1}, 59 μm = 200 cm^{-1}, μm = micrometer, (micron), 10^{-6} meter.

A molecular that contains N atoms has 3N normal vibration modes. These include rotational and translational motions of the entire molecule. For highly symmetrical molecules having very few atoms, the entire infrared and Raman spectrum can be correlated to and explained by the vibrational modes. However, even for most low molecular-weight substances, N is too large for such analysis but useful information can be obtained because some vibrational modes involve localized motions of small groups of atoms and produce absorption bands at frequencies which are characteristic of these groups and the type of motions undergone by them.

In polymers the infrared absorption spectrum is generally very simple, considering the large number of atoms that are involved. This simplicity is due to the fact that many of the normal vibrations have almost the same frequency and so appear in the spectrum as one absorption band and, also from the strict selection rules that avoid many of the vibrations from causing absorptions.

The Raman effect is produced when the frequency of visible light is changed in the scattering process by the absorption or emission of energy produced by changes in molecular vibration and vibration-rotation quantum states.

Experimental Methods

Here we will take up briefly some problems of sample preparation for polymers. The greatest experimental difficulties in infrared spectroscopy of polymers, is in obtaining sufficiently thin samples. Methods of sample preparation are many but compression molding is by far the most widely used, other methods include dissolving the polymer in a solvent, like carbon disulfide or tetrachloroethylene, whose spectrum is relatively free of intense absorption bands: preparing a thin film by microtoming or milling; casting a thin film from solution; and pressing a finely ground mixture of the sample with KBr to form a disc or water.

Detection of Chemical Groups in Typical Spectra

A few of the many chemical linkages or groups that can be detected in polymer spectra, along with the approximate wavelengths at which they occur are shown in Fig. 8.3. The infrared absorption bands of interest in common polymers are those in polyethylene which correspond to C—H stretching (3.4 µm), C—H bending of CH_2 groups (6.8 µm) and CH_3 groups (shoulder at 7.25 µm on an amorphous band at 7.30 µm), and CH_2 rocking of sequences of methylene groups in paraffin structures (13.9 µm). Some other absorption bands of interest are those produced due to C = C at 6.1 µm in natural rubber; carboxyl at 5.8 µm and ether at 8.9 µm in poly methyl methacrylate); aromatic structures at 6.2, 6.7, 13.3, and 14.4 µm in polystyrene; C—Cl at 14.5 µm in poly (vinyl chloride); peptide groups at 3.0, 6.1, and 6.5 µm in nylon; and CF_2 at 8.2 – 8.3 µm in polytetrafluoroethylene.

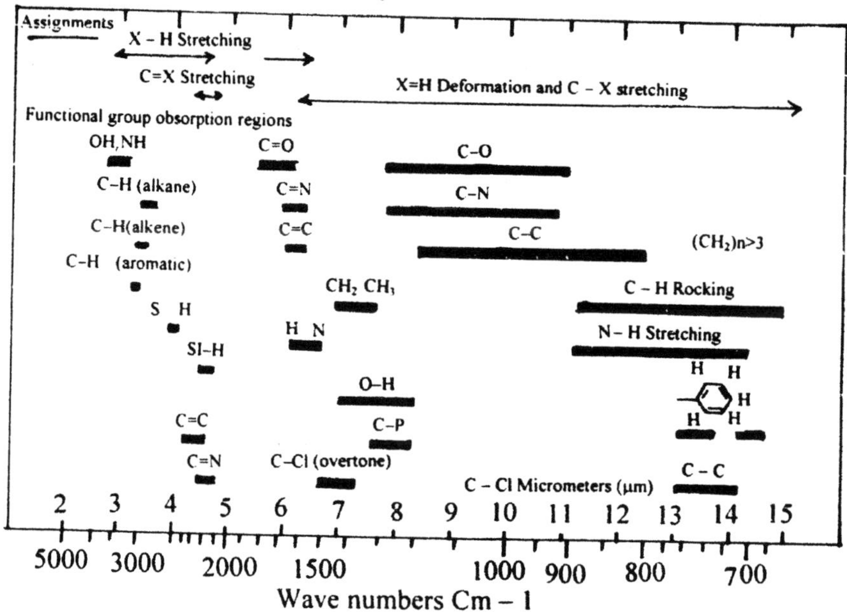

Fig. 8.3: Infrared absorption bands of interest in polymers arranged by approximate wavelength and frequency.

Related Methods

To supplement infrared absorption (in the 2–15 µm wavelength rang) observations in the *far infrared* region (say up to 200 µm) become important. Inspite of serious experimental difficulties some valuable information is available in this region; For some polymers, like polytetrafluorethylene, most of the absorption bands occur above 15 µm (far infrared region).

The applicaticn of *Fourier-transform* infrared spectroscopy of polymers has been reviewed. Since the requirements for activity of a vibration for causing absorption in the infrared and for causing *Raman scattering* are generally different so information from Raman experiments in general supplements information that is obtainable from infrared absorption. Using of lasers as light sources can greatly reduce experimental difficulties in this technique. At present it has become one of the standard tools for polymer analysis.

Dichroism

For any molecular vibration that leads to infrared absorption, there is a periodic change in electric dipole moment. In case the direction of this change is parallel to component of the electric vector of the infrared radiation, absorption takes place; otherwise it does not. In oriented bulk polymers, the dipole-moment change can be confined to specified directions. The use of polarized infrared radiation in such a case leads to absorption which is a function of the orientation of the plane of polarization. The phenomenon is known as *dichroism* and is generally measured as the *dichroic ratio,* the ratio of the optical densities of an absorption band measured with radiation polarized parallel and perpendicular, respectively, to a specified direction in the sample. Dichroic ratios depend both on the degree of orientation and the angle between the direction of the transition moment and the selected direction in the sample. They generally range between 0.1 and 1.0.

Crystallinity

The infrared absorption spectra of the same polymer in the crystalline and amorphous states may differ because of the following two reasons: (i) Specific intermolecular interactions may exist in the crystalline polymer which lead to sharpening or splitting of certain bands; and (ii) Some specific conformations may exist in one but not the other phase, which may lead to bands characteristic exclusively of either crystalline or amorphous material. For example in poly(ethylene terepthalate), the $-OCH_2CH_2O-$ portion of each repeat unit is restricted to the all trans-conformation in the crystal, but can exist in part in the gauche form in the melt. Several bands characteristic of each conformation have been identified. In favourable cases, like 66-nylon percent crystallinity can be determined in absolute terms from infrared absorption data.

Geometric Isomerism

The determination of the various types of geometric isomers associated with unsaturation in polymer chains is of great importance, for example, in the study of the structure of modern synthetic rubbers. In Table 8.1 are listed some of the important infrared absorption bands which arise from olefinic groups. In synthetic "natural" rubber, *cis*-1, 4-polyisoprene, relatively small amounts of 1,2 and 3,4-addition can easily be detected, though it is more difficult to distinguish between the cis and trans-configurations. Nuclear magnetic resonance spectroscopy is also useful for this analysis.

Table 8.1: Absorption Wavelengths of Olefinic Groups

Group Containing C=C	*Wavelength (μm)*
Vinyl, $R_1CH=CH_2$	10.1 and 11.0
trans-$R_1CH=RHR_2$	10.4
Vinylidene, $R_1R_2C=CH_2$	11.3
$R_1R_2C=CHR_3$	12.0
cis-$R_1CH=CHR_2$	14.2 (variable)

Nuclear Magnetic Resonance Spectroscopy

Since about 1960 nuclear magnetic resonance (NMR) spectroscopy has become an important tool for the study of chain configuration, sequence distribution and microstructure of polymers. Its use started from early broad-line studies of the one-set of molecular motion in solid polymers and passed through the solution studies of proton NMR, to the application of the more difficult but more powerful carbon-13 NMR methods to both liquids and solids.

Some important aspects of NMR spectroscopy are discussed here.

Experimental Methods

The NMR technique makes use of the property of spin (angular momentum and its associated magnetic moment) possessed by nuclei whose atomic number and mass number are not both even. Such nuclei include the isotopes of hydrogen and ^{13}C, ^{15}N, ^{17}O, and ^{19}F. Application of a strong magnetic field to material containing such nuclei splits the energy level into two, representing states with spin parallel and antiparal'el to the field. Transitions beiween the states lead to absorption or emission of an energy

$$E = h\nu_0 = 2\mu H_0 \qquad ...(8.1)$$

where the frequency ν_0 is in the microwave region for fields of strength H_0 of the order of 10,000 gauss and up, and μ is the magnetic moment of the nucleus. The energy change is observed as a resonance peak or line in experiments where either H or ν_0 is charged, while the other kept constant.

With an assembly of nuclei, the field on any one is modified by the presence of the others:

$$h\nu_0 = 2\mu(H_0 + H_L) \qquad ...(8.2)$$

where H_L is the local field having a strength of 5—10 gauss. A distribution of local fields generally exists and resonances line gets broadened. This broadening can be studied as a function of temperature to indicate the temperature of the onset of molecular motion. However, with ample molecular motion present (as with liquids and solutions) quite narrow lines can be observed. The positions of these lines on the scale of frequency or magnetic field depend on the local fields, that in turn arises from the nature and location of the atomic groups in the vicinity of the protons. The displacements in the resonance known as *chemical shifts* can be measured in parts pcr million in frequency (or the equivalent field strength) on a scale labelled δ. The zero of the δ scale is a reference point provided by the single resonance of the equivalent protons in a substance showing minimum chemical shift, like tetramethyl saline. Tables of chemical shifts arising from group in the neighbourhood of the nucleus being studied are available in any standard text-book of spectroscopy.

In proton NMR, additional complexity and additional information arises from coupling of the resonance of protons on adjacent carbon atoms which result in the splitting of their resonance into $n+1$ peaks, where n is the number of equivalent neighbouring protons. For in the interpretation, two experimental modifications are found useful. One of these is the use of high magnetic field strengths, in the range 60,000–220,000 gauss compared to the few tens of thousand gauss used in early broad-line work. Super conducting magnets are used to field strengths. Fig. 8.4 shows the increased resolution at the higher field strength which helps in the interpretation of the spectra.

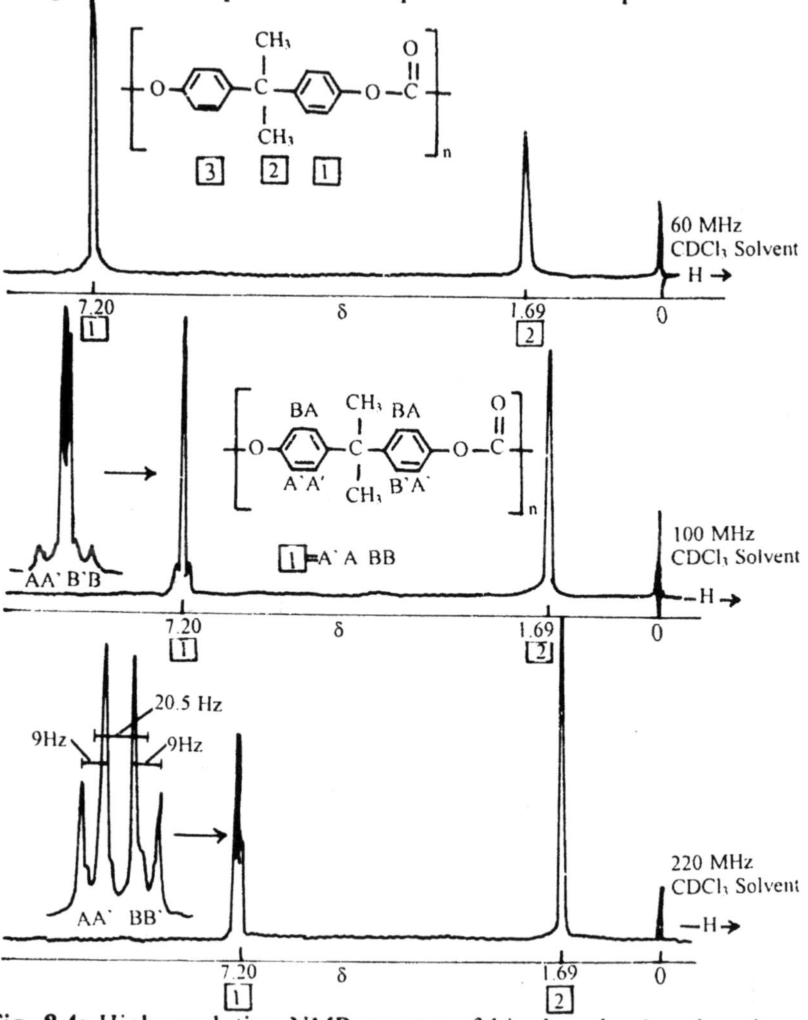

Fig. 8.4: High-resolution NMR spectra of biophenol—A polycarbonate in deutrochloroform solution showing the effect of frequency on resolution.

The other is double resonance or spin decoupling which effects great simplifications in the spectra. A second radiofrequency field is used which has the effect to removing the coupling and collapsing multiplet spectra to much simpler ones (Fig. 8.5). ^{2}H NMr is also used to provide spectra simplified by elimination of proton-proton coupling because the lower frequency of occurrence of the deuteron.

The advantage of ^{13}C NMR over proton NMR is that much greater range of chemical shifts is exhibited by ^{13}C (and most other nuclei), some 200 ppm in contrast to 10 ppm for ^1H. Moreover, there is no carbon–carbon spin coupling because of the low chemical abundance of the ^{13}C nucleus, 1.1% relative to ^{12}C. Finally, the resonances of carbon nuclei are sensitive to configurational as also the chemical environments. However, these advantages are offset by the lower abundance of the ^{13}C nuclei and their lower nuclear magnetic moment about one quarter that of the proton; both effects reduce the intensity of the NMR signal.

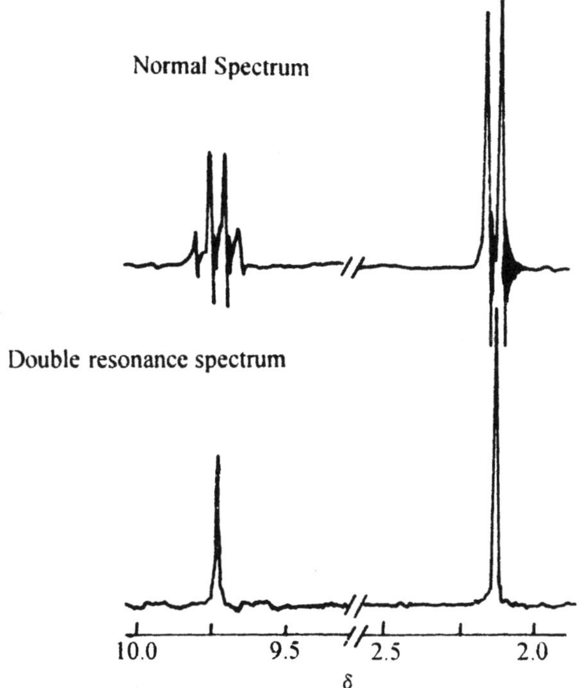

Fig. 8.5: NMR spectra of propylene oxide in carbon tetrachloride solution; (top) conventional spectrum and (bottom) double-resonance spectrum showing simplification resulting from collapse of the splitting due to coupling.

This lower sensitivity can be using pulsed Fourier-transform NMR, wherein a high-power microsecond pulse of radiofrequency energy sets all the carbon nuclei into resonance at once, thus eliminating the need to sweep the frequency or magnetic field. The data are recorded as the subsequent decay of the resonance with time, which is the Fourier transform of the desired spectrum. Records from repeat pulses are summed by computer, that transforms the data to the spectrum desired when the signal-to-noise ratio has reached a satisfactory level.

The local field seen by a pair of like nuclei in a solid sample is given by an isotropic component (the origin of the chemical shift observed in solution) plus a broadening component proportional to 3 $\cos^2\theta - 1$, where θ is the angle between the line joining the nuclei and the direction of the magnetic field. In the random, structure of the polymer solid this component broadens the resonance to a width greater than the total ^{13}C spectrum but when the specimen can be oriented and spun around an axis at $\theta = 54.7°$, for which $3\cos^3\theta - 1 = 0$ – the so-called **"magic angle"** –the line width gets reduced to such an extent that all the carbon resonance can be resolved.

Stereochemical Configuration

Another important application of NMR to polymer systems is the elucidation of the stereochemcial configurations of polymer chains. Poly(methyl methacrylate) was first studied by Bovey in (1960). It is now possible to analyze for the statistical frequency of occurrence of all possible combinations of up to four successive pairs of units (dyads) capable of occurring with either the same (*meso*) or opposite (*racemic*) configurations.

Geometric Isomerism

Both proton ^{13}C NMR are found to be quite useful in supplementing infrared spectroscopy for determining isomerism around the carbon–carbon bond in polymer chains.

Copolymer Sequences

The principles of analyzing short sequences of monomers ir a copolymer are same as those of determining stereochemical configuration. NMR results have contributed to a large extent in this field.

Electron Paramagnetic Resonance Spectroscopy

Electron paramagnetic resonance (EPR) and NMR spectroscopy are quite similar in their basic principles and in experimental techniques. They detect different phenomena and thus yield different information. The major use of EPR spectroscopy is in the detection of free radicals which are uniquely characterized by their magnetic moment that arises from the presence of an unpaired electron. Measurement of a magnetic property of a material containing free radicals, like its magnetic susceptibility, provides the concentration of free radicals, but it lacks sensitivity and cannot reveal the structure of the radicals. Electron paramagnetic resonance spectroscopy is essentially free from these defects.

Experimental Method

As in NMR spectroscopy due to the action of a strong magnetic field on a material containing free radicals the degeneracy of their groundstate energy levels is removed. For low radical concentrations the new energy levels are given by two terms. The first is

$$E = h\nu_0 = g\beta\mu_0 H_0 \quad \text{...(8.3)}$$

where g is a *tensor* relating the field direction and the symmetry directions in the radical, β is *magnetic moment* of the electron spin, and μ_0 is the *magnetic permeability* a vacuum. The second term represents coupling of the electron spin with the nuclear spins in the molecule. This coupling leads to the splitting of the resonance line into a symmetrical group of lines whose positions and amplitudes are found to depend on, the structure of the radical. So it provides information about the structure of the radical.

Applications to Polymers

The investigation of free radicals formed by high energy irradiation of polytetrafluoroethylene is an example of an early application of EPR to polymer radicals.

The EPR spectrum of irradiated polytetrafluoroethylene can be interpreted as arising from radicals of the type shown below:

$$\text{—}(CF_2)_x CF(CF_2)_y\text{—}$$

which remain trapped in the polymer after a C—F bond has been broken and the fluorine atom has diffused away. The spectrum shows fine structure because of coupling between the unpaired electron and neighbouring ^{10}F nuclei. The fine structure is lost at a temperature below 270 K because the motion of the ^{16}F nuclei gets slowed down. The spectrum does not change at higher temperature which shows that the radical is stable to at least 550 K. Secondary products, formed by the reaction of the primary radical with such substances as O_2 and NO, have also been studied.

It may be noted that the strength of the C—F bond and the tendency of polytetrafluoroethylene to degrade rather than crosslink on irradiation suggests the breaking of a C—C rather than a C—F bond as a likely source of radicals in this polymer.

8.6. Thermal Analysis

With introduction of simple, inexpensive instruments for several types of thermal measurements it became possible to carry out thermal analysis of polymers about 20 years ago thermal analysis now includes many equipments for thermogravimetric analysis and thermomechanical analysis, electrical thermal analysis, and effluent gas analysis. We can study the enthalpy changes associated with heating, annealing crystallizing, or otherwise thermally treating polymers, as also can study a wide variety of responses of the system to temperature which include polymerization, degradation, or other chemcial changes.

Differential Scanning Calorimetry

Experimental Methods

Measurements of specific heat and enthalpies of transition are now usually carried out on quite small samples in a *differential scanning calorimeter (DSC).* DSC is applied to two different moles of analysis, of these the one is more closely related to traditional calorimetry and is described here. In DSC an average-temperature circuit measures and controls the temperature of sample and reference holders to conform to a predetermined time-temperature programme. This temperature is plotted on one axis of an $x-y$ recorder. Simultaneously a temperature-difference circuit compares the temperatures of the sample and reference holders and proportions power to the heater in each holder so that the temperatures remain equal. When the sample undergoes a thermal transition, the power to the two heaters is adjusted to maintain their temperatures, and a signal proportional to the power difference is plotted on the second axis of the recorder. The area under the resulting curve gives a direct measure of the heat of transition.

Though DSC is less accurate than a good adiabatic calorimeter (1–2% versus 0.1%), but its advantages of speed and low cost makes it outstanding instrument for most modern calorimetry.

Application to Polymers

Fig. 8.6 shows the specific heat temperature curves obtained (by adiabatic calorimetry) on heating quenched (amorphous) specimens of poly(ethylene terephthalate). Each curve rises linearly with temperature at low temperatures and then rises more steeply at the glass transition, 60–80°C. With the onset of mobility of the molecular chains above this transition, crystallization occurs, as shown by the sharp drop in the specific heat curve. At still higher temperature, 220–270°C, the crystals melt with a corresponding rise in the specific heat curve.

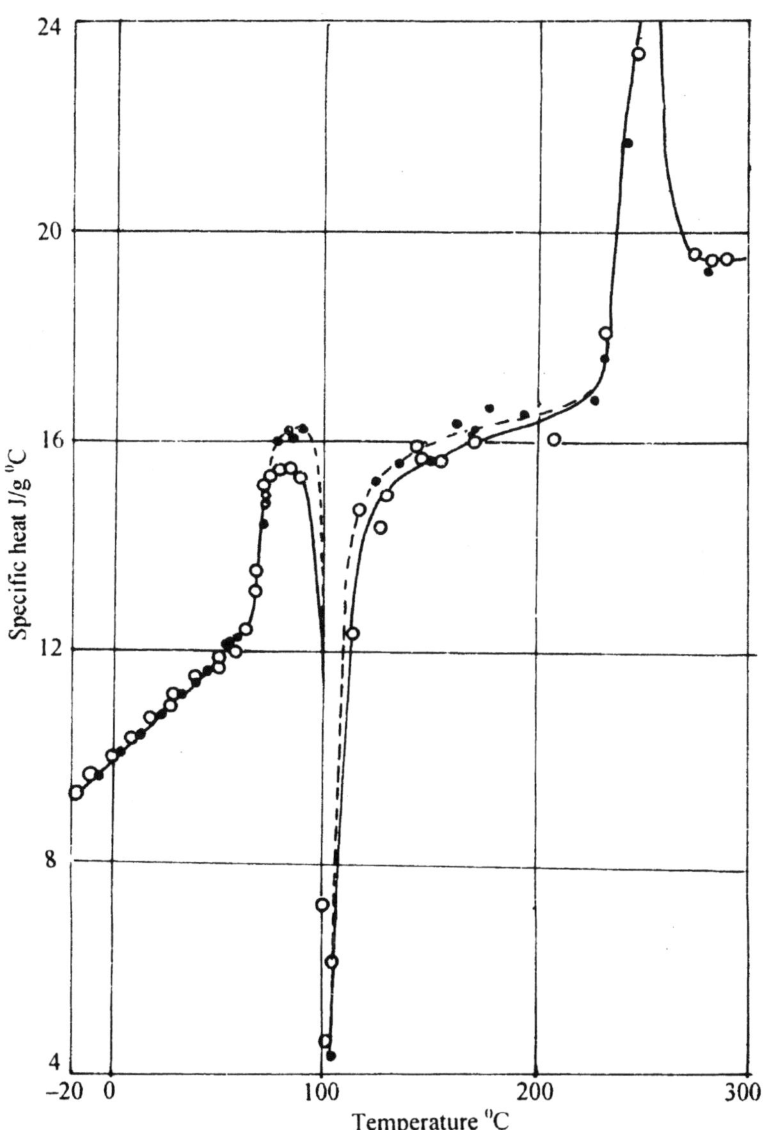

Fig. 8.6: Curves of specific heat as a function of increasing temperature for quenched (amorphous) poly(ethylene terephthalate).

Differential Thermal Analysis

Experimental Methods

In differential thermal analysis (DTA) the sample and an inert reference substance, undergoing no thermal transition in the temperature range under study, are heated at the same rate. The temperature difference between sample and reference is measured and plotted as a function of sample temperature. The temperature difference is finite only when heat is being evolved or absorbed because of exothermic or endothermic activity in the sample, or when the heat capacity of the sample changes abruptly. As the temperature difference is directly proportional to the heat capacity so the curves are similar to specific heat curves, but are inverted because, by convention, heat evolution is registered as an upward peak and heat absorption as a downward peak.

Application to Polymers

Fig. 8.7 shows the differential thermal analysis curve for poly (ethylene terephthalate) which shows features that may readily be identified by comparison with Fig. 8.5. The lower crystalline melting range in the specimen of Fig. 8.7 can be attributed to impurities present in the polymer.

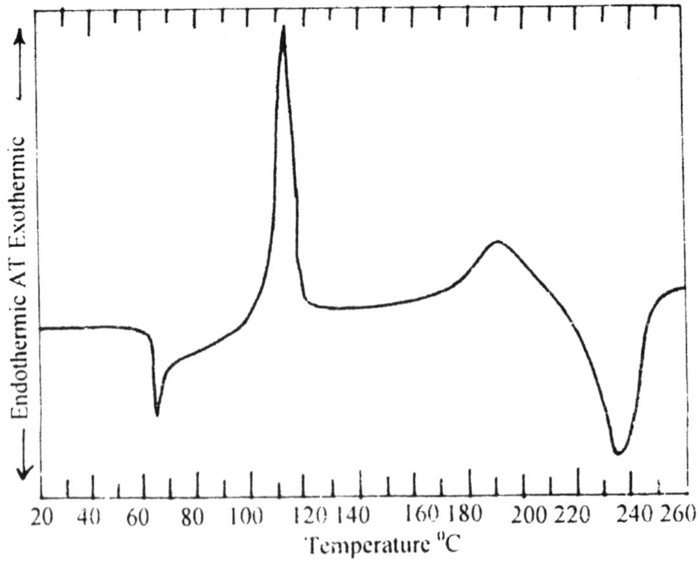

Fig. 8.7: Differential thermal analysis curve for amorphous poly(ethylene terephthalate).

Other Thermal Methods

Thermogravimetric Analysis

In thermogravimetric analysis (TGA) a sensitive balance is used to follow the weight change of the sample as a function of temperature. Its applications include the assessment of thermal stability and decomposition temperature (Fig. 8.8), extent of cure in condensation polymers, composition and some information on sequence distribution in copolymers, and composition of filled polymers, among many others.

Interpretation of these curves show that Poly(vinyl chloride) (PVC) first loses HCl; later the mixture of unsaturated carbon-carbon backbone and unchanged poly(vinyl chloride) partly degrades to small fragments. Poly (methyl methacrylate) (PMMA), branched polyethylene (HPPE), and polytetra fluorethylene (PTFE) degrade completely to volatile fragments, while a polyimide (PI) partially decomposes, forming a char above 800°C.

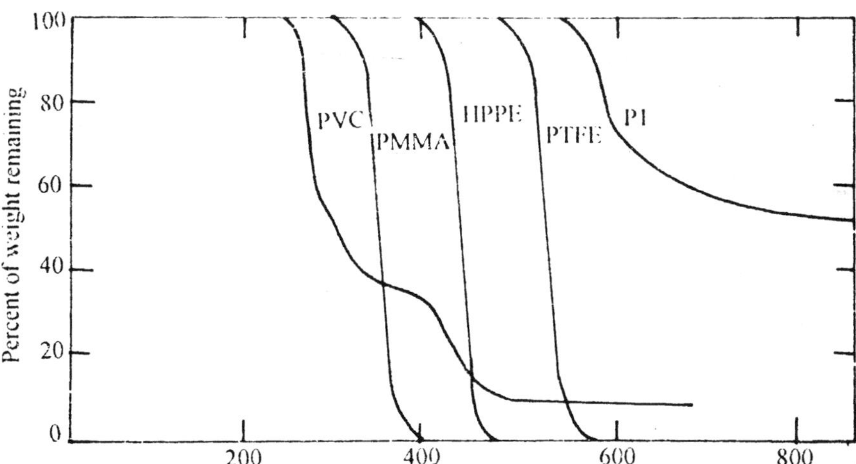

Fig. 8.8: Relative thermal stability of polymers as determined by weight loss on heating at 5°C/min in nitrogen in thermogravimetric analysis.

Thermomechanical Analysis

Thermomechanical analysis (TMA) helps to measure the mechanical response of a polymer system with the change of temperature. These measurements include *dilatometry, penetration* or *heat deflection, torsion modulus,* and *stress-strain behaviour*.

8.7. Molecular Weight Determination

The molecular weights of polymers can be determined by chemical or physical methods of functional group analysis, by *measurement of colligative properties, light scattering* or *ultra-centrifugation* or by *measurement of dilute solution viscosity.* The range in which each of these methods is applicable is given below:

Method		*Molecular weight*
End group analysis	....	upto 3×10^4
Ebullioscopy, cryoscopy and Isothermal distribution	....	upto 3×10^4
Osometry	....	10^4 to 10^6
Light scattering	....	10^4 to 10^7
Sedimentation in ultracentrifuge and diffusion	....	10^4 to 10^7

Except the method of end group analysis all the above methods are based on the determination of properties of polymer solution. The molecular weight of the polymer is related to its chain length or to the degree of polymerisation (DP) defined as the ratio between the molecular weight of polymer and that of the monomer.

For determination of molecular weights by these methods we need complicated apparatus and the methods are time consuming. In laboratory generally the viscometric method is used.

The properties such as *b.p., m.p., v.p.,* etc., are dependent on the size and shape of macro molecules and also on the interaction between individual segments of molecules. They are also found to depend on conformations of macromolecules and thus on the structure of molecules.

1. End Group Analysis Method

It is the simplest method for the determination of molecular weights of the polymers. Determinations by this methods require that the polymer contain a known number of determinable groups per molecule. The long chain nature of polymers limits such groups to end groups. Thus the method is known as *end-group analysis.* The determination by this method yield *number average molecular weight* ($\overline{M}_n$) for poly-disperse materials.

Because different types of end groups are found in addition

polymers and condensation polymers the methods is different in the two cases as discussed below:

The end-group analysis method can be carried out by using the following chemical and physical methods:

(i) **Chemical methods.** By using classical analysis, titration etc.
(ii) **Physical methods.**
(a) Spectrophotometric method.
(b) Tracer analysis using radioisotope-labelled substances.
(iii) Dye techniques.

Molecular Weight Determination of Condensation Polymers

End-group analysis in condensation polymers usually involves chemical methods of analysis for functional groups. Carboxyl groups in polyesters and in polyamides are usually titrated directly with base in an alcoholic or phenolic solvent, while amino groups in polyamides may be titrated with acid under similar conditions. Hydroxyl groups are usually determined by reacting them with a titrable reagent, but infrared spectrosocpy has been used. The chemical methods are often limited by insolubility of the polymer in solvents. Hydroxyl group can be determined by using a dry technique developed by Professor S.R. Palit. The chemical methods are often limited by insolubility of the polymer in solvent suitable the titrations.

Determination of Molecular Weight of a Polyester

Consider a polyester prepared the polycondensation of a hydroxy acid. From the structure of the polymer it is known that this polyester has a carboxyl group for each molecule and when titrated one mole of sodium hydroxide is used per mole of polyester. To calculate the molecular weight of the polyester, we have only to calculate how much polyester (in grams) gets neutralized by one mole of sodium hydroxide. Hence if the titration of 1.5 g of polyester needs up 0.75 ml. of 0.1 (N) sodium hydroxide (=0.003 g of sodium hydroxide), then we need 1.5 × 40/0.003 g of the polyester to use up 1 mole = 40 g of NaOH. Thus the molecular weight of the polyester is given by

$$\overline{M}_n = \frac{1.5 \times 40}{0.003} = 20,000$$

Acid Number (AN)

It is generally used in case of a polyester and is defined as *the number of milligrams of KOH required for titration of 1g polyester.* In the above numerical example 1.0 g of polyester need 0.50 ml of 0.1 N KOH, the acid number is 2.8 (0.5 × 56 × 0.1 = 2.8 g of KOH reqd. for 1 g of polyester).

Thus the molecular weight of the polyester can be calculated as under using the general formula for the number-average weight of the polyester given below:

$$\overline{M}_n(\text{polyester}) = \frac{56,000}{AN}$$

Using this formula for the numerical example considered above, we get

$$\overline{M}_n = \frac{56\times1000}{2.8} = 20,000$$

If the polyester is formed by the condensation of dicarboxylic acids and glycols, the same method can be used for the determination of the molecular weight of the polyesters.

If the dicarboxylic acid has been present in excess in the reaction mixture, then the resulting polyester uses up more sodium hydroxide for the same molecular weight of the polyester which has been prepared by the complete equivalencies of the monomers. If the amount of dicarboxylic acid during the synthesis of the ester is very high and if all the molecules of the polymers contain carboxyl groups at both ends, then the consumption of sodium hydroxide will be twice than in case of the first category of polyesters. In such a case, the molecular weight of the polyester may be calculated by using the equation give below:

$$\overline{M}_n = \frac{n\,q\,E}{a}$$

when n = number of groups that can be determined per polymer molecule

q = equivalent weight of the reagent expressed in gm that is used for the determination

E = weight of the polymer in grams.

a = amount of the reagent consumed in grams.

From the analytical method, the fraction of the groups that have reacted can be calculated as under

We know that,

$$\text{The degree of polymerization, DP} = \frac{1}{1-p}$$

where $$p = \frac{\text{number of reacted functional groups}}{\text{number of functioanl groups present}}$$

Consider the formation of polyester using five molecules of ω-hydroxy acid.

$$\text{HO—R—COOH} + \text{HO—R—COOH} + \text{HO—R—COOH} + \text{HOR—COOH} + \text{HO—R—COOH}$$

$$\downarrow$$

$$\text{HO—R—COO—R—COO—R—COOR—COOR—COOH} + 4HO_2$$

In this case, we have

Total number of functional groups present = 10

Number of functional groups that has reacted = 8

Number of unreacted groups present = 2

Thus the degree of polymerization,

$$\overline{\text{DP}} = \frac{\text{Total number of functional groups present}}{\text{The number of unreacted functional groups}}$$

$$= \frac{10}{2} = 5.$$

Hence if we know the degree of polymerization it is possible to calculate the molecular weight of the polymer easily.

Thus the end-group analysis method is quite useful for the determination of the molecular weight of the polymers that possesses end groups which can be determined precisely by some analytical reaction. In addition to polyamides and polyesters, where the end group analysis method is commonly used, the method can also be used for polysaccharides.

2. Dye Technique

Dye technique is found useful is case of the polymers that have ionic groups or these which can be transferred to ionic groups. The two kinds of dye-techniques are:

(i) Dye partition technique.

(ii) Dye interaction technique.

Dye techniques are very sensitive and are characterized by simplicity and rapidity of operation. For these only common equipment are needed. Sometimes, the dye technique is the only suitable method for the determination of OH end group.

3. Physical Method

There are a number of physical methods available but of all the physical methods, tracer analysis using C^{14}, H^3 or S^{35-} labelled initiators or other substances has received considerable attention. However, this method is incapable of giving results specific to the actual functional group and is highly susceptible to interference by absorbed radioactive impurities. It is moreover hazardous, expensive and needs sophisticated apparatus.

Molecular Weight Determination of Addition Polymers

No general procedures can be outlined for end-group analysis in addition polymers due to the variety of type and origin of the end groups. When the polymerization kinetics is known analysis can be done for initiator fragments containing identifiable functional groups, elements, or radioactive atoms, or for end groups arising due to transfer reactions with solvent or for unsaturated end groups in linear polyethylene and poly-α-olefins, as in the infrared spectroscopic analysis for vinyl groups.

Molecular Weight Determination by Colligative Property Measurements

Methods for the determination of molecular weight based on colligative property are vapour-pressure lowering, boilinpoint elevation (*ebulliometry*), freezing-point depression (*cryoscopy*), and the osmotic pressure (*osmometry*).

Following equations are used for determination of molecular weights by these measurements:

$$\lim_{c \to o} \frac{\Delta T_0}{c} = \frac{RT^2}{p\Delta H_r} \frac{1}{\overline{M}_n}$$

$$\lim_{c \to o} \frac{\Delta T_f}{c} = \frac{RT^2}{p\Delta H_r} \frac{1}{\overline{M}_n} \quad ...(8.4)$$

$$\lim_{c \to 0} \frac{\pi}{c} = \frac{RT}{\overline{M}_n}$$

where ΔT_f, ΔT_f, and π are the boiling-point elevation, freezing-point depression, and osmotic pressure, respectively; p is the density of the solvent; ΔH_r, and ΔH_f are the enthalpies of vaporization and fusion, respectively, of the solvent per gram; and c is the solute concentration in grams per cubic centimeter. The *number-average molecular weight* $\overline{M}_n$, has been inserted to make the equation applicable to polydisperse solutes.

1. Molecular Weight Determination by Osmometry

Osmotic pressure is one of the colligative properties of solutions containing both low-molecular weight compounds and high polymers. The major difficulty faced in the study of the behaviour of low molecular weight compounds in solution by the osmotic pressure measurement method is the selection of a suitable semipermeable membrane.

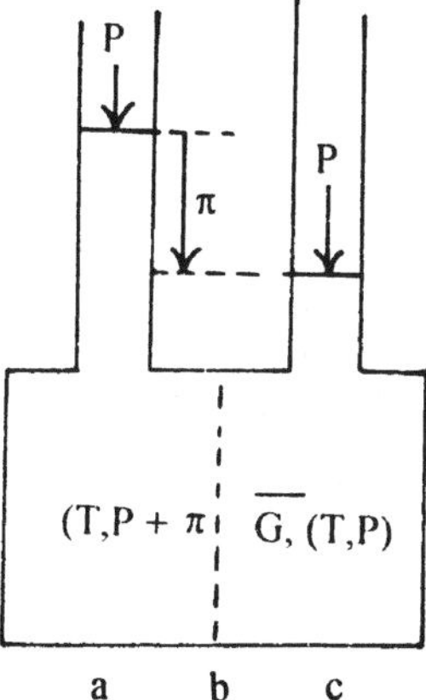

Fig. 8.9: Schematic representation of the phenomenon of osmosis (a) solution (b) membrane, (c) solvent.

In case of solutions of high molecular weight compounds, the selection of semipermeable membrane is easier, because the solvent and the solute molecules are quite different in their size. The relationship between the osmotic pressure of solution of a macromolecular compound and the molecular weight is widely used for determination of molecular weights and in the study of the interaction between the solvent and the solute molecules in the solution.

The osmometric method is very useful for the determination of the molecular weights in the range of 30,000 to 15,00,000. The lower limit of molecular weights that can be determined by this method is only determined by the possibility of obtaining suitable semipermeable membrane.

The number average molecular weight ($\overline{M}_n$) of the polymer can be determined osmometrically by taking into consideration the physical principle that *The chemical potential of a solvent in solution is lower than that of the pure solvent when the solution is separated from the pure solvent by a semipermeable membrane.* The molecules of the solvent can pass into the solution (or vice versa) until the osmotic pressures of the solvent and the solution become equal.

Let the chemical potential of the solvent molecules at temperature T and pressure P in the solvent compartment be equal to G_0, (T, P). When the osmotic pressure equilibrium is attained, the chemical potential of the solvent molecules in the solution compartment becomes equal to μ_0(T, P + π). Hence at the osmotic pressure equilibrium.

$$\mu_0(T, P+\pi) = \overline{G}_0(T,P) \qquad ...(8.5)$$

$$\text{But } \mu_0(T, P+\pi) = \mu_0(T, P) + \left(\frac{\partial \mu_i}{\partial P}\right)_T \pi = \mu_0(T, P)+V_0\pi \qquad ...(8.6)$$

Thus,

$$\overline{V}_0 = \overline{G}_0(T,P) - \mu_0(T,P) = -\Delta\mu_0$$

For ideal solutions,

$$\mu_0(T, P) = \overline{G}_0(T,P)+RT\ ln\ x_o \qquad ...(8.7)$$

and

$$-\Delta\mu_0 = -RT\ ln\ x_0 = -RT\ ln\ (1-x) \qquad ...(8.8)$$

where x_0 and x are the mole fraction of the solvent in solution and the mole fraction of the solute in solution respectively.

If the molar concentration is very low, then

$$-ln(1-x) = x = \frac{n}{n_0}$$

where n_0 and n are the mole faction of the solvent and of the solute in solution, respectively.

Again, $n_0\overline{V}_0 = N$, where V is the volume of the solvent, therefore,

$$\pi = -\frac{\Delta\mu_0}{V_0} = \frac{n}{V}RT = \frac{RT}{M}.C \qquad ...(8.9)$$

where C is the concentration of the solution in g/ml.

However in actual practice polymer solutions are not ideal due to the interaction between the polymer and the solvent molecules and the large difference in size between the different molecules in solution.

Hence,

$$\Delta H \neq O \text{ and } \Delta S \neq -R[x_0 + x \, ln \, x]$$

For this reason, the relationship between the reduced osmotic pressure π/C and the concentration is generally expressed in the form of virial equation as given below:

$$\pi / C = RT\left(\frac{1}{M} + A_2C + A_3C^2 + ...\right) \quad ...(8.10)$$

where A_2 and A_3 are the second and third virial coefficients respectively.

Huggins and Flory proposed the following equation for the variation of the osmotic pressure with concentration,

$$\pi / C = RT\left[\frac{1}{\overline{\overline{M}}_n} + \left(\frac{1}{2} - \mu\right)\frac{v^{-2}}{\overline{\overline{V}}_0}C + \frac{1}{3}\frac{v^{-2}}{\overline{\overline{V}}_0}C + ...\right] \quad ....(8.11)$$

where v is the partial differential of the specific volume of the polymer in solution, $\overline{V}_0$ is the volume of one gram-equivalent of the solvent, and μ is the parameter characteristic of the polymer-solvent system.

Comparing equation (8.10) and (8.11), we get,

$$A_2 = \left(\frac{1}{2} - \mu\right)\frac{\overline{v}^2}{\overline{\overline{V}}_0} \quad ...(8.12)$$

$$\left(\frac{\pi}{C}\right)_{C \to O} = \frac{RT}{\overline{M}_n}$$

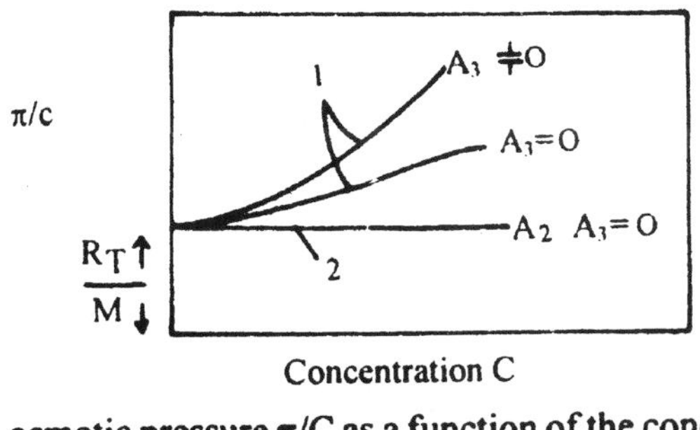

Fig. 8.10: Reduced osmotic pressure π/C as a function of the concentration of the solution C (1) polymer solution, (2) ideal solution.

Hence, the osmotic pressure of a polymer solution is determined at various concentrations and a graph is plotted between π/C as a function of 'C' and Fig. (8.10). The curve can extrapolated to infinite dilution and the extrapolation gives the value of π/C at $C = 0$; the number-average molecular weight $\overline{M}_n$ can be is then calculated by using the formula.

The virial coefficient A_2, A_3 etc. is a measure of the resultant interaction between the polymer chains, and between the molecules of the polymer and the solvent. These virial coefficients depend on the degree and nature of salvation of the polymer by the solvent and also on the state of the polymer in solution. As the accuracy of the determinations of osmotic pressure is limited the value of A_3 is just approximate. The second virial coefficient A_2 reflects the nature of the interaction between the polymer and the solvent molecules.

In a good solvent, the randomly distributed polymer molecules are loosened, and therefore the repulsive forces between the macromolecules predominate. In such a case A_2 is positive. When a poor solvent is added, the macromolecules of the polymer get coiled up due to which the mutual interaction increases; A_2 therefore decreases. As the precipitation point, A_2 approach zero (Fig. 8.11). Hence there is mutual repulsion due to salvation and the excluded volume effect, gets compensated by the interaction between the chains.

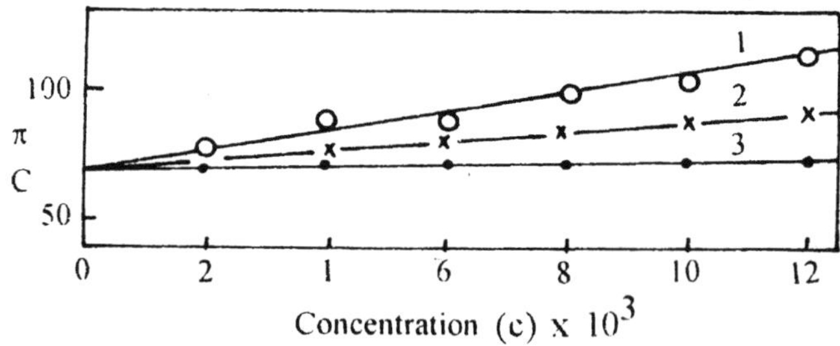

Fig. 8.11: Reduced osmotic pressure of polystyrene solutions in mixed-solvents as a function of the concentration.

(1) Pure methyl ethyl ketone, (2) methyl ethyl ketone + methanol mixture (95 : 5) (3) methyl ethyl ketone + methanol mixture (90 : 10).

It is possible to estimate the influence of temperature on A_2 from the change in the state of the polymer. The value of A_2 depends on the molecular weight and also on the molecular weight distribution of the polymers. As the molecular weight increases, the value of A_2

decreases. Flory has equated the second virial coefficient to the excluded volume by the equation; $A_2 = \frac{Nu}{2M}$ where N is Avogadro's number, u is the excluded volume. and M is molecular weight of the polymer.

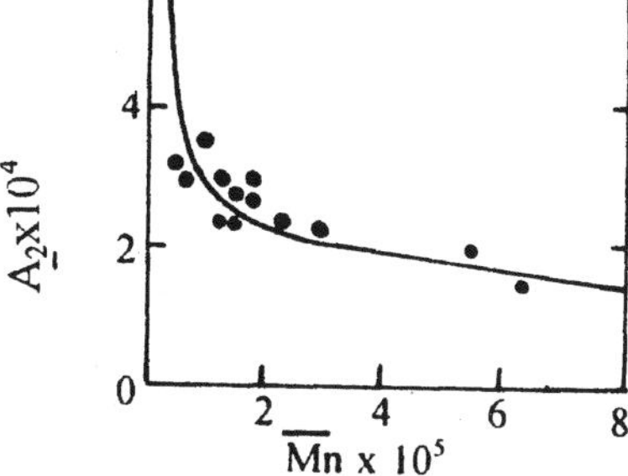

Fig. 8.12: Second virial coefficients as a function of molecular weight for a solution of poly (methyl methacrylate) in acetone.

By determining the osmotic pressure of a polymer solution at only one concentration, and at various temperatures, the fundamental thermodynamic parameters of the solution (heat of dilution $\Delta\overline{H}_0$ and entrophy of dilution $\Delta\overline{S}_0$) can be calculated by using the following relationship

$$-\overline{V}_0\pi = \Delta\mu_0 = \Delta\overline{H}_0 - T\Delta\overline{S}_0$$

therefore,

$$\frac{\partial\left(\frac{\Delta\mu_0}{T}\right)}{\partial\left(\frac{1}{T}\right)} = \Delta\overline{H}; \qquad \frac{\partial\Delta\mu_0}{\partial T} = -\Delta\overline{S}. \qquad \text{...(8.13)}$$

The values of $\Delta\overline{H}_0$ and $\Delta\overline{S}_0$ are then found from the slopes of the curves giving $\frac{\overline{V}_0\pi}{T}$ as a function of $\frac{1}{T}$ and $\overline{V}_0\pi$ as a function of T.

Osmometers

In the past few years various types of osmometers have been developed and used. They differ generally in the details of their technical construction. These osmometers are based on two types of cells:

(i) Cell with a horizontal membrane: G.V. Schulz.

(ii) Cell with a vertical membrane: R.M. Russ and D.J. Mead.

Osmometers with Horizontal Membrane

The simplest types of osmometer contains an osmotic cell with a horizontal (Fig. 8.13). This type of osmometer was first proposed by Schulz. Several modification of this type of osmometer have subsequently been ınade.

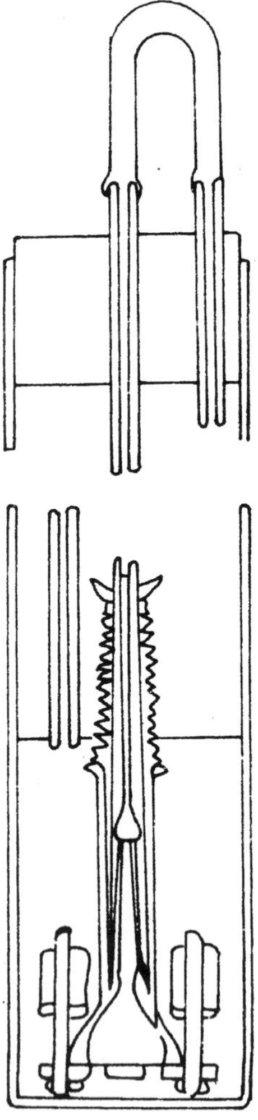

Fig. 8.13: Schulz osmometer.

The polymer solution is put in the cell that is joined to the capillary using a ground joint and a mercury seal. The osmometer is completely immersed in the solvent. The height of the column in the working capillary can be determined with a **cathetometer.** This height is equal to the height of the column and corresponds to the osmotic pressure of the solution plus the height of its capillary rise. The capillary rise of the solution can be determined by filling both the cell and outer vessel with the solvent. Because the determination of the osmotic pressure involves only very dilute solutions, the capillary rise of the solution is quite close to the capillary rise of the solvent. Many types of osmometers have a reference capillary fixed to the working capiliary. The height difference of the liquid columns in these two capillaries corresponds to the osmotic pressure of the solution.

The membrane area in osmometers having a horizontal membrane is generally very small and the time for equilibration is, very long. Nevertheless, osmometers of this type are now widely used because of their simple construction.

Osmometers with Vertical Membrane

Such an osmometer was first designed by Herzog and was further modified by Fuoss and Mead. This type of osmometers are widely used for the determination of molecular weight.

In such an osmometer the lateral walls on the depth of the half cells are made up of perforated brass plates the width and the depth of each groove in the plate are; 1.5 mm and the distance between two adjacent groove is also 1.5 mm. The diameter of osmometer cell is 11.5 cm. The semipermeable membrane is clamped between the two half-cells. The solution is placed in the glass-tube having a needle-type stopcock and is fitted with pure solvent. The volume of the osmometer cell is about 7 ml. The assembled osmometer is put in a double-walled air thermometer. The temperature fluctuations in the thermostate are ±0.05°C.

The osmotic pressure equilibrium in this type of osmometers is attained quick and very small quantities of solvent and solution are needed in this case. Sometimes mixed solvents may be used in this type of osmometer. The disadvantages of the osmometer are its complicated design and high price. It is difficult to fix the membrane tightly, and to detect the removal of the air bubbles from the osmometer cell.

Zimm-Meyerson Osmometer

More popular and widely used these days are relatively small and simple osmometers based on the Zimm-Meyerson design (Zimm 1946), in which two membranes are held against a glass solution cell by means of perforated metal plates, as shown in Fig. 8.14.

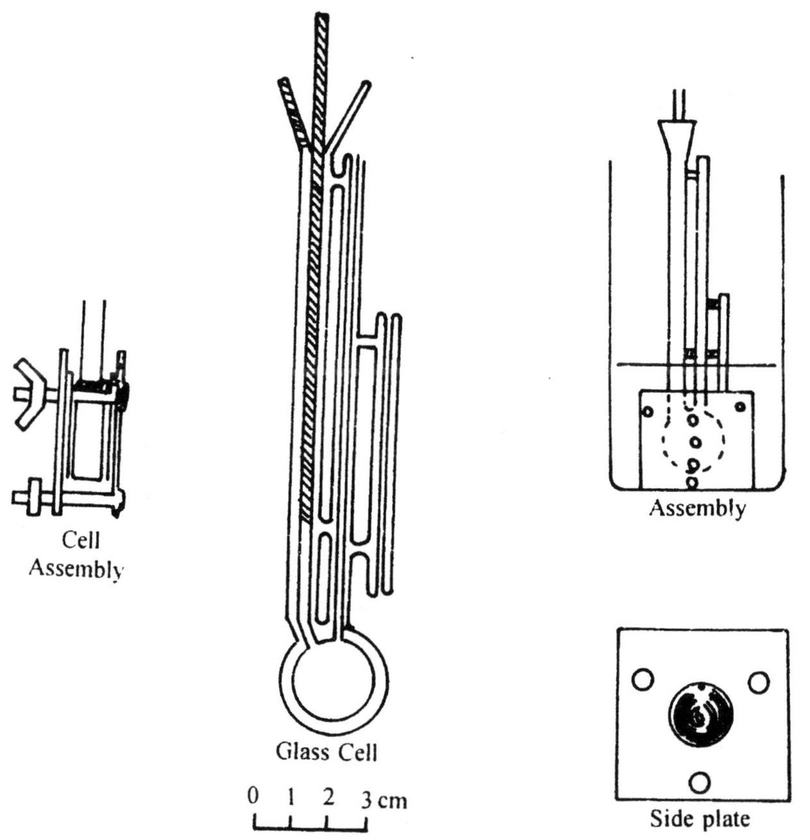

Fig. 8.14: Diagram of the Zimm-osmometer (Zimm 1946). A typical diameter for the measuring and reference (solvent cell) capillary is 0.5—1 mm. The closure of the filling tube is a 2-mm metal rod. A mercury seal is used at the top to ensure tightness.

The osmotic cell is a thick glass tube, of 2 cm, inner diameter, joined to the capillary of 0.5 mm in diameter and 10 cm long. The bottom part of the cell is joined to the U-shaped capillary, whose inner diameter is 1-2 mm. The solution is introduced through this capillary.

When the osmometer is in operation, a nichrome wire of a suitable diameter is inserted into the capillary (the diameter of the wire is almost equal to the inner diameter of the capillary). The wire serves for adjusting the liquid level in the measuring capillary and is held in the capillary with the aid of a cork.

The lateral surfaces of the osmometer cell are carefully ground with fine abrasive powders. It is very important that the wall surfaces of the cell be parallel. The semipermeable membranes are applied to these surfaces and are tightly clamped with two ground brass plates, perforated with 1-2 mm holes. The reference capillary shows the level of the solvent; for this reason the inner diameters of two capillaries should be equal.

Working of Zimm-Meyerson Osmometer

First of all the solution is introduced into the cell and then the nichrome wire is inserted into capillary-4 through funnel 5; mercury is now poured into the funnel, which provides a mercury seal. The osmometer is now placed in a tall glass beaker with a lid 5 cm inner diameter. Solvent is poured into the beaker in such a way that some of it moves into the capillary. The difference between the liquid levels in the capillaries 2 and 3 gives the osmotic pressure of the solution. Advantage of this osmometer is its simplicity in use. In this system, only a small volume of solvent is needed. The disadvantage of this osmometer is that relatively large volume of the solvent must be used; there is no membrane holder, so that the membrane may be easily distored.

Recording Osmometer

Melabs Model (CSM-2) the best recording osmometer in use at present. It consists of a stainless stell suitably equipped to operator at constant temperature. The cell contains two membrane-separated compartments, a flexible stainless steel diaphragm and a strain gauze. This is connected to a 0.1 mv DC stripchart recorder. The sensitivity of the strain gauze out put can be adjusted to accommodate four preset pressure ranges of 0 to 5, 0 to 10, 0 to 50 and 0 to 100 cm. The hydrostatic head can be set constant level using a light source-photo-electric cell micro-ammeter system that is designed to indicate a change in amperage as the miniscus passes the fixed pencil of light being transmitted throught the solution-inlet to tube.

The cell is made up of one glass tube and three metal valves. The valves are open when they are in vertical position; and are closed when in the horizontal position; and allow maximum flow at the 45 position. When the three valves are closed, the liquid automatically flows into the solution compartment of the cell; when the right-hand valve (solvent-inlet valve) is open, the liquid will be directed from the solution in-let tube into the solvent compartment of the cell, by passing the solution compartment. The left hand valve is the solvent drain valve, and the near valve, the solution-drain valve.

Membrane Osmometry

In membrane osmometry the two compartments of an *osmometer* are separated by a *semipermeable membrane* only solvent molecule can penetrate through the semipermeable membrane (Fig. 8.15) which is closed except for capillary tubes. The polymer solute remains confined to one side of the osmometer and the activity of the solvent is different in the two compartments. Because of the thermodynamic drive toward equilibrium a difference in liquid level in the two capillaries results.

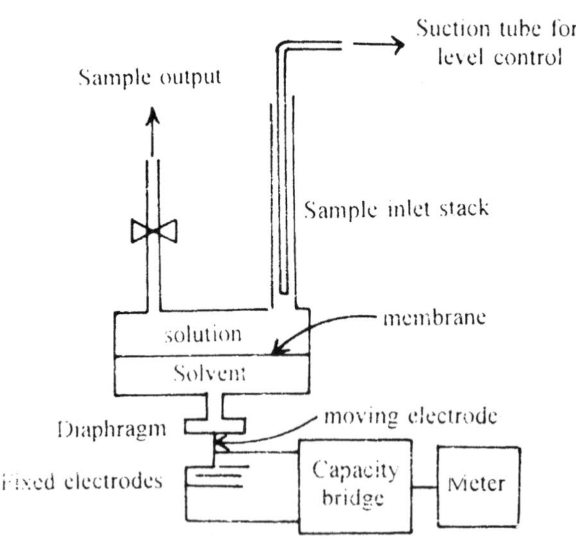

Fig. 8.15: Schematic of a modern rapid automatic osmometer. Osmotic pressure is directly after suitable calibration) as the change in capacitance of a condenser caused by the deflection of the flexible diaphragm.

Increase hydrostatic pressure results in an increase in the solvent activity on the solution side until. the applied pressure becomes equal to the osmotic pressure when equilibrium is attained.

Membranes

The substances generally used as osmotic membranes include collodion (nitrocellulose of 11—13.5% nitrogen); regenerated cellulose, obtained by denitration of collodion; gel cellophane that has never permitted to dry after manufacture; bacterial cellulose, obtained by the action of certain strains of bacteria; rubber, poly(vinyl alcohol); polyurethances; poly(vinyl butyral); and polychlorotrifluoroethylene. At present gel cellophane is most widely used.

Permeability of Solute Through the Membrane

For the success of the osmotic experiment the availability of a membrane through which solvent but not solute molecules can pass freely is essential. Existing membranes can only be considered as of approximate ideal semipermeability. The major limitation of the osmotic method is the diffusion of low-molecular-weight species through the membrane. Thus the method is reliable and accurate only for experiments where in diffusion is absent, such as measurement of unfractionated polymers with $\overline{M}_n > 50,000$, or polymers from which low-molecular-weight species have been removed by fractionation or extraction with $\overline{M}_n > 20,000$. The upper limit to molecular weights measurable by the osmotic method is about 10^6. This limit has been set by the precision with which small osmotic heights can be read. When diffusion is present, the apparent osmotic pressure is always less than the true osmotic pressure and falls with time. By extrapolation to zero time we can get too low an osmotic pressure and hence too high a molecular weight. The magnitude of the error is due to the solute diffusion and depends on the type of measurement, Table 8.2 lists some data to illustrate it.

Table 8.2: Apparent Molecular Weights of Branched Polyethylene by Membrane Osmometry

Method	*Apparent* $\overline{M}_n$ *of Sample*		
	75	*76*	*77*
Average of cryoscopy, ebulliometry, and vapour-phase osmometry (Table 8.3)	11,000	15,300	18,500
Rapid membrane osmometer, first observable values (5-7 min after sample introduction)	19,900	22,100	27,900
Rapid membrane osmometer, 36 min after sample introduction	29,400	25,000	32,800
Conventional membrane osmometer 1000 min after sample introduction	39,500	3,3000	36,400

Measurement of Molecular Weight

To measure molecular weight prepare a solution of polystyrene log/litre and from this solution, 10 ml of solutions having concentrations 8, 6, 4, 2 g/litre respectively were prepared by diluting with toluene. The osmotic pressure of various solutions were measured.

A graph is then plotted between the value of π/C (with C in g/litre and π in cm) against C. The value of $(\pi/C)_0$ is determined from this plot. The number average molecular weight of the polymer can then be determined using the following expression:

$$\left(\frac{\pi}{C}\right)_0 = RT\left(\frac{1}{\overline{M}_n}\right)$$

the second virial coefficient is obtained by using the following equation

$$\frac{\pi}{C} = RT\left(\frac{1}{\overline{M}_n} + A_2C\right)$$

and the excluded volume effect can be found by using the following equation

$$A_2 = \frac{N_u}{2M}$$

Vapour-Phase Osmometry

Vapour phase osmometry is found satiable for determining M_n of low molecular weight samples. This method involves an indirect measurement of vapour pressure lowering by a membrane osmometer. A "vapour-pressure osmometer" (Fig. 8.16), used to measure vapour pressure lowering which is the small temperature difference arising due to different rates of solvent evaporation from the condensation onto droplets of pure solvent and polymer solution maintained in an atmosphere of solvent vapour. This temperature difference is proportional to the vapour-pressure lowering of the polymer solution at equilibrium and thus to the number-average molecular weight: Due to heat losses, the full temperature difference expected from theory can not be attained. Measurements have to be made at different concentrations and extrapolated to $c = 0$. Like ebulliometry and cryoscopy, the method is calibrated with low-molecular-weight standards and is useful for values of M_n at least up to 40,000. It is rapid and needs only a few milligrams of sample. The lower limit of M_n measurable by this method is one at which the solute becomes appreciably volatile. As the method does not measure equilibrium vapour-pressure lowering but depends on the development of quasi-steadystate phenomena so care must be taken to standardize variables such as time measurement and drop size between calibration and sample measurement.

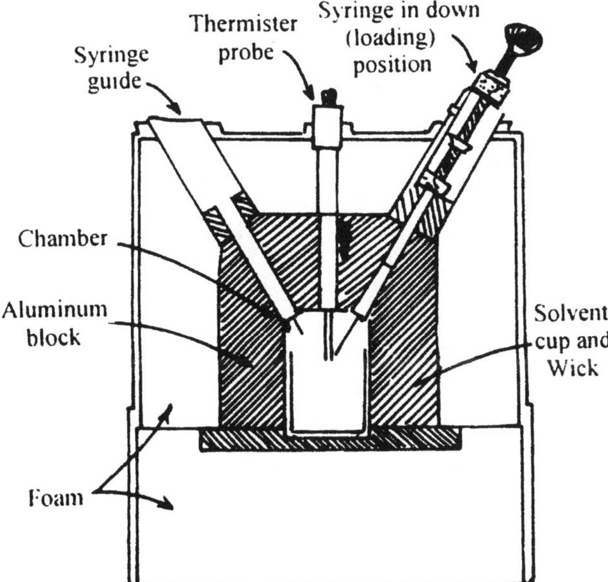

Fig. 8.16: Measurement chamber of a vapour-phase osmometer.

Machrolab Vapour pressure osmometer (VPO) model 30/A is composed of a sample-chamber assembly and a control unit which houses a wheatstone bridge, a null indicator, and a heater-control circuit.

The sample chamber of the osmometer shown in Fig. 8.16 consists of a foaminsulated thermal block containing solvent. The chamber is machined such that the syringes can be lowered in order to apply a drop of solution to one thermistor and a drop of solvent to the other without any need to open the system. The syringe tips and the thermistors can be viewed through the minar viewing path located on the side of the sample chamber.

Vapour pressure osmometer is a variation of the *isopiestic* or of *the isothermal distillation techniques* by which a solvent and a solution in that solvent are placed side by side in a closed container. It measures the difference in temperature created by the condensation of solvent on a sensitive thermistor containing a solution of the solute whose molecular weight is to be determined. *The isopiestic method measures a difference in vapour pressure while the isothermal distillation technique depends upon a difference in volume.* Despite the specific changes being measured in the techniques each change is proportional to the colligative property of the solution – the lowering of the vapour pressure.

The vapour pressure osmometer method is more acceptable of all the methods involving measurement of colligative properties because of the sensitivity of the detector. For ideal solvent–solvents with a low heat of vapourization, the differential thermistors of the VPO can detect differences in temperature of the order of 0.001°C; this sensitivity determines the molecular weight of the samples upto 20,000.

Since the polymer-solvent interaction changes with concentration, molecular weight measurements by VPO, must be conducted within controlled concentration range. If the concentration is too high, significant condensation will take place on the solution droplet, thus reducing the difference in vapour pressure between the solution and the solvent.

The changes in a colligative property of a polymer solution with concentration can be expressed by a virial expansional given below:

$$\frac{\Delta R}{KC} = \frac{1}{M_n}[1 + T_2C + gT_2^2C^2] \qquad ...(8.14)$$

where ΔR is the difference in resistance caused by the difference in vapour pressure between the solvent and the solution, K is a calibration constant. T_2 is the second virial coefficient, and g is a polymer-solvent interaction function. The last term in equation (8.14) may be neglected at very low concentration ranges. Hence, rearranging equation (8.14), we can get

$$\frac{\Delta R}{C} = \frac{K}{M_n} + \frac{KT_2}{E_n}C \qquad ...(8.15)$$

Hence a plot of $\Delta R/C$ versus C should be linear and $\overline{M}_n$ will be equal to the quotient of K divided by the intercept of the line, provided K is given in the dimensions of ohmslitres/g and C in the dimensions of g/litre.

When the ideal value for a polymer in good solvent is used then equation (6.25) could be converted to

$$\frac{\Delta R}{C} = \frac{K}{M_n}\left[1 + \frac{T_2}{2}C\right] \qquad ...(8.16)$$

It is preferable to plot $\left(\frac{\Delta R}{C}\right)^{\frac{1}{2}}$ versus C if $\Delta R/C$ is greater than twice $\left(\frac{\Delta R}{C}\right)_0 \left(\frac{\Delta R}{C} \text{ when } C = 0\right)$. The use of the square root function helps to keep the relation linear over a larger concentration range.

Another correction that must be made is to account for the decrease with time of the value of $\Delta R/C$ caused due to the change in concentration of the solution droplet resulting from the condensation of solvent on the sample thermistor. If we assume that the concentration time dependency is a linear function, then

$$-dc/dt = kc \qquad ...(8.17)$$

Integrating equation (8.17), when $C=C_0$ and $t=t_0$, we get

$$C = C_0 \, exp{-kt} \qquad ...(8.18)$$

By substituting equation (8.15) into equation (8.18) and neglecting second term in the latter equation, we get

$$\Delta R = \frac{K}{\overline{M}_n} C_0 \, exp - kt \qquad ...(8.19)$$

Equation (8.19) is found to be valid at high kt values; however, with solvents where the k value is relatively small.

Now equation (8.19) may be expressed in a linear form, as under:

$$\Delta R = \frac{K}{\overline{M}_n} C_0 - \frac{K}{M_n} C_0 \, kt \qquad ...(8.20)$$

A plot either of ΔR versus 't' or of log ΔR versus 't' should be a straight line whose intercept $(\Delta R)_0$ gives the difference in the resistance for the solution concentration as prepared.

2. Ebulliometry

In ebulliometry the boiling point of the polymer solution is compared directly with that of the (condensing) pure solvent in a vessel called an ebulliometer. Sensing devices include differential thermometers and multiplication thermocouples or thermistors which are arranged in a Wheatstone bridge circuit. The ebulliometer is generally callibrated with a substance of known molecular weight, for example, octacosane (M = 396) or triseterian (M = 892).

Making careful use of an ebulliometer we can make an accurate measurement of molecular weights of substances having a number average molecular weight $\overline{M}_n$ upto about 30,000 or more. The ebulliometric experiment sometimes suffers from a limitation in the tendency of polymer solutions to foam on boiling and this may lead to unstable operation, as also the polymer may concentrate in the foam due to its greater surface, thus rendering uncertain the actual concentration of the solution. No equipment suitable for ebulliometry in the high-polymer range is commercially available, and this method remains largely in the category of a reference technique.

3. Cryoscopy

The freezing-point depression or cryoscopic method is quite similar to the ebulliometric method in many respects. The most preferred temperature-sensing element is the thermistor which is used in a bridge circuit. The freezing points of solvent and solution are often compared sequentially. Calibration is done with a substance of known molecular weight. Although the limitations of the method seem to be somewhat less as compared to those of ebulliometry proper care must be taken and supercooling be controlled. For this a nucleating agent be used which provide controlled crystallization of the solvent and is helpful in this respect. Relibale results can be obtained for molecular weights as high as 30,000. Due to lack of commercial equipment this method relegates to reference status.

4. Determination of Weight Average Molecular Weight by Light Scattering

The dispersion of a beam of light at the same frequency as the incident beam is known as **light scattering. The scattering** from large particles is the **Tyndall scattering** which is characteristic of dust particles and colloidal suspensions. In **Tyndal scattering,** the hallo created from the photodispersion causes the particle to appear larger than they actually are; therefore, the incident beam becomes clearly visible when the light is viewed at right angles.

Raleigh showed in his electromagnetic theory that in scattering light by gas molecules, the molecules act as a source of light, Polarization of molecules due to light creates an instantaneous dipole in the molecules. These instantaneous dipoles are oscillating at the same frequency as that of incident light and these oscillating dipoles act as sources of scattered light.

In 1871 Lord Rayleight applied classical electromagnetic theory to the problem of the scattering of light by the molecules of a gas and showed that the quantity of light scattered, for particles small compared to the wavelength of the light, is inversely proportional to the number of scattering particles per unit volume and to the fourth power of the wavelength; the latter dependence accounts for the blue colour of the sky.

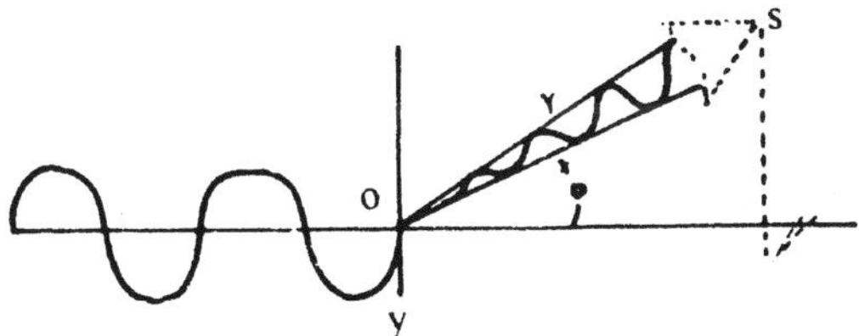

Fig. 8.17: Representation of the scattering of a polarized transverse wave.

Scattering, by particles small compared with the wavelength of light used and having a refractive index different to that of their surroundings is known as **Rayleigh scattering.** Two common manifestations of Rayleigh scattering are exemplified by the whiteness or the opacity of crystalline substance and by the colour of the Sky.

Peter Debye in 1944 further extended the work of Rayleigh and the fluctuation theory of Smoluchowski and Einstein to include the measurement of the scattering of light by macromolecular solutions for determining molecular size.

When the electric field strength of the incident light is E_0, the induced dipole will be $\mu_1 = \alpha E_0$ where α is the optical polarizability. The electric field strength of radiation scattered by the induced dipole E_s, depends on second derivative of μ_1 with respect to time. The useful experimental quantities are intensities of scattered light (i_s) and incident important light (I_s). These are respectively proportional to E_s^2 and E_0^2, averaged over a vibrational period *i.e.*, from time $t = 0$ to λ_0/C, where λ_0 is wavelength of light in vacuum and C is velocity of light. Hence,

$$i_s = \frac{16\pi^2\alpha^2 I_0 \sin^2\theta_1}{\lambda_0^4 r^2} \qquad ...(8.21)$$

Using the relationship $(n^2 - 1) = 4\pi v\alpha$ and changing the number concentration v to weight concentration, we get

$$\alpha = \frac{M}{2\pi N_0}\left(\frac{dn}{dc}\right) \qquad ...(8.22)$$

where n is the refractive index, M is the molecular weight, N_0 the Avogadro's number and $\left(\frac{dn}{dc}\right)$ is the variation of refractive index with concentration. Hence,

$$i = \frac{4\pi^2 M^2 I_0 \sin^2\theta_1}{N_{02}\lambda_0^{2} r^2}\left(\frac{dn}{dc}\right)^2 \qquad ...(8.23)$$

We can determine experimentally the scattering per unit volume and multiplying it by v the number of particles per *cc* can be found, $v = \frac{N_0 c}{M}$, scattering per unit volume is

$$i_s = \frac{4\pi^2 I_0\, cM \sin^2\theta_1}{N_0\lambda_0^4 r^2}\left(\frac{dn}{dc}\right)^2 \qquad ...(8.24)$$

When vertically polarized light is used i_s is independent of angle and $\sin^2\theta_1$ term vanishes from eq. 8.24. For commonly used unpolarized light i_s is given by

$$i_s = \frac{2\pi^2 I_0\, cM}{N_0 \lambda_0^{\,4} r^2}\left(\frac{dn}{dc}\right)^2 (1+\cos^2\theta) \qquad ...(8.25)$$

where θ is the angle of observation with respect to the direction of incident light.

For polymer solutions, scattering due to solvent molecules forms a finite but small amount and a difference between solution and solvent scattering should be considered. This is also known as **excess scattering.** If n_0 is the refractive index of pure solvent n, that of solution, the relation $(n^2 - n_0^2) = 4\pi\nu\alpha$ gives an expression similar to eq. 8.22. Then

$$i_s = \frac{2\pi^2 n_0^{\,2} I_0\, cM}{N_0 \lambda_0^{\,4} r^2}\left(\frac{dn}{dc}\right)^2 (1+\cos^2\theta) \qquad ...(8.26)$$

introducing $\overline{M}_W$ for M it can be written as

$$R_\theta = K\overline{M}_{W}c \qquad ...(8.27)$$

where R_θ is known as the excess Raleigh ratio and

$$R_\theta = i_\theta\, r^2\,(1+\cos^2\theta)/I_0 \qquad ...(8.28)$$

$$K = 2\pi^2 n_0^{\,2}\left(\frac{dn}{dc}\right) / N_0\lambda_0^{\,4} \qquad ...(8.29)$$

$$= K'\left(\frac{dn}{dc}\right)^2 \qquad ...(8.30)$$

The dimensions of Raleigh ratio are cm^{-1}. It is common to measure scattering at an angle of 90° and hence,

$$R_{90} = i_{90}.r^2/I_0.$$

Equation 8.27 forms the basis for determination of molecular weight from light scattering data. Like osmotic pressure measurements, it is essential to consider the nonideality of solutions and the concentration dependence. Following Debye, eq. 8.27 gets modified to

$$Kc/R_\theta = (\overline{M}_W)^{-1} + 2A_2c \qquad ...(8.31)$$

Hence, scattering measurements are needed at different

concentration and the results be extrapolated to $c = 0$. The term turbidity $\tau(=16\pi R_\theta/3)$ is generally used L.H.S. of eq. 8.31 is then replaced by Hc/τ *i.e.*,

$$Hc/\tau = (\overline{M}_W)^{-1} + 2A_2c \qquad ...(8.32)$$

where $$H = 32\pi^3 n_0^{\,2}\left(\frac{dn}{dc}\right)^2 3N_0\lambda_0^{\,4} \qquad ...(8.33)$$

For determination of molecular weight by scattering of light photometers/light scattering photometers are used these described here.

Photometers

Modern light-scattering instruments use either a mercury arc or a laser as a source and detect the scattered light photo-electrically. Most of the photometers are interfaced to computers for control, data handling and computation of results. The essential features of a light-scattering photometer are similar to these of widely used Brice-Phoenix instrument (Fig. 8.18). Light from a mercury-arc source (S) passes through a lens (L), a polarizer (P_1), and monochromatizing filter (F), and then strikes either a calibrated reference standard or a glass cell (C) which hold the polymer solution. On passing through the cell, the primary light beam is absorbed in a light-trap tube (T) and scattered light from cell or standard is viewed by a multiplier phototype (R) after passing through a slit system D_1 — D_2 and polarizer (P_2). The phototube is powered by high voltage from a regulated electronic power supply. Its output signal gets transmitted either to a stripchart recorder or to an analogue-to-digital converter for computer use.

A scattering cell consists of a glass cylinder with flat entrance and exit windows. The scattering cell is centred on the axis of rotation of the receiver phototube.

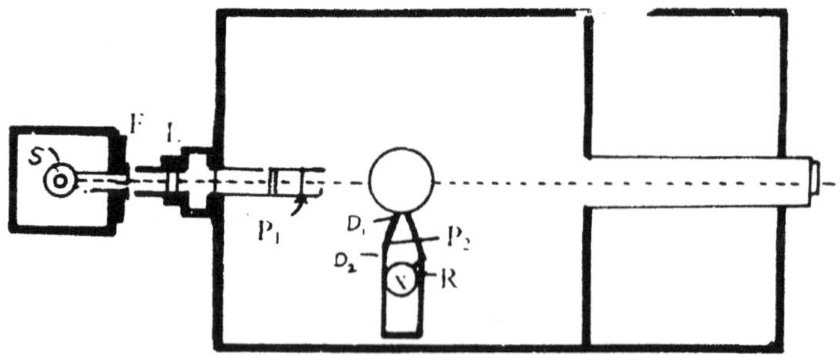

Fig. 8.18: Sketch of the essential components of a light scattering photometer.

Unusually initial adjustments are made using pure solvent. Introducing solution in the cell changes the reading and using one or more sets of neutral filters, the reading can be obtained in the range of 0 to 100 scale divisions. Knowing the constants of filters and calibration of the apparatus, turbidity can be calculated. With proper geometry of the apparatus the ratio of the meter readings G_θ/O_0 would directly give the turbidity τ. Extreme care is needed to prevent stray light, reflections and presence of dust particles. Appropriate precautions must be taken for preparing dust free solvents and solutions. Through cylindrical glass cells can be used for Debye's method special cells of semioctagonal shape are also available.

In addition to light scattering experimental data, it is necessary to know refractive index n_0 of pure solvent and the coefficient $\left(\frac{dn}{dc}\right)$ at the same wavelength as used in light scattering photometer. For this purpose, a sensitive refractometer is needed; generally a differential refractometer is used for rapid and accurate results.

Various steps involved in experimental determination are discussed below:

Sample Preparation

The preparation of a sample for light scattering measurement is the most important step. Polymer molecules are generally small as compared to particles of dust or other extraneous material. The scattering of these extraneous substances, though generally confined to low angles, can easily outweight the scattering from the polymer solution. Thus, solvents and polymer solutions must be clarified by filtration or by ultracentrifugation. Proper clarification, as evidenced by obtaining a rectilinear Zimm plot, is essential for the unambiguous interpretation of light-scattering data.

For accuracy in light-scattering measurement the proper choice of solvent is necessary. The difference in refractive index between polymer and solvent should be as large as possible. Moreover, the solvent should itself have relatively low scattering and the polymer-solvent system must not have too high a second virial coefficient as the extrapolation to zero polymer concentration becomes less certain for high A_2. Mixed solvent should be avoided unless both components have the same refractive index.

Calibration

It is possible to relate observed scattered intensity to turbidity through knowledge of the geometry of the photometer but calibration with substance of known turbidity is common by practice. These substances include reflecting standards, colloidal suspensions, and simple liquids. Tungstosilicic acid, $H_4SiW_{12}O_{40}$, M=2879, is recommended as a primary calibration standard in preference to pure liquids (too sensitive to impurities) or uniform-particle-size latexes (too sensitive to residual polydispersity).

Treatment of Data

The weight-average molecular weight ($\overline{M}_W$) in the inverse of the intercept at $c = 0$ and $\theta = 0$ in the Zimm plot (Fig. 8.21). The second virial coefficient can be calculated from the slope of the lines at constant angle by Eq. 8.34, and to a first approximation is independent of angle. The radius of gyration is

$$K.\frac{C}{\Delta R_{90}} = H.\frac{C}{\Delta\tau} = \frac{1}{M_W P(\theta)} + 2A_2C + \ldots \quad \ldots(8.34)$$

obtained from the slope of the zero-concentration line as a function of angle. The appropriate relationship, is

$$\bar{s}_z^2 = \frac{3\lambda s^2}{16\pi^2} \times \frac{\text{initial slope}}{\text{intercept}} \quad \ldots(8.35)$$

Range of Applicability

Light scattering from solutions can be used to measure molecular weights as low as that of sucrose and as high as those of proteins. In actual practice, polymer molecular weights are of the order of 10,000-10,000,000 and can readily be measured. The method has been used for determination of absolute values $\overline{M}_W$, particularly for proteins and other biological materials, and calibration of intrinsic viscosity-molecular weight relations. However, it is not possible to obtain information on particle shape of polydispersity from light-scattering measurement except under most unusual conditions. The method can be applied to copolymers only with severe restriction because of the difference in refractive index which generally exists between the two types of chain repeat units. Branched polymers can be measured without restriction.

Asymmetric Scattering

Equations 8.31, or 8.32 are found valid only when the size of the scattering solute particle is less than $\frac{\lambda}{20}$ and this condition is seldoms fulfilled by polymer molecules. In polymer solutions, scattering from different segments of the same molecule can interfere and this depend on scattering angle θ. The scattering intensity therefore is now a function of angle θ and it has to be corrected. Debye introduced a correction term P(θ) to account for asymmetric scattering.

$$P(\theta) = \frac{\text{scattering by polymer}}{\text{scattering in absence of interference}}$$

$$= i_\theta / i_\theta^0 \qquad \text{...(8.36)}$$

This correction term, however, in a function of size and shape of the particle and also depends on wavelength of light use. In case of polymer chains, use of r.m.s. radius of gyration $(\bar{s}_0^{\,2})^{-1/2}$ for size parameter is more appropriate. For different shapes, of molecules, the radius of gyration have relationships as under:

(a) Thus rods with lengths L, $s = L / \sqrt{12}$

(b) Spheres of radius R, $s = R / \sqrt{3/5}$

(c) Random coil with end-to-end distance r_0, $\bar{s}_0^{\,2} = \bar{r}_0^{\,2} / 6$

The corresponding correction terms P(θ) will be as under:

(a) For rods

$$P(\theta) = \frac{1}{x}\int_0^{2\pi} \frac{\sin\theta d\theta}{\theta} - \left(\frac{\sin X}{X}\right)^2 \qquad \text{...(8.37)}$$

where $$X = \frac{2\pi}{L} \sin\frac{\theta}{2}$$

(b) From spheres

$$P(\theta) = 3\left(\frac{\sin Z - Z\cos Z}{Z^3}\right)^2 \qquad ...(8.38)$$

where $$Z = 4\pi \frac{R}{\lambda_0^2}.\sin\frac{\theta}{2}$$

(c) For random coils

$$P(\theta) = \frac{2}{Y^2}(Y - 1 + c^{-Y}) \qquad ...(8.39)$$

where $$Y = \frac{8\pi^2}{3}.\frac{r_0^2}{\lambda_2}.\sin^2\frac{\theta}{2}$$

In all cases $\lambda = \lambda_0/n_0$. Expanding by Taylor series eqs. 8.37 – 8.39 gives

$$\frac{1}{P(\theta)} = 1 + \frac{16}{3}\pi^2\frac{s_0^2}{\lambda_2}.\sin^2\frac{\theta}{2} + ... \qquad ...(8.40)$$

On extrapolation to $\theta = 0$ or to $\sin^2\frac{\theta}{2} = 0$, we get $P(\theta) = 1$.

Hence, the expression used to calculate molecular weight corrected for non-ideal behaviour as also for asymmetric scattering will be

$$\frac{Kc}{R_0} = \frac{Hc}{\tau} = \frac{1}{M_w.P(\theta)} + 2A_2c \qquad ...(8.41)$$

The two alternative methods to solve eq. 8.41 for calculation of molecular weight are one due to Debye and other due to Zimm.

Both the Debye method and the Zimm method are discussed here.

Debye's Method

Debye gave tables are also graph for $P(\theta)$ as a function of size factor $(\overline{S}^2_0 / \lambda^2)^{1/2}$. Figure 8.19 shows variation of $P(\theta)$ vs. size factor and Fig. 8.20 shows the plot of dissymmetry coefficient Z_d vs. size factor. Experimentally, scattering data are obtained at any two angles symmetric around 90°C generally at 45° and 135°. Dissymmetry coefficient Z_d is then given by

$$Z_d = \frac{P(\theta_1)}{P(\theta_2)} = \frac{P(45°)}{P(135°)} = \frac{i_{45}^{\ 0}}{i_{135}^{\ 0}}$$

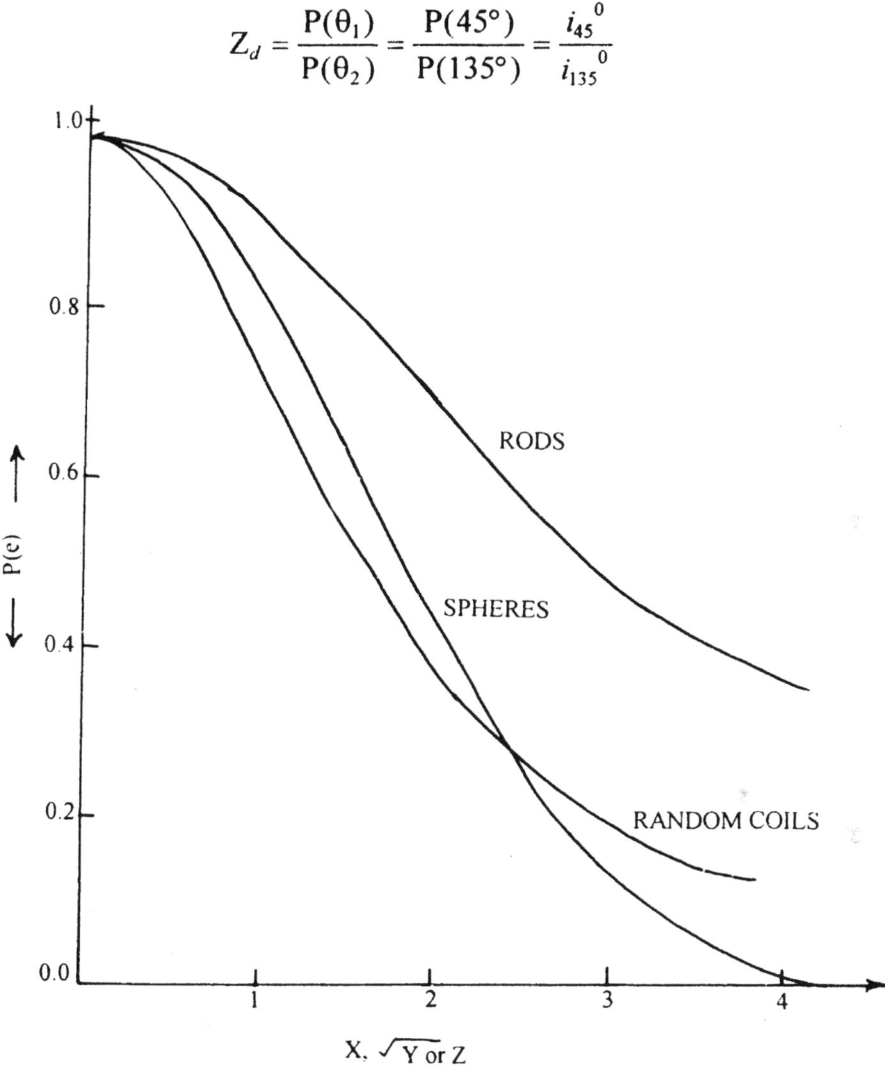

Fig. 8.19: Variation of asymmetric scattering correction with size factor X, $\sqrt{Y}$ or Z.

Referring to Fig. 8.19, plot of Z_d vs. $(\overline{S}^2{}_0 / \lambda^2)^{1/2}$,$(\overline{S}^2{}_0 / \lambda^2)^{1/2}$ can be read for measured value of Z_d. Knowing $(\overline{S}^2{}_0 / \lambda^2)$ now, P(θ) at any desired angle can be either calculated, read from Debye's table (Table 8.3) or read from the graph. An intercept from the plot of $\frac{Kc}{R_\theta}$ vs. c (eq. 8.41) gives $\frac{1}{\overline{M}_W P(\theta)}$ and hence $\overline{M}_W$ can be calculated.

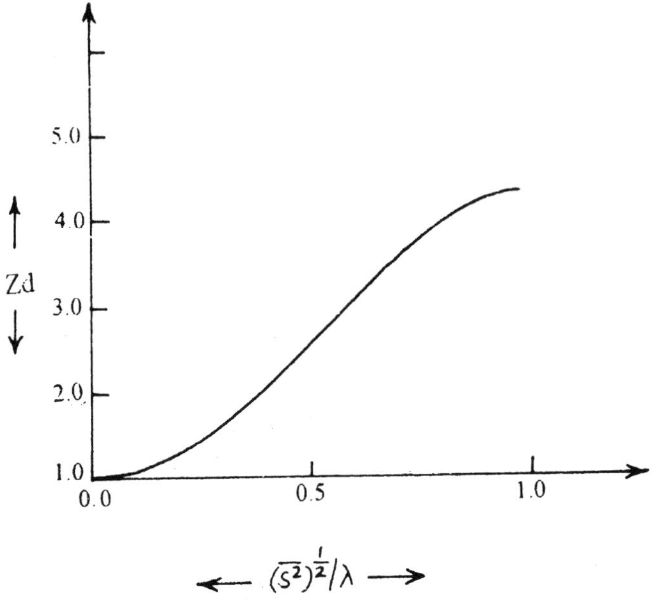

$(\overline{S^2})^{\frac{1}{2}}/\lambda$

Fig. 8.20: Variation of dissymmetry coefficient Z_D with $(S^2)^{1/2}/\lambda$.

Zimm Plot Method

Equation 8.40 can be substituted for $\left(\frac{1}{P(\theta)}\right)$ in eq. 8.41 to get

$$\frac{Kc}{R_0} = \frac{1}{\overline{M}_W}\left[1 + \frac{16\pi^2}{3\lambda^2}\sin^2\frac{\theta}{2}(\overline{S}_0^{\,2})\right] + 2A_2c \qquad ...(8.42)$$

It suggests that evaluation of $\overline{M}_W$ requires two corrections and so two extrapolations of the quantity $\frac{Kc}{R_\theta}$. Therefore

$$\frac{1}{\overline{M}_W} = \lim_{\theta \to 0} \frac{Kc}{R_\theta} \qquad ...(8.43)$$

The extrapolation may be done separately or simultaneously as in case of Zimm plot method. In Zimm plot shown in Fig. 8.21 $\frac{Kc}{R_\theta}$ is plotted against $\sin^2\frac{\theta}{2} + B.c$, where B can be arbitrarily chosen to be 50 or 100 or any suitable numerical value. Double extrapolation to

$c=0$ and $\theta = 0$ gives the intercept that corresponds to $\frac{1}{M_w}$.

Table 8.3: Dissymmetry coefficient Z_d and $[P(90°)]^{-1}$ for different particle shapes

Coil		*Rod*		*Sphere*	
Z_d	$[P(90°)]^{-1}$	Z_d	$[P(90°)]^{-1}$	Z_d	$[P(90°)]^{-1}$
1.02	1.02	1.02	1.01	1.02	1.01
1.12	1.09	1.13	1.09	1.12	1.08
1.21	1.15	1.22	1.15	1.20	1.14
1.31	1.22	1.33	1.24	1.32	1.21
1.62	1.46	1.62	1.48	1.62	1.39
2.03	1.81	2.04	2.01	2.00	1.60
3.05	2.99	–	–	3.02	2.09
4.01	4.99	–	–	3.91	3.44

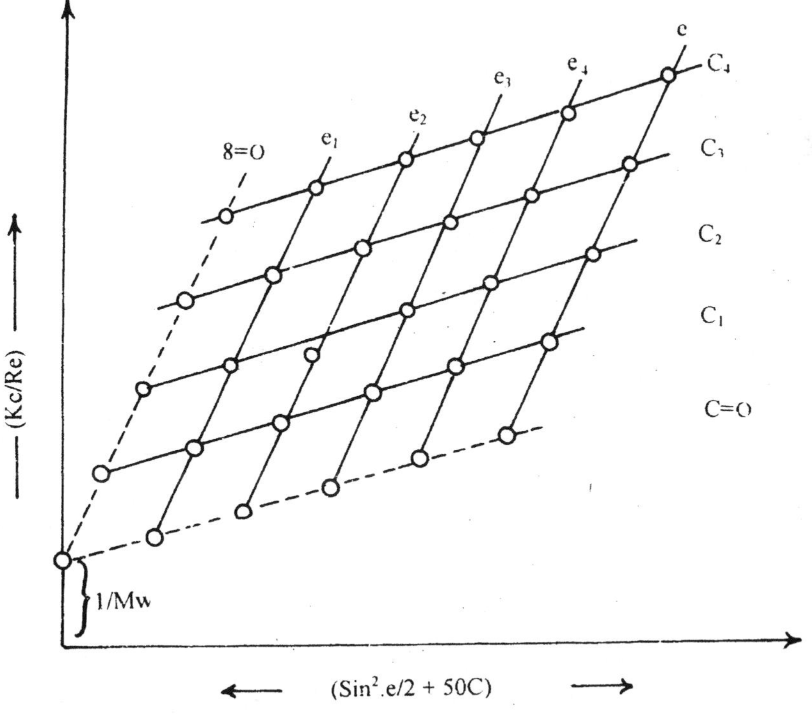

Fig. 8.21: A typical Zimm plot.

In this method, at every concentration, scattering should be measured at a number of different angles and therefore light scattering photometer should have a provision for continuous variation of scattering angle. Using recently developed laser sources, scattering can be accurately measured at angles as 1—2° and in such cases $P(\theta)$ correction can be neglected.

Other Scattering Methods

Dynamic Light Scattering

Dynamic light scattering is applied in a number of related methods wherein the spectrum of scattered light is determined by both static and dynamic means. The most valuable additional information obtained is the rotational divisional coefficient of the polymer.

Neutron Scattering

These experiments which are entirely analogous to light scattering, the angular dependence of scattering at small angles can be observed when a monochromatic collimated beam of coherent "cold' neutrons is directed at a solid polymer sample. Advantage is taken of the much greater scattering power of deuterons as compared to protons to provide the equivalent of the refractive-index difference between polymer and solvent in light scattering. Fully deuterated polymer samples are prepared and studied in mixtures with normal, protonated samples. The major results are in the area of polymer conformations in the solid state.

5. Sedimentation and Ultracentrifuge

The use of an ultracentrifuge allows sedimentation of polymer molecules from solution to take place in a short time, if their density is appreciably different from that of the solvent. Either rate of sedimentation or equilibrium position of solute molecules after a balance between sedimentation and diffusion has been attained, is measured to calculate molecular weight of the solute.

In sedimentation velocity experiments, a dilute solution in high centrifugal field gets separated into a clear pure solvent layer and a solution. The initial clear boundary between the two then spreads because of diffusion. The rate of movement of boundary can be measured by suitable methods. Sedimentation constant s is given by

$$S = \frac{1}{x\omega^2} \cdot \frac{dx}{dt} \qquad ...(8.44)$$

where x is the distance of boundary from the rotor axis $\frac{dx}{dt}$ is the rate of movement of the boundary in the cell and ω is the angular velocity of the rotor. By introducing a diffusion constant D of the solute in the givne solvent, the ratio (D/S) can be extrapolated to infinite dilution and is related to molecular weight as under:

$$\left(\frac{D}{S}\right)_{c=0} = \frac{RT}{M(1-\overline{v}_\rho)} \qquad ...(8.45)$$

where R is the gas constant, v is the partial specific volume of polymer and ρ is density of solvent. The method gives weight average molecular weights. Polydispersed samples show diffused boundaries, and using Schlieren optics, and photographing the boundaries at different times, it also becomes possible to calculate molecular weight distribution. However, diffusion constants for each component must be known.

Alternatively, in sedimentation equilibrium experiments, the ultracentrifuge is operated at slow speeds for longer time to permit the solute molecules to attain equilibrium between sedimentation and diffusion. If the sample is monodisperse, its concentrations c_1, c_2 can be measured at 2 positions x_1, x_2 in the cell. Then M is given by

$$\overline{M} = \frac{4.606.RT\log(c_2/c_1)}{w^2(1-\overline{v})(x_2^2 - x_1^2)} \qquad ...(8.46)$$

For polydisperse sample, situation becomes a bit more complex, but a detailed analysis can yield average molecular weights with different weightings such as $\overline{M}_n$, M_n, M_z which are useful measures of breadth of distribution.

Ultracentrifugation

Ultracentrifugation techniques are the most intricate of the various existing methods used for determining the molecular weights of high polymers. They are found to be more successful in application to compact protein molecules than to random-coil polymers, where extended conformations increase deviations from ideality and the chances of mechanical entanglement of the polymer chains. Inspite of various recent theoretical and experimental developments which have helped to increase the utility of the techniques for random-coil polymer system, the method is used mostly for biological materials.

Experimental Details

The ultracentrifuge is made up of an aluminium rotor several inches in diameter and it is rotated at high speed in an evacuated chamber. The solution to be centrifuged is kept in a small cell within the rotor near its periphery. The rotor can be driven electrically or by an oil or air turbine. The cell is fitted with windows, and the concentration of polymer along its length is determined by optical methods which are based on measurements of refractive index of absorption solvents to be used for ultracentrifugation experiments must be choosen for difference from the polymer in both density and refractive index. An effort be made to avoid mixed solvents. Low solvent viscosity is also desirable.

Preparative Separation

The usefulness of the ultracentrifuge in the preparation of samples rather than in the production of analytic data should not be overlooked. Preparative ultracentrifuges have utility in fractionating polymer samples and in freeing them from easily seaimented impurities.

Sedimentation Equilibrium

In the sedimentation equilibrium experiment, the ultracentrifuge must be operated at a low speed of rotation up to 1 or 2 weeks under constant conditions. A thermodynamic equilibrium is reached where in the polymer gets distributed in the cell solely according to its molecular weight and molecular-weight distribution, the force of sedimentation on each species is just balanced by its tendency to diffuse back against the concentration gradient which results in its movement in the centrifugal field. The force on a particle of mass m at a distance r from the axis of rotation is given by w^2r $(1-\bar{v}_\rho)m$, where w is the angular velocity of rotation, $\bar{v}$ is the partial specific volume of the polymer, and ρ is the density of the solution. Writing the partial molar free energy in terms of this force and using the conditions for equilibrium we get the result, for ideal solutions.

$$\overline{M}_w = \frac{2RT\, ln(c_2/c_1)}{(1-\overline{v}_p)\omega^2(r_2^2 - r_1^2)} \quad ...(8.47)$$

where c_1 and c_2 are the concentrations at two points r_1 and r_2 in cell; r_1 and r_2 should be taken at the meniscus and the cell bottom so as to include all the molecular species in the average. When the data is treated differently, it can be shown that the *z-average molecular weight* $\overline{M}_z$ is obtained:

$$M_z = \frac{\sum_{i=1}^{\infty} N_i M_i^3}{\sum_{i=1}^{\infty} N_i M_i^2} = \frac{\sum_{i=1}^{\infty} w_i M_i^2}{\sum_{i=1}^{\infty} w_i M^2} \quad ...(8.48)$$

To overcome the problem of nonideality the work be carried out at the Θ temperature because in nonideal solutions the apparent molecular weight is a linear function of concentration at temperatures near Θ and the slope depending primarily on the second virial coefficient.

The main disadvantage is the long period required to reach equilibrium, even if the time is shortened by use of short cells or a synthetic boundary cell to reduce the distance the various species must travel as equilibrium is established.

The distribution of molecular weights can be obtained directly from sedimentation-equilibrium measurements. The solution of an integral equation is involved, however, but the analysis remains a difficult one.

Equilibrium Sedimentation in a Density Gradient

In this approach a mixed solvent is chosen so that the relative sedimentation of the two components may give rise to a density gradient. The solute from a band that centres at the point where its effective density is equal to that of the solvent mixture. The band has a Gaussian shape with respect to solute concentration, the half-width is inversely proportional to the molecule weight of solute. A major importance is its sensitivity to small differences in effective density among the solute species.

Approach to Equilibrium

Archibald (1947) showed that measurement of *c* and *dc*/*dr* at the cell boundaries permit molecular weight determination at any stage in the equilibrium process. However when measurements are made early enough, before the molecular species have time to redistribute in the cell, the weight-average molecular weight and second virial coefficient can be evaluated. In practice, measurements can be made about 10–60 min after the start of the experiment.

6. Solution Viscosity and Molecular Size

The usefulness of viscosity as a measure of polymer molecular weight was recognized in the early work of Staudinger (1930). Solution viscosity is a measure of the size or extension in space of polymer molecules. It is empirically related to molecular weight for linear polymers; the simplicity of the measurement and the usefulness of the viscosity-molecular weight correlation are so great that viscosity measurement constitutes an extremely valuable and simple tool for the molecular characterization of polymer.

The viscosity is measured with the help of a viscometer and the method is known as **viscometry.**

Molecular Weight Determination by Viscometry

For the determination of molecular weight of macromolecular compounds, the viscosity method was introduced by H. Staudinger (1930) and is the most commonly used in day-to-day research and development. This may be attributed to the fact that the experimental procedure used in viscosity determinations is quite simple and accurate. Moreover it requires only simple apparatus. However, this method of determining molecular weights is not an absolute method. To find the molecular weight from the viscosity data, it is essential to use empirical equations that give viscosity as a function of the molecular weight. These equations are set up with the aid of any suitable absolute method for the determination of the molecular weights.

The empirical dependence that is established for a polymer of a specified chemical structure is only valid for a given solvent and temperature. Viscosity of polymer solutions are generally higher as compared to those of pure solvent. A number of viscosity designation have been defined for dilute polymer solutions. For the sake of consistency, the more common usage is adopted in present discussion.

Relative Viscosity $\eta_{\gamma el} = t/t_0$ ...(8.49)
(Viscosity ratio)

't_0' is the flow time through a viscometer of a reference liquid, and 't' is the flow time through the same viscometer of a dilute solution of polymer in the reference liquid.

Specific viscosity : $\eta_{sp} \quad \eta_{el}^{-1}$ ...(8.50)

Reduced specific viscosity : $\eta_{sp}/C = (\eta_{\gamma\, el}^{-1})/C$ ...(8.51)
(viscosity number)

Inherent viscosity
(logarithmic number) viscosity $\beta_{inh} = ln\ \eta_{rel}/C$ ...(8.52)

When the polymer solution is dilute,

Thus, $$ln\ \beta_{rel}/C = \frac{ln(1+n_{sp})}{C} \quad ...(8.53)$$

$$= \frac{\eta_{sv}}{C}\left[1 - \frac{1}{2}\eta_{sp} + \frac{1}{3}\eta^2{}_{sp.....}\right] \quad ...(8.54)$$

and Thus,

$$\lim_{C\to 0} ln \frac{n_{rel}}{C} = \lim_{C\to 0} ln \frac{n_{ps}}{C} = [\eta] \quad ...(8.55)$$

The intrinsic viscosity $[\eta]$ can be obtained by drawing a plot between η_{sp}/C *i.e.*, reduced viscosity or $ln\,\eta_{rel}/C$ *i.e.*, inherent viscosity against concentration (*g'dl g*/100 ml) and extrapolating it to the zero concentration. This is shown in Figure 8.22. Hence intrinsic viscosity $[\eta]$ is independent of concentration because of extrapolation to C = 0.

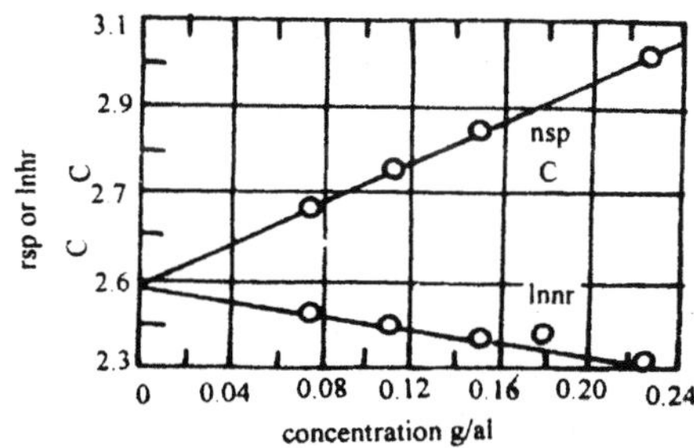

Fig. 8.22: Reduced and inherent viscosity-concentration curves for a polystyrene in benzene.

Empirical Correlations between Intrinsic Viscosity and Molecular Weight for Linear Polymers

Staudinger predicted in 1930 that the reduced viscosity of a polymer is proportional to its molecular weight. It has been modified slightly by substituting intrinsic viscosity for the reduced viscosity. It has also been recognized that the proportionality is to a power of the molecular weight that is somewhat less than 1. The relation can be expressed mathematically as:

$$[\eta] = K' M^a \qquad ...(8.56)$$

where K' and a are constants which can be determined from a double logarithmic plot of intrinsic viscosity and molecular weight (Fig. 8.23). Such plots are generally straight iines with error with in experimental limits over large ranges of the variables. In case of randomly coiled polymers the exponent a changes from 0.5 in a Θ solvent to a maximum of about 1.0. For many systems a lies between 0.6 and 0.8. Mostly values of K' lie between 0.5 and 5×10^{-4}. Both K' and a are functions of the solvent as also of the polymer type. This empirical relation between viscosity and molecular weight is found to be valid only for linear polymers.

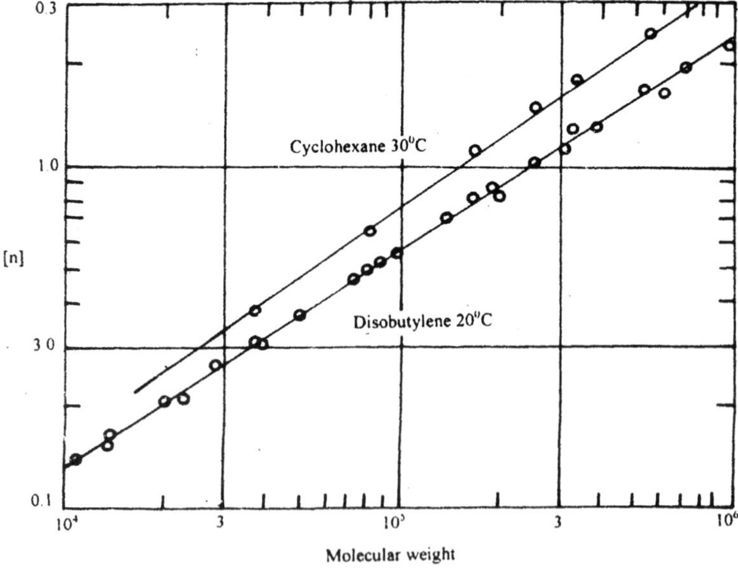

Fig. 8.23: Intrinsic viscosity-molecular weight relationships for polyisobutylene in disobutylene and cyclohexane.

Intrinsic-viscosity measurement leads to the *viscosity-average molecular weight,* which may be defined as under:

$$\overline{M}_v = \left(\sum w_i M_{v^a}\right)^{1/a} \left(\frac{\sum_{i=1}^{\infty} N_i M_i^{1+a}}{\sum_{i=1}^{\infty} N_i M_i}\right)^{1/a} \quad ...(8.57)$$

Thus $\overline{M}_v$ depends on a as also on the distribution of molecular species. For many polymers $\overline{M}_v$ is 10–20% less than $\overline{M}_w$. For $a = 1$, $\overline{M}_v = \overline{M}_w$. In cas of polymers, for which $\overline{M}_n \simeq \overline{M}_v \simeq \overline{M}_w$, any absolute molecular-weight measurement may be combined with viscosity measurement to evaluate the constants in Eq. 8.56. If the molecular-weight distribution of the samples is know in sufficient detail to calculate $\overline{M}_v$ from a molecular-weight average that has been measured, the constants in the equation

$$[\eta] = K'\overline{M}_v^{\,a} \quad ...(8.58)$$

can be found. Because $\overline{M}_v$ is nearer to M_w than to M_n, the weight-average molecular weight is preferred for this calibration. We can show that polymers having fairly broad but well-known distributions are preferred to fractions for this purpose. A less accurate relation. Given below may be used to relate

$$[\eta] = K'\overline{M}_w^{\,a} \quad ...(8.59)$$

weight-average molecular weight directly to intrinsic viscosity for a limited series of samples. Extensive tabulations of K' and a are also available.

Equation 8.59 to becomes inaccurate for molecular weights less than about 50,000, because of deviations from the linear relationship set in. For better results in this region, use of one of several expressions of the form given below is recommended

$$[\eta] = KM^{1/2} + K''M \quad ...(8.60)$$

Here the first term is determined by short-range interactions, as in Eq. 8.63, and the second term by long-range interactions. The form and value of K' vary among the several theories available.

Intrinsic Viscosity and Molecular Size

It has been shown, by theories of the frictional properties of polymer molecules in solution, that the intrinsic viscosity is proportional to the effective hydrodynamic volume of the molecule in solution divided by its molecular weight. The effective volume is proportional to the cube of a linear dimension of the randomly coiling chain. If $(\bar{r}^2)^{1/2}$ in the dimension chosen,

Mathematically

$$[\eta] = \Phi \frac{(\bar{r}^2)^{3/2}}{M} \qquad ...(8.61)$$

where Φ is a universal constant. Substituting $(\bar{r^2})^{1/2}$ for $x(\overline{r_0^2})^{1/2}$ and recalling that $\bar{r}_0^2 / M$ is function of chain structure independent of it surroundings or molecular weight, we get

$$[\eta] = \Phi(\frac{r_0^2}{M})^{3/2} M^{1/2}\alpha^3 = KM^{1/2}\alpha^3 \qquad ...(8.62)$$

where $K = \Phi(\overline{r_0^2})^{3/2}$ is a constant for any given polymer and is independent of solvent and molecular weight. The dimensions $(\overline{r^2},^{1/2}$ and $(\overline{r_0^2})^{1/2}$ and the expansion factor α are generally identified. Remember that Φ is not a universal constant, and that the power of α in Eq. 8.30 is somewhat less than 3.

From the properties of α we have, at $T = \Theta$, $\alpha = 1$ and

$$[\eta]_\Theta = KM^{1/2} \qquad ...(8.63)$$

A lot of experimental evidence is available for the validity of this equation. Values of K are near 1×10^{-3} for a number of polymer systems.

Flory's viscosity theory also furnishes confirmation of the Θ temperature as that in which $a = \frac{1}{2}$, and it permits the determination of the unperturbed dimensions of the polymer chain. Even if a Θ solvent is not available, several extrapolation techniques can be used for the estimating the unperturbed dimensions from viscosity data in good solvents. The simplest of these techniques seems to be that of Stockmayer.

Determination of Chain Branching

The size of a dissolved polymer molecule, that can be estimated from its intrinsic viscosity is found to depend upon a number of factors such as molecular weight, the local chain structure (as reflected in the factor $\overline{r_0^2} / M$), the perturbations (reflected in x), and the gross chain structure, wherein the degree of long-chain branching is an important variable. By evaluating or controlling each of the other factors, viscosity measurements can be used for estimating the degree of long-chain branching of a polymer.

The relation between molecular size and the number and type of branch points has been calculated (Zimm 1949), leading to a relation between the radii of gyration (g) for branched and linear chains of the same numbers of segments:

$$g = \frac{\overline{s^2}(\text{branched})}{\overline{s^2}(\text{linear})} \quad \text{...(8.64)}$$

Various theories have been suggested to relate g to $[\eta]$ (branched)/$[\gamma_1]$ (linear). Recently it has became possible to decide which of these is best fit to experimental data. At present it still appears appropriate to consider the ratio $[\eta]$(branched)/$[\eta]$ (linear) evaluated at constant molecular weight as a qualitative indication of chain branching rather than to attempt to assign numerical values to the degree of chain branching in a given polymer sample.

Viscometers

For molecular weight determination by viscometry we do not need absolute η value, viscosity measurements may be carried out in simple Ostwald Viscometer. Because of (the non-Newtonian behaviour of most macromolecular solutions at high velocity gradients in the capillary, the viscometer dimensions are choosen in such a manner that the viscosity gradient is the smallest possible.

In the Ostwald viscometer, (Fig. 8.24) the solution is introduced through the tube '*a*' and is then sucked through the tube '*b*' until its level is above mark 'C', the efflux time of the solution from mark *c* to mark *d* is then noted.

For determining the viscosities of a polymer solution at different concentrations, it is simplest and most convenient to carry out the successive dilution of the solution directly in the viscometer. A suspended level instrument of the Ubbelohde type (Fig. 8.24) is used for this purpose. The solution is sucked through tube *b*, tube *a* being closed. Before the liquid in tube *b* begins to flow out, tube *a* is opened letting air enter bulb. The liquid flowing out of the capillary is suspended over the air and flows down along the wall, and thus turbulence is avoided. The height of the liquid column is independent of the level of the liquid in tube *a i.e.*, the efflux time is independent of the liquid introduced into tube *a*.

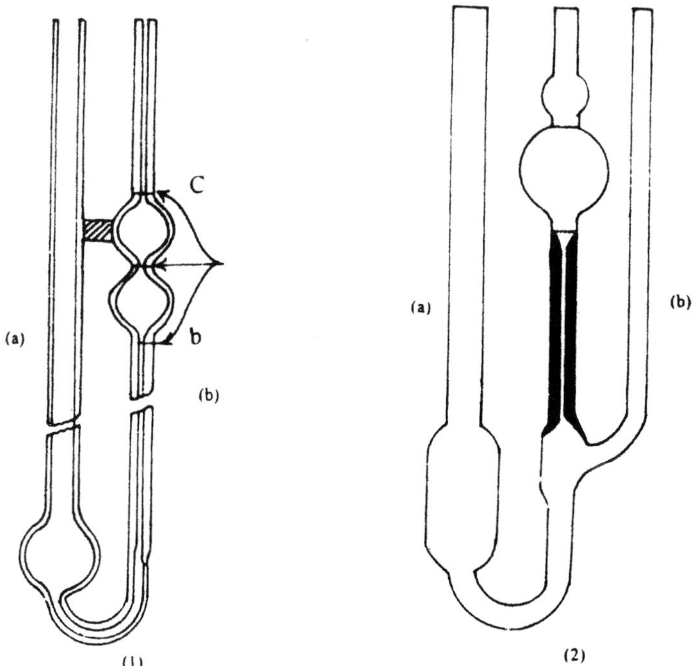

Fig. 8.24: Capillary viscometers commonly used for measurement of polymer solution viscosities: (i) Ostwald-Fenske and (ii) Ubbelohde.

When Ostwald viscometer is used, and equal volume of the solution should be taken for each determination. The error in the determination of the efflux time because of the deflection of the viscometer from the vertical is much larger in Ostwald viscometer than in Ubbelohde viscometer.

7. Gel Permeation Chromatography

Gel permeation chromatography is also known as *size exclusion chromatography* and is a separation method for high polymers. It is similar to *gel filtration* that is carried out by biochemists. It has become a prominent and widely used method for estimating molecular-weight distributions after its discovery just about four decades ago in 1961. The separation occurs in a chromatographic column filled with beads of a rigid porous "gel"; highly cross-linked porous polystyrene and poous glass are preferred as column packing materials. The pores in these gels are of the same size as the dimensions of polymer molecules.

A sample of a dilute polymer solution is introduced into a solvent stream flowing through the column. When the dissolved polymer molecules flow past the porous beads (Fig. 8.25), they may diffuse into the internal pore structure of the gel to an extent that depend on their size and also the pore-size distribution of the gel. Larger molecules can enter only a small fraction of the internal portion of the gel, or are completely excluded; smaller polymer molecules penetrate a larger fraction of the interior of the gel.

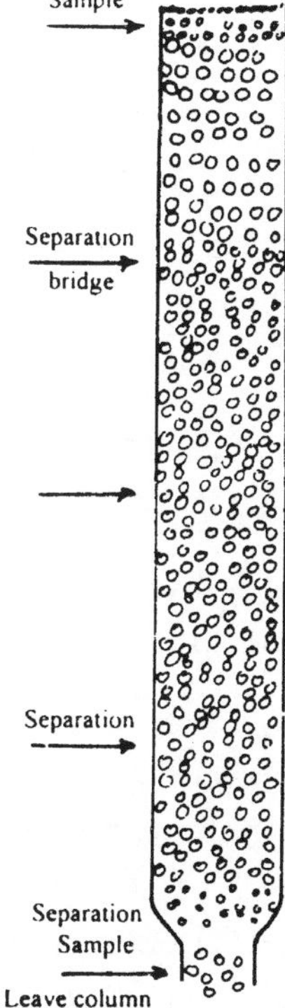

Fig. 8.25: Principle of the separation of molecules according to size by gel permeation chromatography.

Thus larger the molecule, the less time it spends inside the gel, and the sooner it flows through the column. The different molecular species get eluted from the column in order of their molecular size as distinguished from their molecular weight, the largest emerging first.

A complete theory predicting retention times or volumes as function of molecular size has not yet been formulated for gel permeation chromatography. A specific column or set of columns is calibrated empirically to give such a relationship, by means of which a plot of amount of solute versus retention volume (the chromatogram Fig. 8.26) can be changed into a molecular-size-distribution curve Generally commercially available narrow distribution polystyrenes are used. If the calibration is done in terms of a molecular-size parameter, for example, $[\eta]M$ it can be used for a wide variety of both linear and branched polymers (Fig. 8.27).

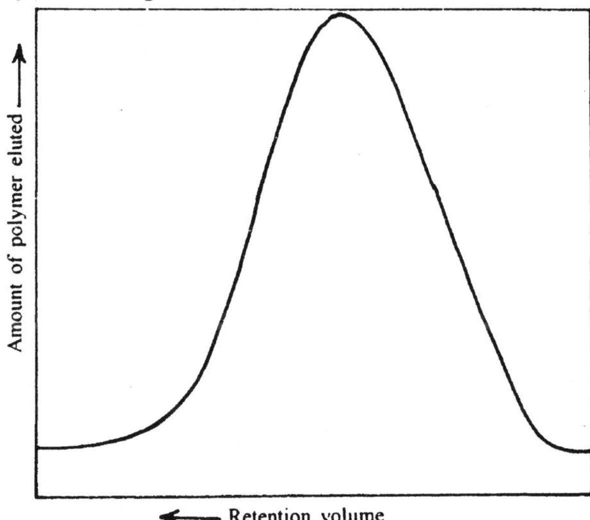

Fig. 8.26: A typical gel permeation chromatogram: polystyrene in tetrahydrofuran, which $M_s/M_w = 2.9$.

Similar to all chromatographic processes the band of solute that emerges from the column can be broadened by a number of processes, including contributions from the apparatus, flow of the solution through the packed bed of gel particles, and the permeation process. Corrections for this zone broadening may be made empirically; it generally becomes unimportant when the sample has $\overline{M}_w /_n \overline{M} > 2$.

Gel permeation chromatography is very valuable for both analytic and preparative work with a wide variety of systems ranging from low to very high molecular weights. The method can be used in a large number of solvents and polymers, depending on the type of gel used. With polystyrene gels, relatively nonpolar polymers can be measured in solvents like tetrahydrofuran, toluene, or (at high temperatures) O-dichloro-benzene; with porous glass gels, more polar systems, including aqueous solvents, can be used. Only a few miliigrams of sample is needed for a analytic work, and the determination is complete in a few minutes using modern high-pressure, high-speed equipment. The technique is known as high performance liquid chromatography (**HPLC**).

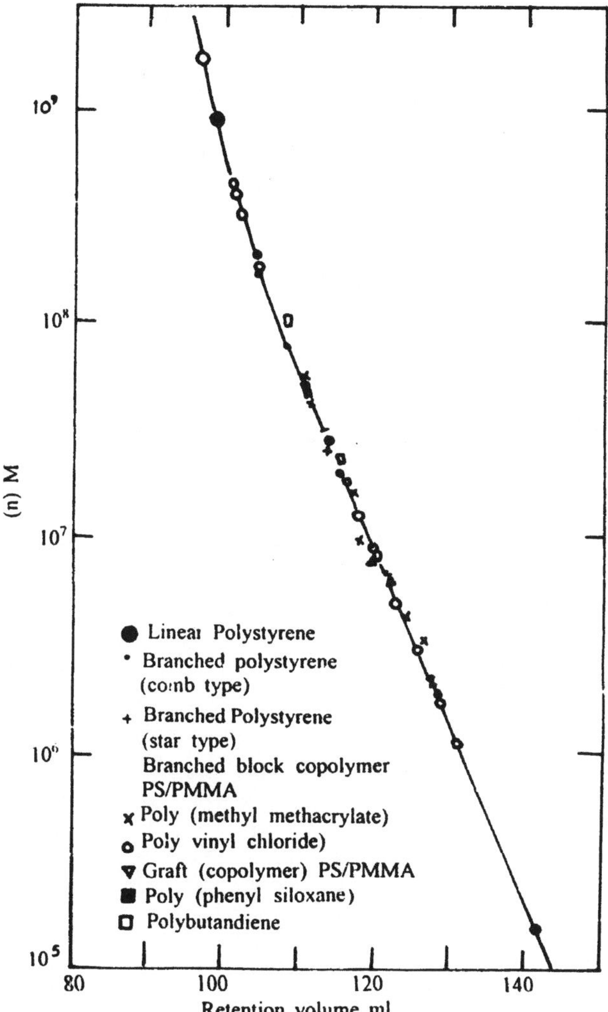

Fig. 8.27: Calibration curve for gel permeation chromatography based on hydrodynamic volume as expressed by the product [η] M. The polymer types shown are linear polystyrene, two types of branched polystyrene (methyl methacrylate). poly(vinyl chloride), polybutadiene poly(phenyl soiloxane), and two types of copolymer.

The results of gel permeation chromatography experiments for molecular-weight distribution are found to agree well with results from other techniques. Fig. 8.28 shows the extent of agreement this method and a solvent-gradient elution fractionation, while Fig. 8.29 shows the degree of agreement between the experiment and a distribution curve calculated from polymerization kinetics.

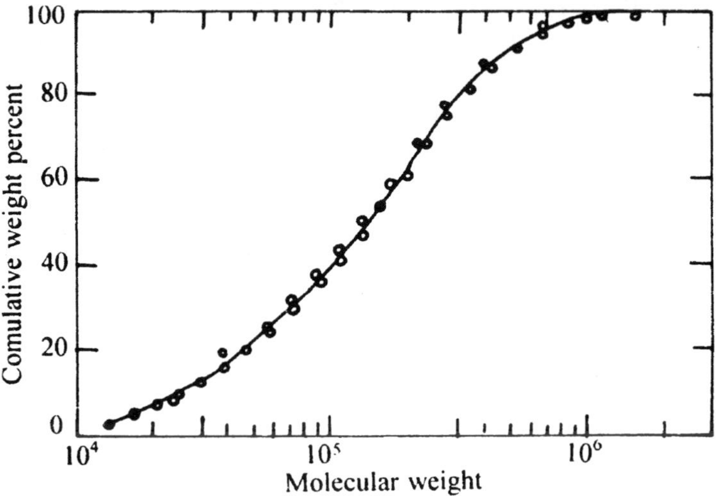

Fig. 8.28: Typical cumulative molecular-weight distribution curve for a sample of polypropylene gradient-elution data (O) and data from gel permeation chromatography.

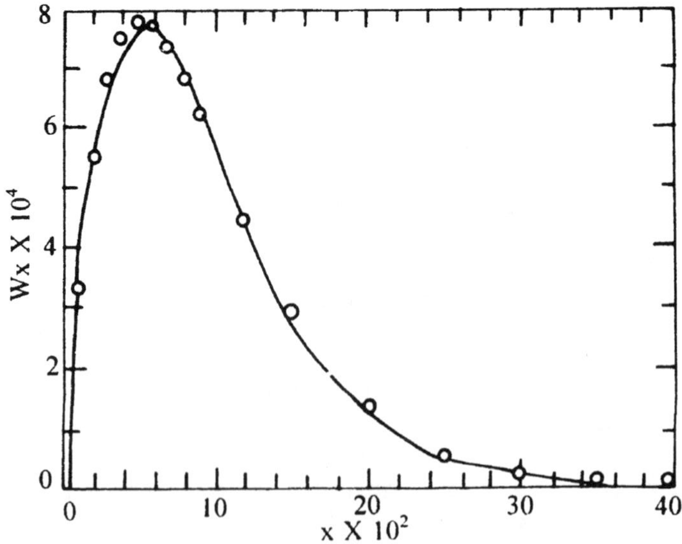

Fig. 8.29: Gel permeation chromatography data for polystyrene to a molecular-weight distribution curve calculated from polymerization kinetics.

Polyelectrolytes

Polymers having ionizable groups along the chain are known as polyelectrolytes. They generally exhibit properties in solution which are quite different from those with nonionizable structures. Examples of polyelectrolytes, includes polyacids like poly (acrylic acid) and hydrolyzed copolymers of maleic anhydride, polybases like poly (vinyl amine) and poly (4-vinyl pyridine), polyphosphates, nucleic acids, and proteins.

When they are soluble in nonionizing solvents (for example, poly (acrylic acid) in dioxane—polyelectrolytes show normal behaviour but in aqueous solution they are ionized, and the mutual repulsion of their charges produces expansions of the chain far beyond those resulting from changes from good to poor solvents with ordinary polymers. The size of the polyelectrolyte random coil is a function of the concentrations of polymer and added salt, and both these influenced the degree ionization.

Moreover the ionization of the electrolyte groups leads to a number of unusual effects in the presence of small amounts of added salt. The intensity of light scattering decreases due to the ordering of the molecules in solution and the osmotic pressure and ultracentrifugation behaviour are determined predominantly by the total charge on the molecule.

Finally, the ionic charges attached to the chains create regions of high local charge density thus affecting the activity coefficients and properties of small ions in these localities. Though these effects cannot be separated completely, the results of chaing expansion are of primary interest for the measurement of molecular weight and size.

High Performance Liquid Chromatography

This technique in commonly used these days and a typical HPLC set up is shown in 8.30. Different types of detectors are used to monitor solute concentration. The most commonly used are refractive inlet dector, *uv* absorbance dector. IR dector etc.

Only a small quantity (only a microlite solution 1%) is needed and the instrument can directly give calculated results as polydispersity, weight fraction of each component. A sample separation is shown in Fig. 8.31.

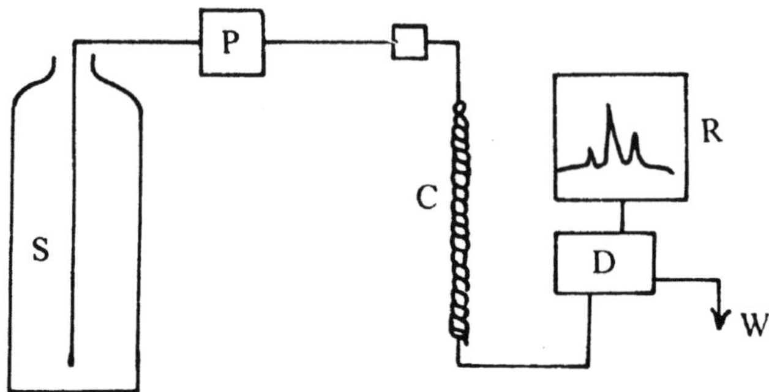

Fig. 8.30: A schematic representation of high performance liquid chromatograph, S—solvent reservior; P—pump; C—column; D—dector, R—recorder; W—drain, I—injector port.

Properties which depend on the size of the chain, for example, viscosity and angular dependence of light scattering, get strongly affected by chain expansion. The viscosity may even increase markedly as polymer concentration *decreases,* the increase in viscosity results in an increase in the degree of ionization of the polymer.

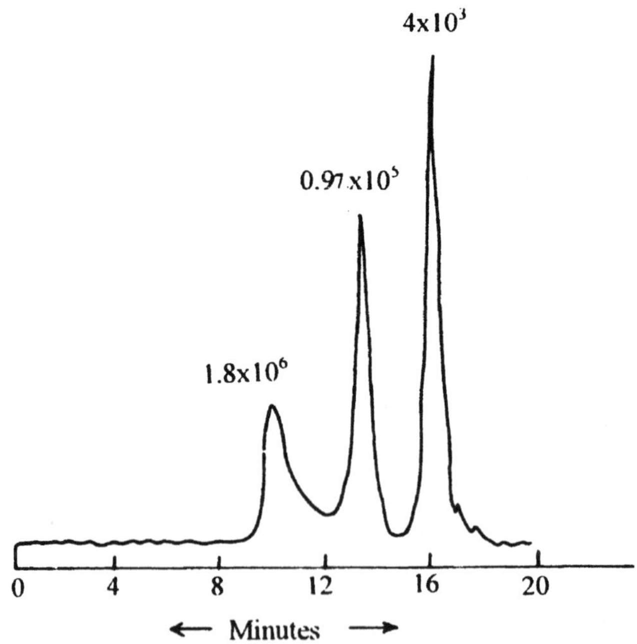

Fig. 8.31: HPLC of a polystyrene sample. This figure give molecular weight of the fractions present in the sample.

When very high chain extensions are reached, the effect gets reversed but it does not disappear at infinite dilution. On the other hand, the addition of low molecular-weight electrolyte (salt) to the aqueous solution increases the ionic strength of the solution outside the polymer coil as compared to that inside, and also reduces the thickness of the layer of "bound" counterions around the chain. Both effects cause the chain to contract, and when the concentration of added salt reaches, about 0.1 M, behaviour again becomes normal. Using some special precautions, molecular weight can be measured by light scattering and equilibrium ultracentrifugation and intrinsic viscosity-molecular weight relations can be established polyelectrolytes in the presence of added salt.

Of the preponderance of small ions, the colligative properties of polyelectrolytes in ionizing solvents measure counterion activities rather than molecular weight. In the presence of added salt, however, correct molecular weights of polyelectrolytes can be measured by membrane osmometry, since the small ions can move across the membrane. The second virial coefficient differs from that previously defined, since it is determined by both ionic and nonionic polymer-solvent interactions.

9

Industrial Polymers

9.1. Introduction

We know that collulose (chief component of the cell walls of a plant), proteins essential constituents of living cells), rubber, leather and natural fibres like silk, wool etc. are all polymers and these are known as **natural polymers.**

The development of synthetic materials as substitutes for naturally occurring polymers led to the growth of polymer science. The term **"High polymers"** is used to represent macromolecules in which more that 100 monomer units are involved. Smaller combinations are referred to as dimer, trimer, tetramer etc. depending on whether the polymer molecule contains 2, 3, 4...etc units of monomers.

Synthetic polymers include a variety of products such as plastic, fibres, elastomers, rubber etc.

An attempt will now be made to discuss various types of synthetic polymers.

1. Polyolefins

These are the polymers which are based on unseparaied aliphatic hydrocatbons containing are double bond in molecule. The important polyolefins are polyethylene, polypropylene, poly(isobutene) and poly (4-methyl l-pentene).

2. Polyethylene or Polythene

Ethylene is the simplest unsaturated hydrocatbon but its polymerisation was exceptionally difficult. Gibson and Fawcett in

1933 polymerised ethylene at 170°C and 2000 atm pressure to a waxy solid in the presence of benzaldehyde. Commercial production of polyethylene started in 1939 to provide electrical insulation for new radar installations. This type of ethylene which is obtained at high pressure is called *low density polyethylene*.

In 1953, Ziegler in Germany, Phillips Petroleum Co. (U.S.A.) and Standard Oil Co. (U.S.A.) were able to discover the process of manufacturing polyethylene by avoiding the use of high pressure. The polyethylene so obtained is termed as the *high density polyethylene*.

However, the bulk of the commercial polyethylene is obtained by using high pressure technique.

Manufacture of Ethylene

Ethylene is a colourless gas having boiling point −14°C. It is obtained by the dehydration of ethanol or hydrogenation of acetylene.

$$C_2H_5OH \xrightarrow{-H_2O} \begin{array}{c} CH_2 \\ \| \\ CH_2 \end{array}$$

$$\begin{array}{c} CH \\ \| \\ CH \end{array} + H_2 \xrightarrow{\text{catalyst}} \begin{array}{c} CH_2 \\ \| \\ CH_2 \end{array}$$

Ethylene is also obtained by cracking ethane or propane obtained from the petroleum.

Different Varieties of Polyethylenes

There are two varieties of polyethylenes, viz., low density and high density. Low density polyethylene is made up of molecules having branches, whereas the high density variety is essentially linear.

Preparation of Low Density Polyethylene

Low density polyethylene (LDPE) can be produced by the high pressure polymerisation of ethylene, making use of oxygen as the initiator. The reaction takes place at pressures as high as 1.5×10^8 pascal (approximately 1500 atmospheres) and the temperature range of 180–250°C Even an extremely small quantity of oxygen, say, 0.1%, is sufficient to initiate the polymerisation. In addition to oxygen, other initiators including peroxides, hydroperoxides and azo compounds can be used. Ethylene can be polymerised by solution or bulk polymerisation techniques. For the solution method, the solvents used are benzene, chlorobenzene etc. The reaction conditions are such that the polymer as also the monomer dissolve in the solvent and the system follows a true solution polymerisation technique.

The reaction temperature is above the critical temperature of ethylene so that the ethylene is in gas phase. High pressures are needed for propagation reaction. Only about 6–25% of ethylene is polymerised. Rest of monomer is recycled. Extensive chain transfer reactions takes place during polymerisation to yield a branched chain polyethylene. In addition to long branches it also contains a large number of short branches of upto 5 carbon atoms produced by intra-molecular chain transfer reactions. A typical molecule of low density polyethylene contains a short branch for about every 50 carbon atoms and one or two long branches per molecule.

As stated earlier LDPE consists of molecules which are branched. The branching occurs during the polymerisation process, either by intermolecular chain transfer reactions or by intermolecular chain transfer as under:

1. $\sim\!\sim\!\sim CH_2-\dot{C}H_2 + \sim\!\sim\!\sim CH_2–CH_2–CH_2–CH_2–R$

$$\downarrow \text{ INTERMOLECULAR CHAIN TRANSFER}$$

$$\sim\!\sim\!\sim CH_2–CH_3 + \sim\!\sim\!\sim CH_2–\dot{C}H–CH_2–CH_2–R$$

2. $\sim\!\sim\!\sim CH_2–CH_2–CH_2–CH_2–CH_2–\dot{C}H_2$

$$\downarrow \text{ INTRAMOLECULAR CHAIN TRANSFER}$$

$$\sim\!\sim\!\sim CH_2–CH_2–CH_2–\dot{C}H–CH_2–CH_3$$

In the first case, the branches could be as long as the backbone chain itself, whereas, in the second case, the branches are much shorter in length.

A new process for LDPE has been developed by Dow Chemical Co. This process can be carried out at much lower pressure in presence of a catalyst system. Considerable reduction in plant cost is possible by this process.

Manufacture of Low Density Polyethylene

It is manufactured by high pressure process by polymerising ethylene at high pressure of 1000–3000 atmosphere and temperature of 80–300°C. The common initiators used for polymerising ethylene are traces of oxygen, azobisisobutyronitrile, and benzoyl peroxide.

For the oxygen-initiated polymerization, nearly 1500 atmosphere pressure and 200°C is used having 0.05–0.1 percentage oxygen. The reaction is highly exothermic and efficient heat dissipation becomes essential. The reaction temperature is above the critical temperature of ethylene so that the ethylene is in the gas phase.

In some processes, a diluent, like benzene or chlorobenzene are used as the solvent. At high pressure and temperature both the polyethylene and the monomers dissolve in these solvents so that the reaction occur in a solution phase. In a typical process, 10–30% of the monomer is converted to polymer per cycle. Rest of monomer is recycled. Extensive chain transfer reactions take place during polymerisation to yield a branched polyethylene. Apart from long branches it is also having a large number of short branches of unto 5 carbon atoms formed by intramolecular chain transfer reactions. A typical molecule of low density polyethylene is having a short branch for about every 50 carbon atoms and one or two long branches per molecule.

Method of Polymerization

Method – I

In this method a stainless steel tube of one inch diameter and 100 feet length is used. The tube is filled with water, ethylene, the initiator and benzene.

Initiator, water or benzene may be added so as to keep the initiator constant. The gas and liquid phases are removed continuously, the polymer gets separated and the ethylene is recycled after purification.

Method – II

This method used the bulk polymerization in a tower-type reactor. Ethylene having trace amount of oxygen is charged to the reactor at the pressure of 1500 atmosphere and 1900°C temperature.

The reaction is maintained at isothermal conditions. The effluent from the reactor is allowed to pass to a separatory vessel in which unconverted ethylene is removed for recycling. The molten polyethylene gets chilled below its crystalline melting point.

A new propose for the manufacture of low density polyethylene has been developed by Doro Chemical Co. This process uses much lower pressure in the presence of a catalyst system. Considerable reduction in plant cost is possible for this process.

Mechanism

Production of free-radical

In the first step the decomposition of the initiator takes place to produce a free radical.

$$I - I - \rightarrow 2\dot{I}$$

(where I = initiator)

The initiator radical reacts with ethylene to yield ethylene radical

$$\dot{I} + CH_2 = CH_2 - \rightarrow I - CH_2 - \dot{C}H_2$$

Propagation

Now this radical is made to react with the etylene in a very rapid process to yield a polymer chain radical as given below:

$$I - CH_2 - \dot{C}H_2 + CH_2 = CH_2 - \rightarrow I\,CH_2 - CH_2 - CH_2 - \dot{C}H_2$$

$$I - (CH_2 - CH_2)_{n-1} - \dot{C}H_2 + CH_2 = CH_2 - \rightarrow$$
$$\rightarrow I - (CH_2 - CH_2)_n - \dot{C}H_2$$

Termination

Termination can be of two types:

(i) *Mutual*

In this two growing polymer radicals combine to yield a dead polymer.

$$...-CH_2-\dot{C}H_2-+\dot{C}H_4-CH_2...- \rightarrow ...-CH_2 + CH_2-CH_2-....$$

(ii) *Disproportionation*

$$-CH_2-CH-+\dot{C}H_2-CH_2\sim - \rightarrow$$
$$...\sim CH_2-CH_2...+ CH_2=CH-...$$

The low density polyethylene is branched and is having 30 groups per 1000 carbon atoms in the chain. The infrared studies reveals that practically all the branches are short, and have an either ethyl or butyl group.

The production of these branches is ascribed to intramolecular transfer *i.e.*, back-bite.

```
         CH2 —CH2                          CH2 — CH2
        /        \                        /      |
... CH            CH2   —→            HC        CH2
     |           /                              /
     H        CH2                          H —CH2
          CH3                                CH3
CH2=CH2    |                  CH2=CH2        |
-----→   (CH2)3               ------→      (CH2)3
           |                                 |
        —CH —CH2 —CH2               ...— CH —CH2 —CH2  etc
            (1)
```

Besides propagation the radical I may also undergo backbite to yield ethyl groups.

```
                      CH3                       CH3
                       |                         |
                      CH2                       CH2
                       |                         |
                CH2 —CH                   H2C—CH
               /       |                  /
       ........CH2     H  —→—HC                 H
          |                   \                /
         H2C   CH2             H2C—CH2
                               CH3
                                |
                               CH2
                                |

CH2 =CH2                          |
----→ ......CH —CH2— CH—CH2 —CH2  etc.
             |
            CH2
             |
            CH3
```

Properties

Low-density polyethylene exists as a partially crystalline solid, melting point 115°C. Its density has been found to range between 0.91–0.94. It is practically insoluble in any solvent at room temperature but is soluble in many solvents at temperature above 100°C. Some of the useful solvents for it are CCl_4, Toluene, xylene, decaline, and trichloroethylene. The dissolved polymer precipitates on cooling to room temperature.

It is having good toughness over a wide temperature range. It is having good electrical properties. It is transluscent due to its cryastallinity.

Chemical Properties

As polyethylene can be considered as a high molecular weight paraffin, it may be expected as an inert material. It is not affected by most acids, alkalis and aqueous solution. However, strong oxidising agents like concentrated nitric acid and concentrated solution of hydrogen peroxide and potassium permanganate are able to oxidise the polymer. Oxidation of polyethylene also takes place on exposure in air to ultraviolet light and/or elevated temperature.

Uses

Low density polyethylene is used in making film and sheeting, injection molding, wire and cable insulator, coating and blow molding.

LDPE films are used for packing and wrapping frozen food, textile products, and so on. The film combines in itself low density, high tear strength, extreme flexibility and chemical and moisture resistance The flexibility of the thin film is the inherent property of the polymer without recourse to a plasticiser. LDEP's inertness to chemicals and resistance to breakage is used in 'squeeze bottles' and in many attractive containers. Pipes made of LDPE are used for both agricultural, irrigation and domestic water line connection. The nonpolar nature of the polymer makes it ideal of providing insulations to electric cables.

High Density Polyethylene

The high-density polyethylene is linear and can be manufactured by (*i*) co-ordination polymerization of monomer by triethyl aluminium and tritanium chloride. (*ii*) polymerization with supported metal oxide catalysts. Such as chromium or molybdenum oxides supported over alumina-silica bases.

High density polyethylene was made by Ziegler in 1953 by using low pressure (2–4 atm.) and temperature in the range of 50 – 75°C. Usually the polymerisation is carried out in the presence of Ziegler-Natta catalysts based on titanium tetrachloride and aluminium alkyl. The catalyst may be either prepared or formed in the reactor. Usually, the polymerisation is carried out in presence of a hydrocarbon solvent. The polymer is insoluble in the solvent. The reaction is terminated by addition of an alcohol and catalyst extracted with alcoholic hydrochloric acid. Catalyst removal is important for electrical insolution used. The polymer chain obtained by this process is essentially linear.

The details of Ziegler method and its mechanism are given in the following paragraphs.

1. Co-ordination Polymerization (Ziegler Process)

In this method, the reaction is performed at low pressure and temperature. Generally the reaction occurs at pressure only slightly above atmospheric pressure namely 2–4 atmosphers and at temperatures between 50–75°C. Generally the polymerization is carried out in the presence of Ziegler-Natta catalyst. For the polyethylene titanium chloride-diethylaluminium chloride catalyst is generally used. The reaction is performed in an inert atmosphere, preferably in the presence of nitrogen because oxygen and water reduce the effectiveness of the catalyst and may even bring about explosive decomposition.

When ethylene and the catalyst and a diluent are fed continuously into the reactor. Polyethylene gets formed as a powder of granules which is insoluble in the reaction mixture. After the reaction gets completed, the catalyst is destroyed by adding water or methanol or ethanol. Finally, the polymer gets centrifuged, dried, extruded and granulated.

Catalyst removal is important for electrical insulation uses. The polymer chain obtained by this process is essentially linear.

Mechanism

The mechanism of the polymerization of ethylene by Ziegler catalyst may be put as follows.

$$(C_2H_5)_3Al + TiCl_4 \dashrightarrow (C_2H_5)_2AlCl + C_2H_5TiCl_3$$

$$C_2H_5TiCl_3 \dashrightarrow C_2H_5 + TiCl_3$$

$$(C_2H_5)_3Al\ TiCl_3 \dashrightarrow (C_2H_5)_2AlCl + C_2H_5TiCl_2$$

$$C_2H_5TiCl_3 \dashrightarrow C_2H_5 + TiCl_2$$

$$(C_2H_5)_3Al + T_2H_5TiCl_3 \dashrightarrow (C_2H_5)_2\ AlCl + (C_2H_5)_2\ TiCl_2$$

$$(C_2H_5)_3\ TiCl_2 \dashrightarrow C_2H_5 + C_2H_5TiCl_2$$

The active component of the catalyst mixture is a complex which gets formed between titanium trichloride and triethylaluminium. The structure of the complex may be put as follows:

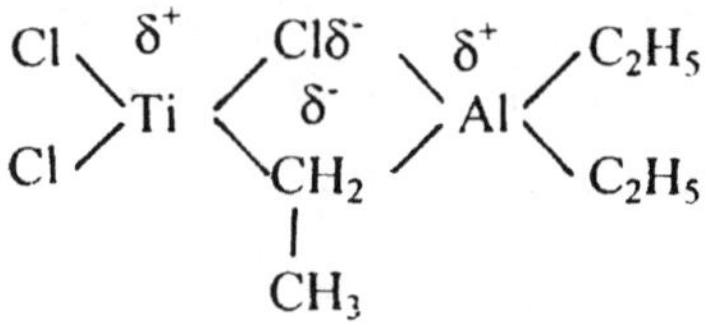

$$\begin{matrix} Cl & \delta^+ & Cl\delta^- & \delta^+ & C_2H_5 \\ & Ti & \delta^- & Al & \\ Cl & & CH_2 & & C_2H_5 \\ & & | & & \\ & & CH_3 & & \end{matrix}$$

In the complex given above the titanium is having unfilled 3d-orbitals to which are co-ordinated π-electrons from the double bond of the vinyl monomer.

Thus, the initiation, propagation and termination steps may be as follows:

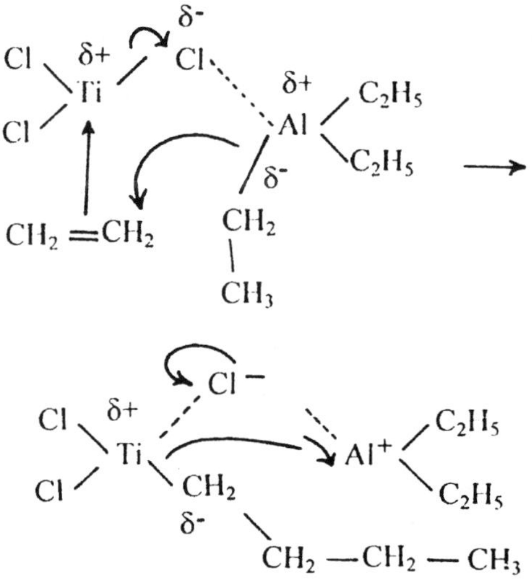

$$\begin{matrix} & & \delta^- & & \\ Cl & \delta+ & Cl & \delta+ & C_2H_5 \\ & Ti & & Al & \\ Cl & \uparrow & & \delta^- & C_2H_5 \\ CH_2{=}CH_2 & & CH_2 & & \\ & & | & & \\ & & CH_3 & & \end{matrix} \longrightarrow$$

$$\begin{matrix} & & Cl^- & & \\ Cl & \delta+ & & & C_2H_5 \\ & Ti & & Al^+ & \\ Cl & & CH_2 & & C_2H_5 \\ & & \delta^- & & \\ & & & CH_2-CH_2-CH_3 & \end{matrix}$$

Propagation

$$\longrightarrow \begin{matrix} & & \delta^- & & \\ & & Cl & & \\ Cl & \delta+ & & \delta+ & C_2H_5 \\ & Ti & & Al & \\ Cl & \uparrow & \delta^- & & C_2H_5 \\ & & CH_2 & & \\ CH_2{=}CH_2 & & & & \\ & & CH_2-CH_2-CH_3 & & \end{matrix}$$

$$\xrightarrow{CH_2=CH_2} \begin{matrix} & & \delta^- & & \\ Cl & \delta+ & Cl & \delta+ & C_2H_5 \\ & Ti & \delta^- & Al & \\ Cl & & CH_2 & & C_2H_5 \\ & & | & & \\ & & CH_2-CH_2-CH_2-CH_2-CH_3 & & \end{matrix}$$

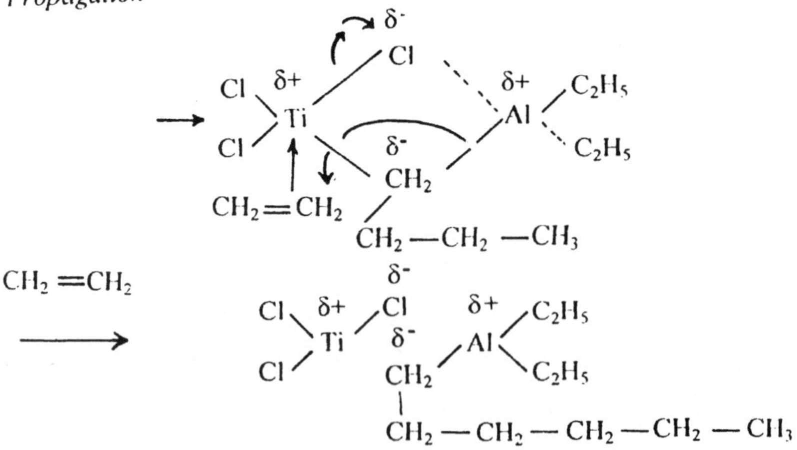

The special aspect of this mechanism is that monomers get added up in a step-wise manner into polarised titanium carbon bond and the polymer is given out of the active centre.

As the propagation end of the polymer chain is negatively charged, the reaction may be considered as an **anionic polymerization reaction.**

Termination

The two types of chain termination are

(i) Internal hydride transfer
$Cal–CH_2–CH_2–[–CH_2–CH_2–]_n–CH_2–CH_3 -\rightarrow$
$Cal–H+CH_2=CH–[–CH_2–CH_2–]_n–CH_2 | CH_3$

(ii) Transfer to monomer

$Cat– CH_2–CH_2 –[–CH_2–CH_2–]_n–CH_2–CH_3–CH_2 =$
$CH_2 -\rightarrow cat–CH_2–CH_3+CH_2= CH–$
$[–CH_2–CH_2]–_n–CH_2–CH_3$

2. Polymerisation with Supported Metal Oxide Catalyst

Two catalyst systems were developed by Standard Oil and Phillps petroleum. Standard Oil process uses metal catalyst such as molybdenum trioxide on supports like alumina or titanium or zirconium dioxide. The process is carried out at 200–300°C at 40–100 atmosphere pressure. Phillips process uses chromium trioxide an silica or silica-alumina support, the process is carried out at 90–160°C and 30–40 atmospheres. Linear polymer is obatined by both these processes.

Properties

The linear polyolefins have been stiffer than the branched material. They are having higher crystalline melting point and greater tensile strength and hardness.

The branched polymer has lower density, melting point, stiffness, surface hardness etc. These properties are due to branching. The polymer possesses excellent electrical insulating property. The polymer can be considered as a high molecule weight paraffin and is inert. At room temperature, it is insoluble in all solvents. At high temperature it dissolves in hydrocarbon and alogenates hydrocarbons.

Polyethylene remains unaffected an action with most acids, alkalis and aqueous solution. Strong oxidising agents like nitric acid and hydrogen peroxide etc. cause deterioration, deterioration also take place with oxygen in presence of ultraviolet light hence anti oxidants are added to the polymer. Polyethylene may be cross linked by exposure to X-rays, Y-rays, fast electrons or by treatment with peroxides.

Uses

HDPE is used in the manufacture of toys and other household articles. Because of its high tensile strength and stiffness polyethylene is also for many other purposes.

The cross linked polyethylene retains shape upto 140°C and is used for cable insulation. Polyethylene is the most widely used plastic due to its low cost and ease of processing. It is used in kitchen ware, as film for packaging, heavy films for protecting grains, crops machinery etc. as insulator for submarine cables, television aerial leads, telephone lead lines, electronics apparatus, as tubing used for transport of water in fields and chemical plant.

Chlorosulphonated Polyethylene

Chlorination of polyethylene (15–40% Cl_2) produces a rubbery polymer which has lost most of the crystallinity. With 50% chlorine content the polymer becomes hard.

Chlorosulphonated polyethylene is obtained by reacting low density polythylene with chlorine in presence of sulphur dioxide using carbon tetrachloride as solvent. The product contains 30% chlorine and 1.5% sulphur. This is a sticky rubbery material and is soluble in chlorinated hydrocarbon solvents. It can be vulcanised by heating with metal oxides like litharge or magnesium oxide in presence of water. The crosslinked product is found to be resistant to chemical attack and is used in gaskets, hoses etc.

Copolymers of Ethylene

Ethylene can be copolymerised with several monomers like propylene, l-butene, vinyl acetate, ethyl acrylate etc.

Ethylene-l-butene (5%) copolymers possess low density, softening point and these are used for making bottles.

Ethylene-vinyl acetate copolymers is used in making soft clear films for food packaging.

Ethylene methacrylic acid (1 to 10%) copolymers are known as

ionomers Some of the COOH groups are esterified in the commercial product.

Applications

The linear polyolefins find use in blow molding, injection molding and manufacture of film, sheet and wire and cable insulation, extrusion coating, and pipe.

3. Polypropylene

Natta synthesised polypropylene in 1954. The first commercial production was done by Montacatini in 1957.

High pressure free radical process is not suitable for propylene due to extensive hydrogen transfer to free radical which results in resonance stabilised alkyl radicals with reduced tendency to react with monomer molecule. Commercial production of polypropylene is largely carried out by Ziegler-Natta process. Slung process is preferred.

Polypropylene is produced from the monomer propylene by using the Ziegler-type catalyst as in case of polyethylene.

Monomer

In case of polypropylene some *atactic polymer* also gets formed in addition to the required *isotactic polymer* ; but much of this atactic material is soluble in the diluent so that the product isolated would be largely isotactic polymer.

Polypropylene is characterised by *isotactic index* which is *percentage of polymer not dissolving in boiling. n-haptane.* Commercial polypropylenes are having isotactic index of 95–98%.

Properties

The isotactic polypropylene is an essentially linear, highly crystalline polymer. The density of polypropylene is 0.905. It has high tensile strength, stiffness and hardness due to its high crystaline character.

It is having the increased softening point and consequently higher maximum service temperature. It is sufficiently stiff for a strain-free article to retain its shape at 140°C. At the room temperature, the impact to strengh of polypropylene has been comparable to that of high density polyethylene, but the impact strenght of polypropylene decreases markedly as the temperature gets reduced while polyethylene exhibits a little change.

It is having excellent electrical properties and the chemical inertness and moisture resistance typical of hydrocarbon polymers.

The solubility properties of polypropylene have been similar to that of polyethylene. Although polypropylene is insoluble at room temperature, yet it is soluble in hydrocarbons and chlorinated hydrocarbons at temperature above 80°C.

Each alternate carbon in the polymer chain of polypropylene has been a tertiary hydrogen atom which is relatively labile. Thus compared to polyethylene it becomes more prone to attack by oxidising agents.

When polypropylene is exposed to high energy radiation or gets heated with a peroxide, cross-linking takes places in polypropylene to give a useful material.

Uses

Components made of polypropylene are used in appliances such as refrigerators, radios and TVs. It is also used for producing films, pipes, storage tanks, seat covers, monofilaments and ropes.

Polypropylene is also used in the manufacture of filament, injection molding and film. Polypropylene is widely used for manufacturing ropes and seat covers. Due to its low cost and low density, polypropylene could ultimately compete heavily with cotton and rayon fibres.

Polypropylene is used in sterilisable medical and chemical equipment machine parts, food packs, book coverings, packing meat and cheese, washable wall paper yarn to make socking, etc.

Comparison of Properties of Polypropylene and Polyethylene

Most properties of polypropylene are similar to polyethylene but polypropylene has higher softening point and at 140°C polypropylene still retains its shape. Polypropylene is more susceptible to oxidation by air at higher temperature. Crosslinking, chlorination and other reactions lead to degradation of polymer chain and are not very useful commercially.

Ethylene-propylene (30-60 mole %) Copolymers produce

substances which are rubbery in nature. They are prepared by using Ziegler catalysts based on vanadium oxychloride/aluminium trihexyl by solution process at 40°C using chlorobenzene or pentane as a solvent. These can be vulcanised with peroxides. Ethylene-propylene-hexa 1, 4-diene terpolymers are rubbers which can be vulcanised with sulphur.

4. Polyisobutene

Polymers having average molecular weight of 15000 are sticky viscous liquids but those with molecular weight 1,00,000 to 2,00,000 are rubber like. It is soluble in hydrocarbon and halogenated hydrocarbon solvents at room temperature. It is largely used as adhesive, for fibre and paper coating etc. Polymerisation is highly exothermic and is accomplished at –100 to –0°C in presence of $AlCl_3$ or BF_3 as catalyst.

The most important limitation of polyisobutene is its tendency to cold flow because of which it cannot be used in self-supporting forms. This defect can be overcome by copolymerisation with isoprene and vulcanisation of product with sulphur. The product is called **butyl rubber**. This product was first marketed in 1943 in USA.

Preparation

A solution of isobutene and isoprene (1-3 mole %) in methyl chloride is cooled to –100°C and introduced along with a solution of $AlCl_3$ in CH_3Cl in the reactor. The product obtained as a slurry is continuously removed from reactor and treated with hot water. To the slurry a lubricant such as zinc stearate and an antioxidant such as phenyl-2-naphthylamine are added. The product is then separated by screening, drying, extrusion or milling.

Properties and Uses

Vulcanised butyl rubber is very similar to vulcanised natural rubber in various physical characteristics but has better resistance to oxidation and has low permeability to gases. Hence it is widely used in tubes for cycles, scooters, motor cars etc. It is also used as rubber in many other applications.

5. Polystyrene

Polyvinyl benzene or styrene is the simplest aromatic

hydrocarbon which can be polymerised. Styrene was obtained by steam distillation of resin from the tree *styrax officinalis*. In 1920's Staudinger gave the name styrene. Patent for polymerisation of styrene was taken out in 1911 by Matthews.

During 1930's the commercial interest of polystyrene started due to its good electrical insulation characteristics. During the second world war two companies, namely, I.G. Farben-industries (Germany) and Dow Chemical Company (USA) started the large scale production of polystyrene.

During the second world war enormous quantities of styrene got produced in U.S.A. for use in the manufacture of synthetic rubber.

Styrene was obtained by catalytic dehydrogenation of ethyl benzene. Styrene gets readily polymerised by light, heat or catalyst by means of all four techniques of polymerisation.

$$C_6H_6 + C_2H_4 \rightarrow C_6H_5C_2H_5 \xrightarrow[600\text{-}650^0C]{\text{metal oxide}} C_6H_5CH{=}CH_2 + H_2$$

Styrene

Bulk and suspension polymerisation are the most commonly used techniques. In **bulk polymerisation** styrene is heated to 80°C for about 2 days to get a viscous solution of polymer in styrene. The solution is then fed to a tower wherein polymerisation is completed at 100°, 150° and 180° stagewise. **In suspension process** styrene is suspended in dimineralised water in presence of suspending agent and initiator like benzoyl peroxide and heated to 20°C. The product is washed with acid, water and dried.

Raw material

The main raw material is styrene which is prepared as follows:

(i) Ethyl benzene is first of all prepared from benezene and ethylene in the presence of $AlCl_3$ at 90°C as shown below:

$$C_6H_6 + CH_2{=}CH_2 \xrightarrow[90^0C]{AlCl_3} C_6H_5CH_2-CH_3$$

ethyl benzene

(ii) When ethyl benzene is passed over an iron oxide or magnesium oxide catalyst at about 600–650°C, styrene is produced which is purified by distillation:

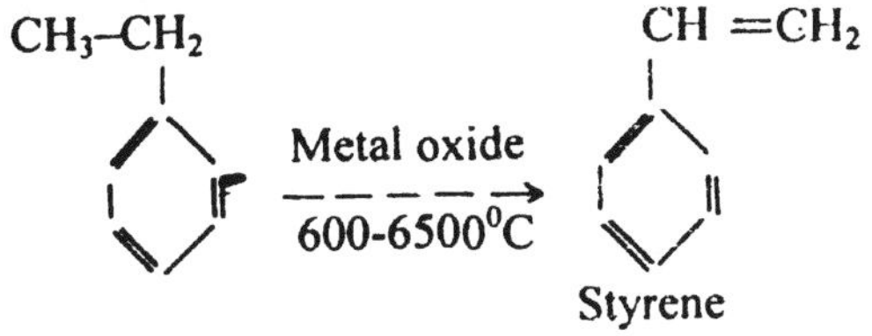

$CH_3–CH_2$ (on benzene ring) —— Metal oxide, 600-6500°C ——→ $CH=CH_2$ (on benzene ring)

Styrene

Manufacture

Polystyrene is manufactured by polymerising styrene by involving the following four methods:

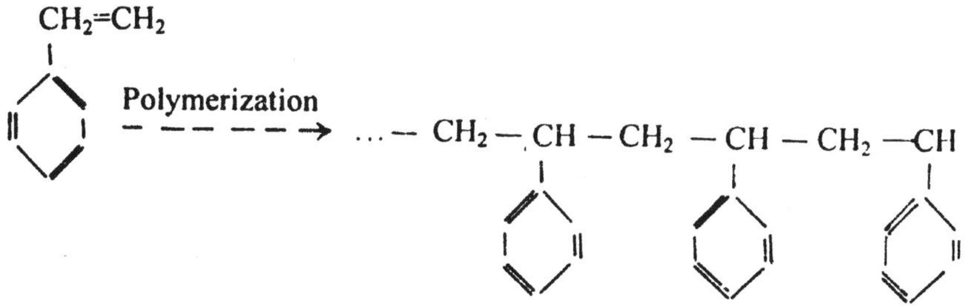

$CH_2=CH_2$ (on benzene ring) —— Polymerization ——→ …– $CH_2 – CH – CH_2 – CH – CH_2 – CH$ (each CH bearing a benzene ring)

(i) *Bulk polymerization*

In bulk polymerization no initiator is required. Styrene gets partially polymerized batch-wise by heating the monomer in large vessels at 80°C for two days until there occurs about 35% conversion. Then the syrupy mixture is allowed to fed continuously into the top of the tower which is twenty-five feet high. The top of the tower is kept at a temperature of about 100°C, the centre at 150°C and the bottom at 180°C.

As the feed material gets passed the temperature gradient, polymerization occurs and the fully, polymerized material emerges from the base of the tower. The reaction process gets controlled by a complex array of heating and cooling jackets and coils.

The molten material is finally allowed to feed into a filament and extruder, extruded is then cooled and chopped into granules.

(ii) *Solution polymerization*

In solution polymerization, the reaction is performed in the presence of a solvent such as ethyl benezene or toluene.

Although solution polymerization is able to control and retards auto acceleration, the solvent is rarely intent and the product of lower molecular weight gets obtained by chain transfer with the solvent.

Finally, the solvent must be removed at the end of the polymerization. However, solvent removal is difficult, dangerous, and expensive.

In a typical polymerizing experiment, a mixture of monomer, solvent and initiator is allowed to feed into a chain of three polymerizing reactors. The first reactor is having three heating zones.

In the first zone, the solution is heating for starting polymerization, but because of the exothermic reaction, cooling becomes necessary in the second and third zones. The polymer solution is then extruded as fine strands into a devolatilizing vessel. This vessel is heated to a temperature of 225°C and then the solvent and unreacted monomer is removed. The molten material is fed into an extruder, then cooled and finally chopped.

(iii) *Suspension polymerization*

This method is extensively used for polymerising styrene. This method is very simple and unlike the solution methods, does not involve solvent removal and recovery.

The polymerization is performed batch-wise in a stirred reactor which is having arrangements for heating and cooling. The suspending agents generally used include tricalcium phosphate and dodecylbenzene sulphonate. Benzoyl peroxide finds use as an initiator for the polymerization of styrene. The reaction is performed at 90°C in the form of a slurry. When the polymerization gets completed, it is repeatedly washed with hydrochloric acid and water to remove the suspending agent. Then, it is dried in warm air, extruded and chopped.

(iv) *Emulsion polymerization*

It is a very good process which is used for the preparation of polystyrene. Emulsion polymerization which is mainly used in the production of polystyrene latex used in water-based surface coating.

As polystyrene obtained by free radical polymerization technique is atactic it is therefore non-crystalline. The isotactic polystyrene is obtained by the use of Ziegler-Natta catalysts and *n*-butyl lithium. Isotactic polystyrene is having a high crystalline melting point of 250°C.

It is transparent. It is more brittle than the atactic polymer.

Properties

The linear polystyrene exists as a hard, rigid and brittle material. It is having a relatively low softening point. Polystyrene is highly transparent and is able to transmit about 90% of visible light.

It is also having good electrical insulation characteristics. It is having low moisture absorption and the electrical properties are maintained in humid conditions. Due to the low cost, good mouldability together with its transparency and colourability this material is having widespread application.

Polystyrene can be moulded, cast or extruded in sheets, rods and tubes. It can be easily pigmented. It finds wide application in making household goods, lenses, plastic moulds, toys, wood laminates, films, battery boxes and electrical components. Expanded polystyrene is used as thermal insulating material and packaging. It is also used in life jackets and floats.

Polystyrene is brittle. Rubber (5–15%) is added to improve this property. This is known as *impact polystyrene*. It is obtained by polymerising styrene in the presence a rubber. This impact polystyrene is having reduced clarity, softening point and tensile strength but better impact strength.

Commercially available polystyrene is mostly of the atactic variety and, hence, is amorphous in nature. Polystyrene consists generally of linear molecules and is chemically inert. Acids, alkalis, oxidising or reducing agents have little effect of it.

It swells or dissolves in several organic solvents, is a very good electrical insulator and becomes yellow on prolonged exposure to sunlight.

Sulphonated styrene divinylbenzene cross-linked polymers are used as cation exchange resins.

Disadvantages

Thermal depolymerisation of polystyrene is possible to given back the styrene monomer along with a number of low molecular weight compounds.

It has a low heat distortion temperature (85°C) and, hence, articles made of it cannot be sterilised with steam, yellowing and cracking of the polymer on outdoor exposure is another main drawback.

Expanded Polystyrene

Preparation

The expanded polystyrene can be made by impregnating small porous beads of the polymer with pentane and heating in steam. The volume increases many a times. Expanded beads are kept in warm air for 24-36 hours and moulded with steam into blocks.

Copolymers

Due to its cheapness, transparency and rigidity attempts were made to develop copolymers with better heat resistance and toughness. Styrene-acrylonitrile copolymers containing 20-30% acrylonitrile have higher softening point and better impact strength. They are transparent and resistant to hydrocarbon and oils.

Styrene-Acrylonitrile Copolymer (SAN)

This copolymer is having 20-30% of acrylonitrite. This copolymer is obtained by the solution polymerization technique, by reacting the monomer (styrene and acrylonitrile) and initiator in the presence of a solvent like toluene or ethlyl benzene.

The copolymer is having a higher softening point and improved impact strenght than the linear polystyrene. Dur to the polar nature of acrylonitrile, the copolymers are some what more resistant to hydrocarbons and oils than polystyrene.

The heat resistance impact strength and chemical resistance gets increased with increasing acrylonitrile content Styrene-acrylonitrile copolymers find use in the manufacture of housewares like beakers and judge and industrial mouldings.

Acrylonitrile-Butadiene-Styrene Copolymers (*ABS Copolymers*)

The copolymer obtained from acrylonitrile, butadiene and styrene is termed as an ABS copolymer. It is a terpolymer. Two methods are used for preparing ABS copolymers such as blending and grafting. These two process produce the copolymers which are different in nature.

Blending

This method involves the mechanical blending of styrene-acrylonitrile copolymers and acrylonitrile-butadiene rubbers. Many products are possible depending on the composition of each copolymer and the relative amounts employed.

In an example 70 parts (70 : 30 styrene acrylonitrile-copolymer) gets blended with 40 parts (35 : 65 acrylonitrile butadiene rubber). After it gets blended, the coagulation of the polymer is brought about by adding an acid or salt.

Grafting

By this method the ABS copolymers are obtained by polymerizing acrylonitrile and styrene in the presence of polybutadiene latex at 50°C in the presence of initiator and transfer agent.

ABS copolymers obtained by this method consists of a mixture of polybutadiene, polybutadiene grafted with acrylonitrile and styrene and styrene-acrylonitrile copolymer. The graft copolymer gets cross linked.

Properties

The range of ABS polymers is very large, because the ratio of the monomers and the manner in which they react could get varied Therefore, commercial ABS polymers are available which are having large difference in properties and they get characterized by high impact strength and softening points. In contracts to polystyrene. ABS copolymers have been opaque.

ABS copolymers have high strength and softening points. They are used extensively in cars, panels, switches, shoe heels, domestic appliances, etc. In recent years, their manufacture has started in India.

Copolymerisation of Polystyrene with P-divinylbezene

To increase heat and impact resistance without affecting other useful qualities, polystyrene is copolymerised with a small amount of P-divinylbenzene $p\text{-}CH_2 = CHC_6H_4CH = CH_2$ when cross-linking between styrene chains occur

$$-CH_2-CH-CH_2-CH-CH_2-CH-CH_2-CH-$$

$$-CH_2-CH-CH_2-C-CH_2-CH-CH_2-CH-$$

The copolymer produced in this way has been found to be

insoluble and infusible but is still chemically reactive. On heating the polymer with a mixture of sulphur trioxide and sulphuric acid, the sulphonic acid group gets introduced into the aromatic rings. The resin so formed is known as a *cation exchanger.*

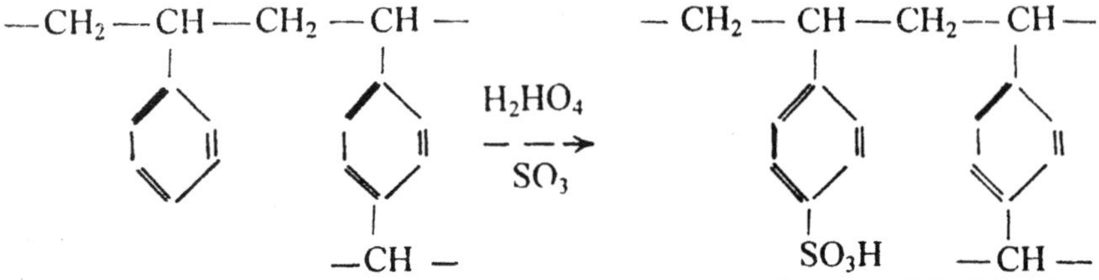

—CH₂—CH—CH₂—CH— →(H_2HO_4 / SO_3) —CH₂—CH—CH₂—CH—

—CH— SO_3H —CH—

(In short RZ, SO_3H)

Cation exchange resins are mainly used in softening hard water by replacing Ca^{2+} and Mg^{2+} ions present in it by Na^+ ions. They are also used in analytical and preparative chemistry.

$$RZSO_3H + M^+ \rightleftharpoons RZSO_3M + H^+$$

$$RZSO_3M + M_1^+ \rightleftharpoons RZSO_3M_1 + M^+$$

When the polymer is treated with formaldehyde and hydrogen chloride in the presence of zinc chloride catalyst (chloromethylation) a resin is formed which when treated with trimethylamine gives *anion exchanger resin.*

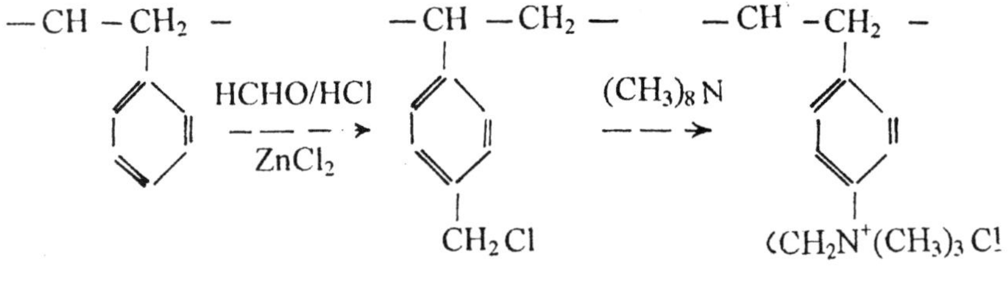

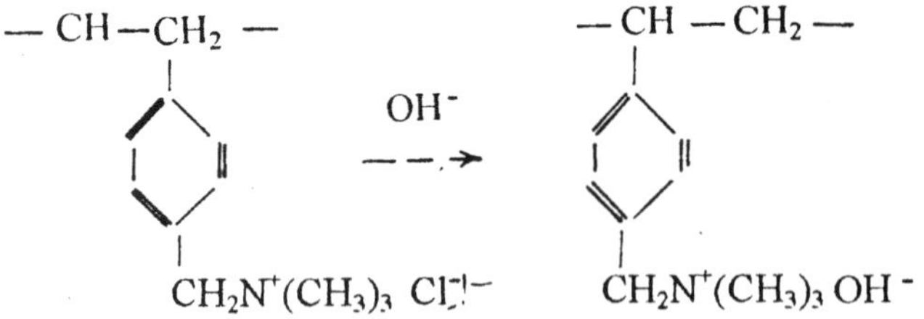

—CH—CH₂— →(HCHO/HCl / $ZnCl_2$) —CH—CH₂— →($(CH_3)_3N$) —CH—CH₂—

CH_2Cl $(CH_2N^+(CH_3)_3Cl$

—CH—CH₂— →(OH^-) —CH—CH₂—

$CH_2N^+(CH_3)_3\ Cl^-$ $CH_2N^+(CH_3)_3\ OH^-$

Anion exchangers are also used for softening hard water by removing undesirable anions.

$$RZ^+OH^- + A^- \rightleftharpoons RZ^+A^- + OH^-$$

The total deionisation of mineral water can be achieved by a combination of a cation exchanger resin and an anion exchanger resin. This has found application in the deionisation of boiler waters to avoid the formation or scales of the dissolved salts on the walls of boilers and pipes. This is also used in the deionisation of water to be used in domestic steam irons.

9.2. Resins

1. Phenol-Formaldehyde Resin

These are produced by the polycondensation of a phenol or a mixture of phenol with formaldehyde. This condensation reaction is catalysed by acids or alkalies. The nature of the product formed by this condensation reactions depends upon the type of catalyst and the mole ratio of the reactants.

Novalak Resins

These are produced by the condensation of a molar excess of phenol with formaldehyde (phenol: formaldehyde –1.25 : 1 ratio, in the presence of an acid which acts as a catalyst.

Preparation

A mixture of phenol and formaldehyde is taken in the distillation vessel. To this hydrochloric acid or sulphuric acid or oxalic acid is added. Now the reaction mixture is heated under reflex at about 100–120°C for 2-4 hours. Water is a by products which is distilled off at the atmospheric pressure. The resin to obtained is having the melting point 65–75°C.

Mechanism

The various steps involved is the reaction between phenol and formaldehyde in the acidic conditions may be outlined as follows:

(i) In the *initial first* step, there occurs the protonation of formaldehyde.

$$CH_2{=}O + \overset{\oplus}{H} \rightleftharpoons \overset{\oplus}{CH_2}{-}OH$$

(ii) In the *second* step, the phenol molecule undergoes electrophilic substitution, thereby forming o- and p-methylol groups.

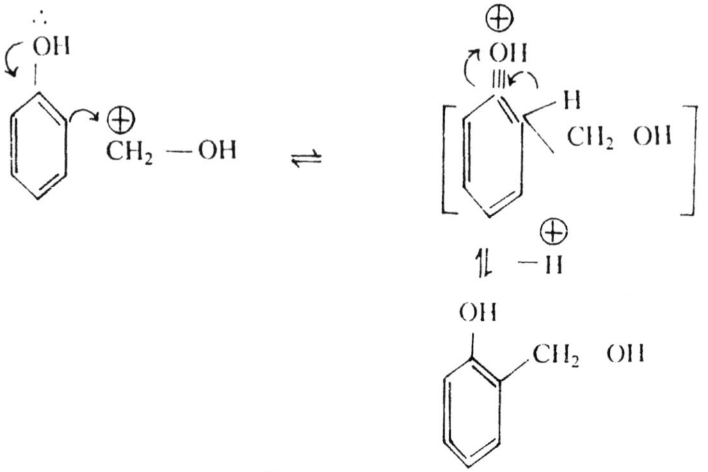

or

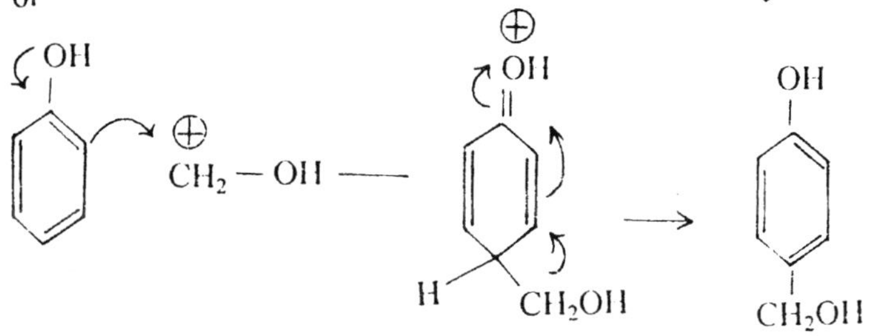

(iii) In the *third* step, methylol phenol reacts with another molecule of phenol to yield dihydroxy diphenylmethanes.

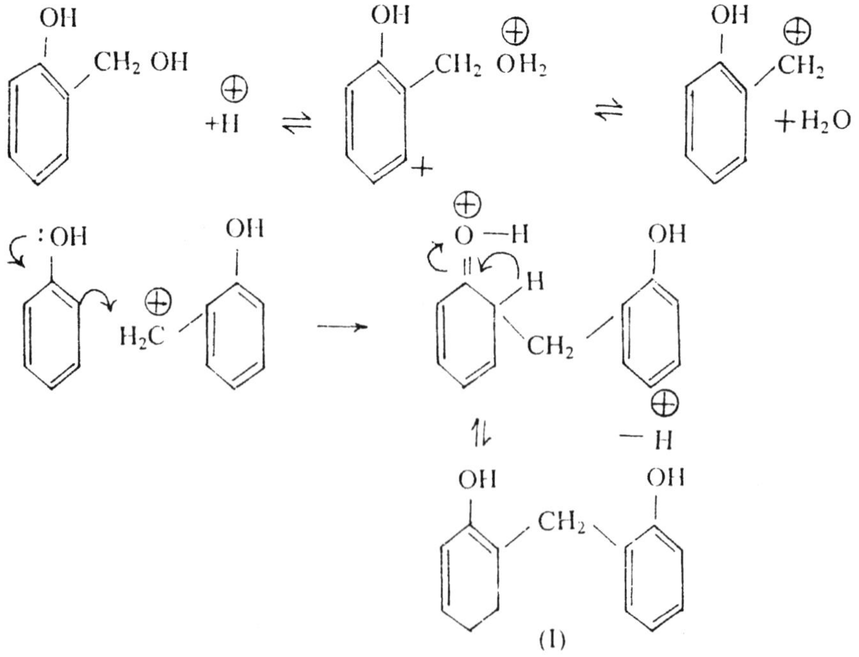

(I)

The structures of other two dihydroxy-diphenylmethanes may be put as follows:

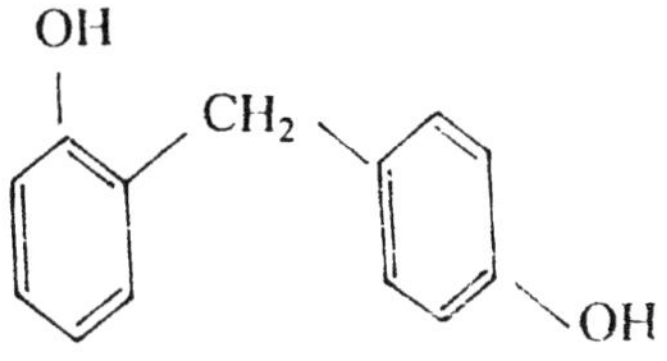

(II) 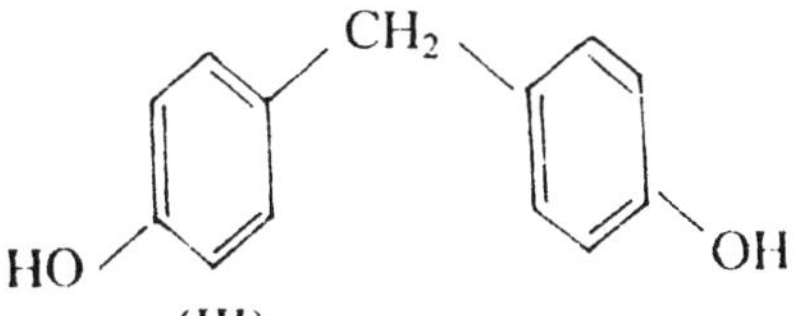 (III)

(iv) When further condensation of dihydroxy-diphenylmethanes with phenol and formaldehyde takes place, there occurs the formation of the polynuclear phenol called **novalak**.

The main feature of novalak resin is that they denote a complete reaction and themselves are having no ability of continue increasing in average molecular weight. The resins therefore become permanently fusible and there occurs no danger of gelation during the production.

Resol Resins

These are produced by reacting phenol with a molar excess of formaldehyde in the ratio of 1: 5, 1.5–2 under alkaline conditions.

Proceduce

A mixture of phenol, formaldehyde and ammonia (nearly 1–3% on the weight of phenol) is made heat under reflux at about 100°C for 25–60 minutes and water is separated by distillation under reduced pressure. After removing water, the resin is obtained as a hard and brittle solid.

Mechanism

The mechanism of the condensation reaction of phenol, formaldehyde in the alkaline medium may be put as follows:

As the resulting o– and p– methylophenols have been more reactive towards formaldehyde than the original phenol, the react – ion with formaldehyde giving rise to the formation of di – and trimethylol derivatives as depicted below:

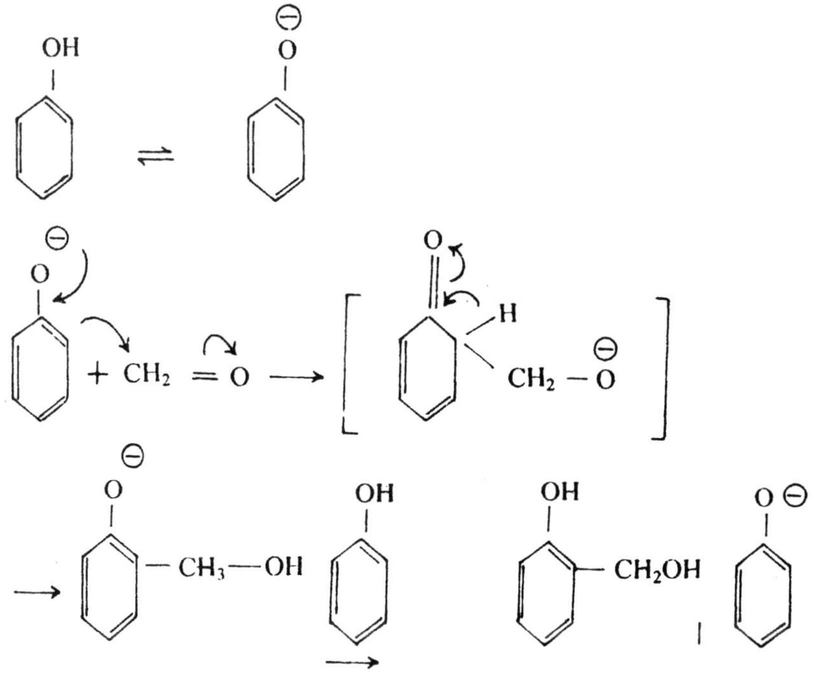

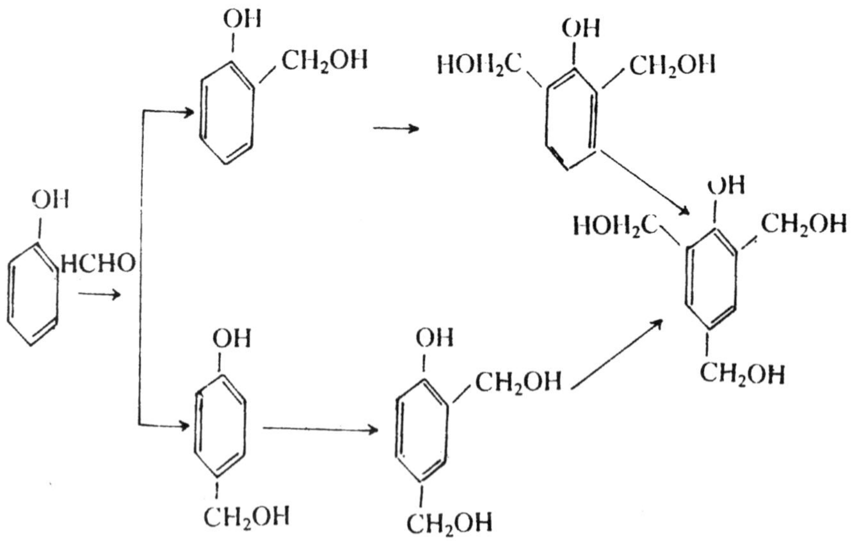

The above reaction gets repeated thereby forming the trinuclear phenols from the dinuclear phenols. Therefore, the product formed by the reaction of phenol with formaldehyde in the presence of an alkali would be a complex mixture of mono and polynuclear phenols in which the methylene groups link the phenolic nuclei. However, the general structure of a resol may be put as follows:

$$HOH_2C-C_6H_2(OH)(CH_2OH)-[-CH_2-C_6H_3(OH)-]_n-CH_2-C_6H_2(OH)(CH_2OH)-CH_2OH$$

Properties and Application

Phenolic resins find numerous application. The main use of phenolic resin is in molding applications. They are used for automotive, radio and television and electrical appliance parts.

In order to impregnate paper, wood and other fillers, nearly 10–15% of the phenolic resins is produced as alcoholic solutions. These find use for decorative purposes for counter tops and wall coverings and industrial laminates of electrical parts.

Phenolic resins also find use in varnishes, electrical insulation, and in other protective coatings. Heat-settings adhesives which are based on phenolics find use in producing plywood. These also find use in the production of ionexchange resins having amine, sulfonic acid, hydroxy or phosphoric acid functional groups.

2. Melamine-Formaldehyde Resin

Introduction

Melamine was isolated by Liebig in 1834 by separating reaction mixture obtained by heating ammonium thiocyanate. It was produced on a commercial scale in 1939.

Manufacture

It involves the following steps:

(a) *Preparation of Melamine*

Melamine was first isolated by Liebig in 1834 by separating reaction mixture obtained by heating ammonium thiocyanate. It was produced on commercial scale in 1939.

When calcium carbinde is made to react with nitrogen; calcium cyanamide is produced which on acid hydrolysis yields cyanamide. Cyanamide on heating to 100°C yields dicyanamide which on heating above its melting point yields melamine.

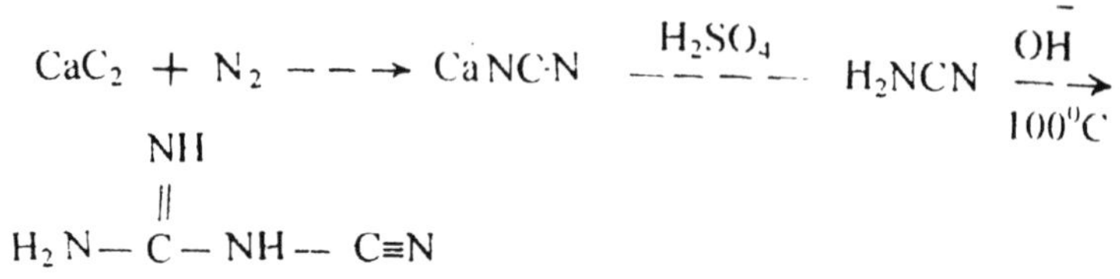

$$CaC_2 + N_2 \dashrightarrow CaNCN \xrightarrow{H_2SO_4} H_2NCN \xrightarrow[100^{\circ}C]{\bar{O}H}$$

$$H_2N-\overset{\overset{NH}{\|}}{C}-NH-C\equiv N$$

Dicyanamide

$$2H_2N-C\begin{matrix}\nearrow NH \\ \searrow NH-C\equiv N\end{matrix} \dashrightarrow \text{melamine } (H_2N-C_3N_3(NH_2)_2)$$

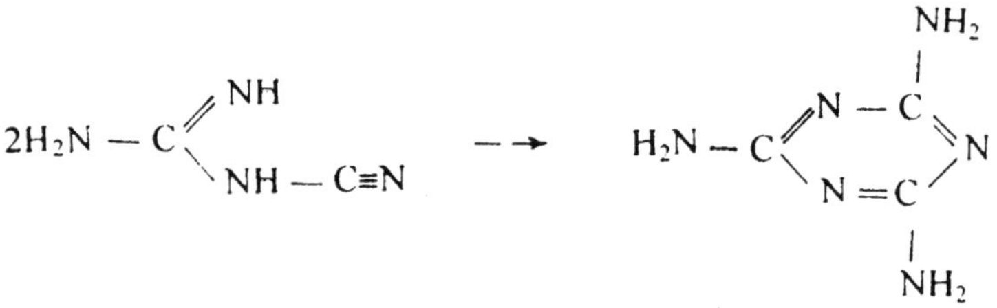

Melamine is also prepared by heating urea in the presence of ammonia at 250–350°C and 40–200 atmosphere pressure.

$$2NH_2CONH_2 \dashrightarrow NH_2CN + NH_2COONH_4$$

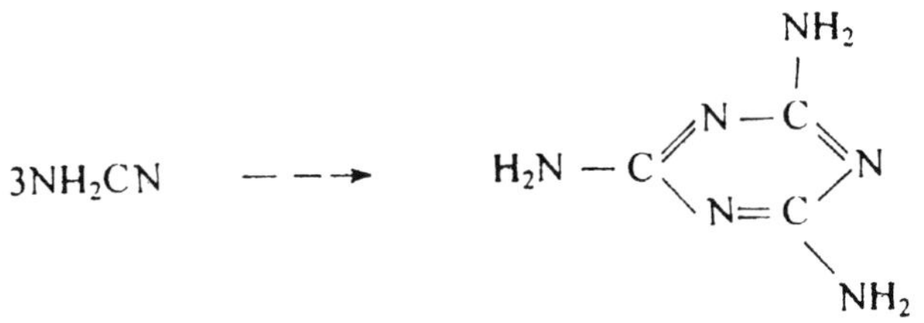

$$3NH_2CN \dashrightarrow \text{melamine } (H_2N-C_3N_3(NH_2)_2)$$

(b) *Preparation of Resin*

Formalin is first of all made alkaline (pH 7.5-8.5) by adding sodium carbonate. Now to this alkaline solution, melamine is added. The melamine and formaldehyde ratio is kept 1 : 3. The mixture is made to heat at 80°C for 1-2 hours. The resulting syrup is stabilized by adding borax and is then used without carrying out further processing.

The reaction between melamine and formaldehyde takes place in two stages which are as follows:

(i) *Formation of methylol melamine.* Melamine reacts with formaldehyde in slightly alkaline conditions to yield methylol derivatives as given below:

```
         NH2                                   NH —CH2OH
          |                                     |
          C                                     C
         / \\                                  / \\
        N    N                                N    N
        ||   |                                ||   |
  HN—C       C — NH —CH2OH                    C    C
   |     \  //                               / \  // \
  CH2OH   N                      HO H2C  HN     N   NH  CH2OH
     Dimethylol                              Trimethylo!
      melamine                                melamine

                    HO H2C        CH2OH
                          \      /
                             N
                             |
                             C
                            / \\
        HOH2C             N     N              CH2OH
             \            ||    |             /
               N—C              C—N
             /     \      //       \
        HOH2C         N              CH2O
                  Hexamethylol melamine
```

(ii) *Condensation of methylol melamine.* When methylol amines are heated a resinous product is obtained. The rate of recertification is strongly dependent on pH. The mechanism of resinification may be put as follows:

(a) $RNH\,CH_2OH + H_2N\text{–}R \rightarrow R\text{–}NH\text{–}CH_2\text{–}NH\text{–}R + H_2O$

(b) $RNH\,CH_2OH + R\text{–}NH\text{–}CH_2OH \dashrightarrow$

$$\begin{array}{l} R\text{–}NH\text{–}CH_2 \\ \quad\quad\quad\quad |\; + H_2O \\ R\text{–}N\text{–}CH_2OH \end{array}$$

(c) $R - NH\text{–}CH_2OH + HOH_2C\text{–}HN\text{–}R$

$\rightarrow RNH\text{–}CH_2\text{–}O\text{–}CH_2\text{–}NH\text{–}R + H_2O$

$R \rightarrow$ melamine residue.

Applications

The tensile strength and hardness of the amino resins have been somewhat better than those of the phenolic resins. The melamine resins are having better hardness, heat resistance and moisture resistance than the urea-resins. The melamine resins find main use as adhesives, largely for plywood and furniture.

Melamine resins also are used for the production of decorative laminates. The amino resins are able to modify textiles like cotton and rayon by imparting crease resistance, stiffness, shrinkage control, fire retardance and water repellency. They are also used to improve the wet strength, rub resistance and bursting strength of paper.

3. Phenol Furfural Resins

Although most phenolic resins are made using formaldehyde, but other aldehydes have also been used. The only one of commercial importance is **furfural** which produces a resin having a high tensile and impact strength.

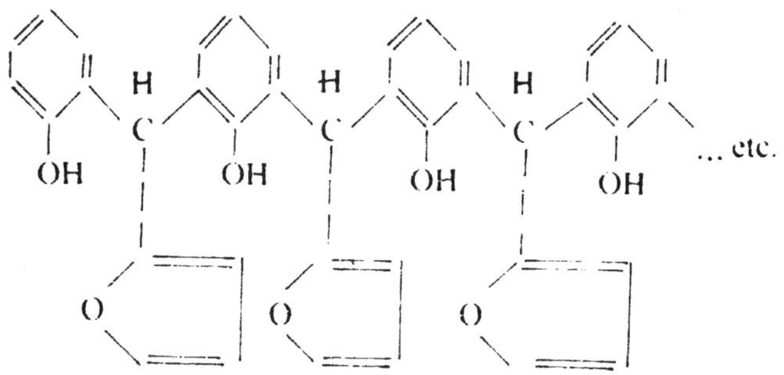

H H H
OH OH OH OH ... etc.
O O O

Phenol-furfural (thermoplastic stage)

Phenol-furfural resins find limited application in the molding of thick parts such as television cabinets, storage battery cases, etc.

4. Amino Resins

Mainly, two types of amino resins are known:

(i) Amino resins obtained from urea and formaldehyde.

(ii) Amino resins obtained from melamine and formaldehyde.

Amino Resins obtained from Urea and Formaldehyde (Urea-Formaldehyde Resins)

It was observed as early as 1897 that urea reacted with formaldehyde to give resins called urea-formaldehyde resins. By early 1920's it was found that waterproof adhesives or glass-like tough plastics could be made from urea-formaldehyde resins. These resins were marketed as synthetic glass in 1929 but because of weathering they lost transparency and developed works. By early 1930's attempts to make these glass-like resins were abandoned. Charleton Ellis experimenting with urea-formaldehyde resins reported that adding a small amount of hydroscoping filler like wood pulp reduced crazing and dulling and stable moulds could be obtained. Urea-formaldehyde resins could be coloured with a large range of pigments giving unlimited colour range brilliance and translascency.

Preparation

First of all sodium hydroxide is added to formalin to make its pH about 8. To this urea is added. Now this mixture is refluxed for about 15 minutes. Then, it is acidified with formic acid and boiled for a further 5-20 minutes. The product is now neutralised with sodium hydroxide and evaporated under reduced pressure to get solid resin.

Mechanism

The reaction between urea and formaldehyde takes place in two stages:

Step I. *Formation of methylol ureas.*

Urea undergoes reaction with formaldehyde in mild alkaline condition to yield monomethylol urea, dimethylol urea and trimethylol urea-respectively, as depicted below:

$$\begin{matrix} NH_2 \\ | \\ CO \\ | \\ NH_2 \end{matrix} \underset{}{\overset{CH_2O}{\rightleftharpoons}} \begin{matrix} NH-CH_2OH \\ | \\ CO \\ | \\ NH_2 \end{matrix} \overset{CH_2O}{\rightleftharpoons} \begin{matrix} NH \quad CH_2OH \\ | \\ CO \\ | \\ NH-CH_2OH \end{matrix}$$

monomethylol urea

dimethylol urea

$$\overset{CH_2O}{\rightleftharpoons} \begin{matrix} N(CH_2OH)_2 \\ | \\ CO \\ | \\ NH-CH_2OH \end{matrix}$$

trimethylol urea

Step II. *Condensation of methylol ureas*

In the second stage of the reaction, there occurs **the condensation** between the methylol groups and the amido hydrogen **in acidic medium** to yield the methylene linkage as given below:

$$-CH_2OH + \overset{+}{H} \rightleftharpoons -CH_2\overset{+}{O}H_2 = -CH_2 + H_2O$$

$$-\overset{+}{C}H_2 + H_2N \dashrightarrow -CH_2-NH_2 \dashrightarrow -CH_2-NH- + \overset{+}{H}$$

The mode of condensation may be depicted as follows:

$$\begin{array}{lccccc}
NH\quad CH_2OH & & NH_2 & -H_2O & NH-CH_2- & NH \\
| & & | & \dashrightarrow & | & | \\
CO & + & CO & & CO & CO \\
| & & | & & | & | \\
NH_2 & & NH_2 & & NH_2 & NH_2
\end{array}$$

$$\begin{array}{cccccccc}
NH-CH_2 & -NH & & NH_2 & -H_2O & NH-CH_2- & NH & \\
| & | & & | & \dashrightarrow & | & | & \\
CO & CO & + & CO & & CO & CO & \\
 & & & & & | & | & \\
NH_2 & NH_2 & & NHCH_2OH & & NHCH_2OH & NHCH_2OH &
\end{array}$$

As the above product is having methylol and amido groups. The condensation may continue to yield polymeric methylene compounds.

As a molar excess of formaldehyde is used to produce resins, the polymers will get terminated predominantly by methyl groups. The general structure of the methylene-containing polymers may therefore by put as follows:

$$HO\,CH_2-[-NH-CO-NH\,CH_2-]_n-NH\,CO\,NH-CH_2OH$$

Uses

The urea formaldehyde resins are used for domestic electrical fittings, bottle caps. These also find use for wood adhesives, surface coatings and textile finishings.

Cross-Linked Resins

It is possible to convert the low molecular weight urea-formaldehyde resin into the high molecular weight cross-linked resins by heating the former in acidic conditions.

In the initial stage of the reaction, the imido groups in the chain undergo reaction with free formaldehyde present in the reaction mixture to form the methylol groups. Then the reaction occurs between methylol groups and imido hydrogen and those involving self-condensation of methylol groups.

$$\begin{array}{lcll} -N- & & ...N... & \\ | & & | & \\ H_2OH & \longrightarrow & CH_2 & +H_2O \\ + & & | & \\ -CO-NH-CH_2- & & CO-N-CH_2.... & \end{array}$$

$$\begin{array}{lcll} -N- & & -N- & \\ | & & | & \\ CH_2OH & & & \\ + & \longrightarrow & CH_2 & +CH_2O+H_2O \\ CH_2-OH & & | & \\ | & & -N- & \\ -N- & & & \end{array}$$

The above reaction will give rise to the formation of a cross-linked polymer as follows:

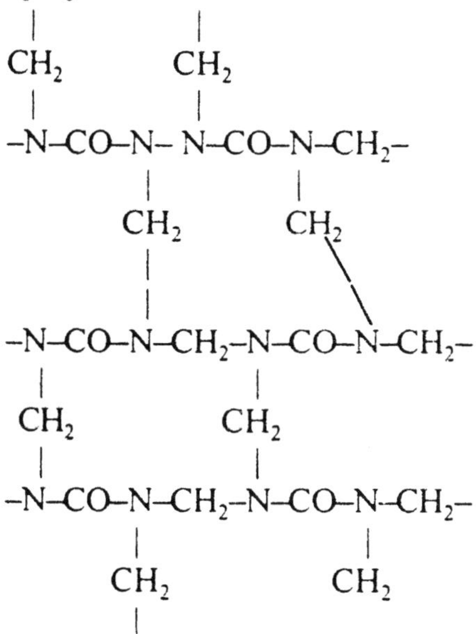

Cross-Linked Urea Formaldehyde Resin

Properties

The cross-linked resins have been righd, infusible and insoluble. They show unusual surface hardness, and may be obtained in a wide range of colours. Cross-linked resins have been found to be very resistant to most organic reagents.

5. Epoxy Polymers

These are crosslinked polymers in which crosslinking is achieved by reaction of Epoxy groups. The epoxy adhesive were developed by Ciba Company in 1943. Devoe and Reynold Company developed epoxies for surface coatings in 1950.

The epoxy polymers are basically polyethers. One type of epoxy polymer (or epoxy resins) are prepared from epichlorohydrin and bisphenol-A. The reaction is carried out with excess of epichlorohydrin. The various reactions for the preparation of an epoxy polymer are given in the following discussion.

Raw Materials

Important raw materials are *bisphenol* A (2,2-bis (4-hydroxyphenyl propane) and *epichlorohydrin.*

(i) Bisphenol A is made by reacting phenol (4 moles) with acetone (1 mole) in presence of hydrogen chloride at 50.70°C. Unreacted phenol is recovered and product crystallised from ethanol:

$$2\,C_6H_5OH + CH_3COCH_3 \xrightarrow{H^+} HO-C_6H_4-C(CH_3)_2-C_6H_4-OH$$

Bisphenol A

(ii) Epichlorohydrin ($ClCH_2-CH=CH_2$) is prepared from propylene by converting it to allyl chloride with chlorine at 500°C and 2 atm pressure. Reaction with hypochlorous acid given dichlorohydrin which on reaction with lime gives epichlorohydrin.

$$CH_3-CH=CH_2 \longrightarrow CH_2Cl-CH=CH_2 \longrightarrow$$

$$CH_2Cl-CHOH-CH_2Cl \xrightarrow{CaO} CH_2Cl-\underset{\text{O}}{CH-CH}$$

Dichlorohydrin Epichlorohydrin

(iii) For preparation of the resin a mixture of bisphenol A (1 mole) is refluxed with epichlorohydrin (4 mole) in presence of sodium hydroxide. The resin is washed with water to remove sodium chloride and dried to remove water. By varying amount of bisphenol A, resins with varying molecular weights can be prepared. Most of these epoxy resins are polymers with low molecular weight (about 4000). Both ends of the polymer molecule have epoxide units while there are hydroxyl groups along the chain. These epoxide and hydroxyl groups are involved in crosslinking reactions with di- or polyamines or di- or polybasic acids or tertiary amines:

$$n\left[\underset{\diagdown O \diagup}{H_2C - CH} - CH_2Cl\right]_n + n\left[HO-C_6H_4-\underset{CH_3}{\overset{CH_3}{C}}-C_6H_4-OH\right]$$

$$\downarrow NaOH$$

$$2\,\underset{\diagdown O \diagup}{CH_2-CH}-CH_2Cl + HO-C_6H_4-\underset{CH_3}{\overset{CH_3}{C}}-C_6H_4-OH$$

$$\downarrow$$

$$ClCH_2\underset{OH}{CH}CH_2O-C_6H_4-\underset{CH_3}{\overset{CH_3}{C}}-C_6H_4-OCH_2\underset{OH}{CH}CH_2Cl$$

$$\downarrow NaOH\ (1-NaCl)$$

$$\underset{\diagdown O \diagup}{CH_2-CH}-OCH_2-C_6H_4-\underset{CH_3}{\overset{CH_3}{C}}-C_6H_4-OCH_2-\underset{\diagdown O \diagup}{CH\ CH_2}$$

$$\downarrow + \text{Bisphenol} - \text{A (Excess)}$$

$$X-\left[C_6H_4-\underset{CH_3}{\overset{CH_3}{C}}-C_6H_4-O-CH_2-\underset{OH}{CH}\ OH\ CH_2-O\right]_n-Y$$

Where $X = \underset{\diagdown O \diagup}{CH_2-CH}\ CH_2-O-$ and

$$Y = -C_6H_4-\underset{CH_3}{\overset{CH_3}{C}}-C_6H_4-O-CH_2-\underset{\diagdown O \diagup}{CH\ CH_2}$$

Instead or bisphenol-A, many other compounds with hydroxyl groups (such as glycols, glycerols and resorcinols) can also be used. The epoxy resins obtained through these reactions will be either highly viscous liquids or solids with high melting points.

Tertiary amines such as benzyldimethylamine, 2-(dimethyl aminoethyl) phenol etc. form an anion with epoxide which leads to their polymerisation. These are used mainly for adhesives and surface coating applications.

Aliaphatic and aromatic amines with at least three active hydrogen atoms present are used as curing agent for epoxy resins *e.g.*, diethylenetramine, triethylenetramine, *m*-phenylenediamine, 4,4'-diaminediphenylmethane, 4,4'-diaminodiphenylsulphone etc. Their reaction with epoxy end group leads to crosslinking at room temperature. They are used in adhesives, casting and laminating. Their main disadvantage is toxicity as they cause severe irritation, rashes or asthmatic response. Several addition products of polyfunctional amines are used commercially to reduce toxicity *e.g.*, glycidyl ether adducts, ethylene oxide adducts or acrylonitrile adducts.

Cyclic acid anhydrides such as maleic, dodecylsuccinic, hexahydrophthalic, phthalic, phyromellitic etc. are widely employed as curing agents for epoxy resins. They form esters with epoxy resins. These resins have better thermal stability and good electrical insulation and chemical resistance expect to alkalis. They are used in castings and laminated.

The cured resins are stable at 150–200°C depending on crosslinking agents used. They are resistant to chemical attack and are flexible and strongly adhesive. They are used as surface coatings and yield an excellent enamel after esterification which is used for floors, walls, tanks, domestic equipment etc.

As adhesives they are very widely used both as cold setting and thermo curing. They can bond together a large variety of surfaces including metals. **Araldite resins** of this type are used widely in aircraft construction. They are also used for bonding concrete in dams and joining teflon film to stainless steel. They are used to encapsulate delineate electrical parts to protect them from shock and moisture. Such parts are used in components of rockets and satellites.

Several other epoxy resins have been made. Many contain glycidyl ether group while others are cyclic aliphatic epoxies and acrylic aliphatic epoxies.

Epoxy resins find a large number of uses because of their remarkable chemical resistance and good adhesion. Epoxy resins are excellent structural adhesives. When properly cured, epoxy resins can yield very tough materials. They are used in industrial floorings, foams, potting materials for electrical insulations, etc. One of the principal constituents in many of the fibre-reinforced plastics (FRP) is an epoxy polymer.

Resin Preparation

(i) *Liquid Epoxy resin.* In order to prepare liquid epoxy resins, a mixture of bisphenol A and epichlorohydrin (1 :4 molar ratio) is heated to about 60°C with stirring. To this reaction mixture solid sodium hydroxide (2 mole per mole of bisphenol A) is added slowly. As the reaction is exothermic cooling becomes necessary to maintain the temperature of the reaction mixture at 60°C. Excess of epichlorhydrin can be removed by vacuum distillation. The reaction mixture is having the resin and sodium chloride.

To the reaction mixture a small amount of toluene is added and the mixture is filtered off. Toluene is removed by distillation under reduced pressure and the resin is made to heat at 150°C and 5 m.m. Hg pressure to remove traces of volatile matter. Finally, the resin is clarified by passing through a fine filter.

(ii) *Solid Epoxy resin.* In order to prepare solid epoxy resin, a mixture of bisphenol A and epichlorohydrin is made to heat at 100°C and aqueous sodium hydroxide is added strongly with vigorous stirring.

When the reaction is completed, the agitation is stopped and a taffy (which is an emulsion of about 30% water in resin) rises to the top of the reaction mixture. The lower layer of brine is removed, the resinous layer gets coagulated and washed with hot water, clarified by passing through a filter and finally allowed to solidify.

It is possible to explain the formation of epoxy resins from bisphenol A and epichlorohydrin according to the following scheme.

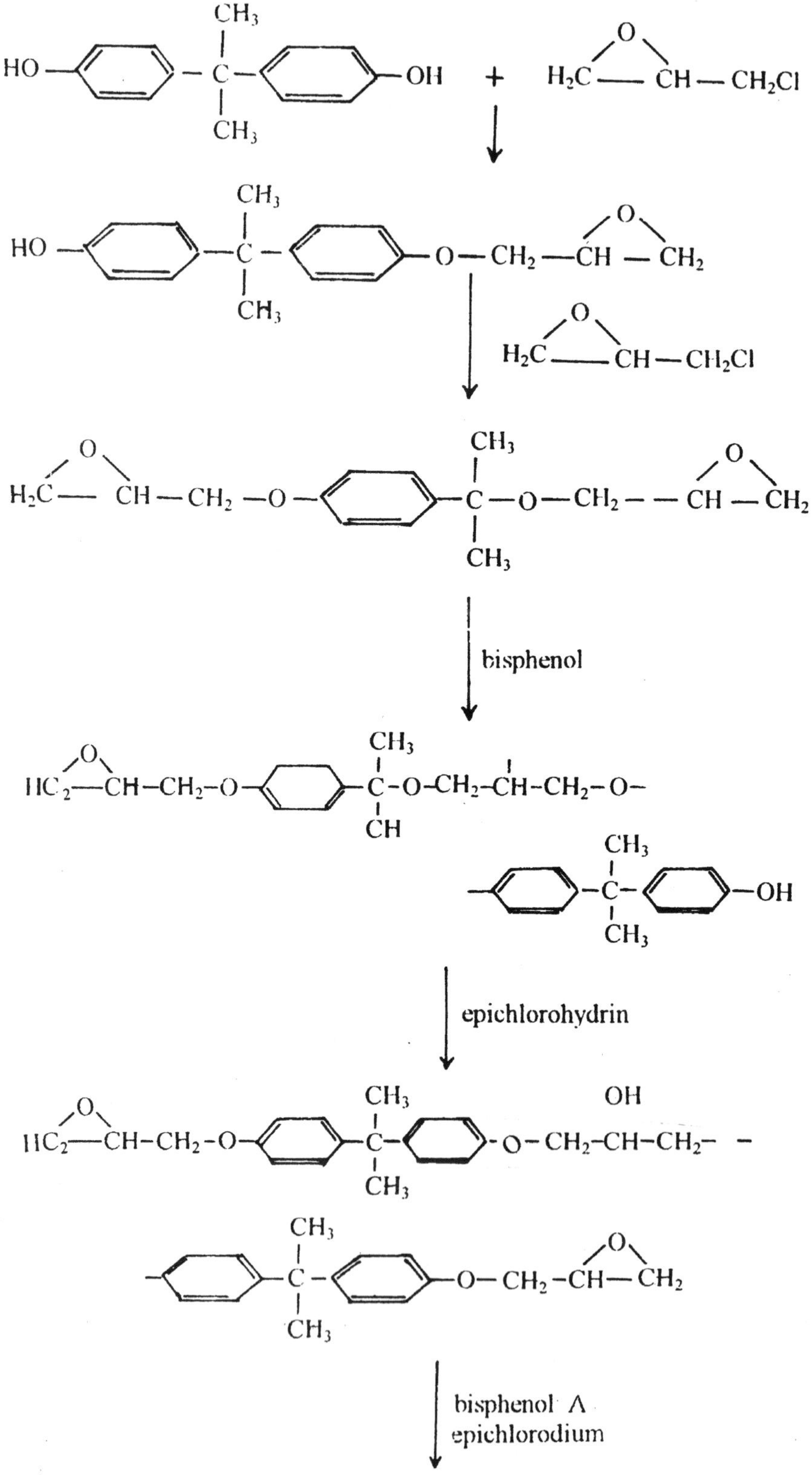

$$HO-C_6H_4-C(CH_3)_2-C_6H_4-OH + H_2C\overset{O}{—}CH-CH_2Cl$$

↓

$$HO-C_6H_4-C(CH_3)_2-C_6H_4-O-CH_2-CH\overset{O}{—}CH_2$$

↓ $H_2C\overset{O}{—}CH-CH_2Cl$

$$H_2C\overset{O}{—}CH-CH_2-O-C_6H_4-C(CH_3)_2-O-CH_2-CH\overset{O}{—}CH_2$$

↓ bisphenol

$$HC_2\overset{O}{—}CH-CH_2-O-C_6H_4-C(CH_3)(CH)-O-CH_2-CH-CH_2-O-C_6H_4-C(CH_3)_2-C_6H_4-OH$$

↓ epichlorohydrin

$$HC_2\overset{O}{—}CH-CH_2-O-C_6H_4-C(CH_3)_2-C_6H_4-O-CH_2-CH(OH)-CH_2-\;-C_6H_4-C(CH_3)_2-C_6H_4-O-CH_2-CH\overset{O}{—}CH_2$$

↓ bisphenol A
epichlorodium

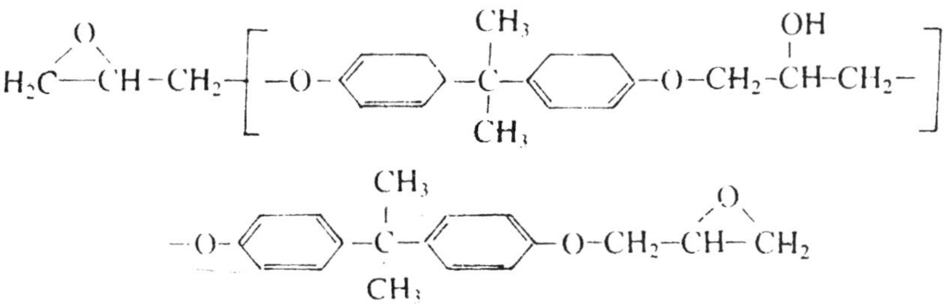

$$H_2C\overset{O}{\frown}CH-CH_2-\left[-O-\langle\bigcirc\rangle-\underset{CH_3}{\overset{CH_3}{\underset{|}{\overset{|}{C}}}}-\langle\bigcirc\rangle-O-CH_2-\overset{OH}{\overset{|}{C}H}-CH_2-\right]$$

$$-O-\langle\bigcirc\rangle-\underset{CH_3}{\overset{CH_3}{\underset{|}{\overset{|}{C}}}}-\langle\bigcirc\rangle-O-CH_2-CH\overset{O}{\frown}CH_2$$

epoxy resin

Applications

Inspite of their high costs, the epoxy resins find many important applications. The resins are used in both molding and laminating technique for making glass fibre-reinforced articles which are having better mechanical strength, chemical resistance and electrical insulating properties.

They find use in casting, potting, encapsulating and embedment in the electrical and tooling industries. The general important used are industrial flooring, adhesive, foams, highway surfacing patching materials and stabilizers for vinyl resins.

9.3 Acrylic Polymers

Theses polymers are made from acrylic acid, its homologues and their derivatives. Glass like resins were made from esters of acrylic acid in 1877 by Fitting and Peter by Kahlbaum. In 1928 Rohm and Hass, a German Company started commercial development of methacrylic esters. Limited production started in 1933. The rapidly expanding air-force used this plastic in place of glass in the aeroplanes. Most of the early production of "**Plexiglass**" was used up by air-force planes. In 1936. ICI marketed methyl methacrylate sheets as "**Perspex**".

The various acrylic polymers are as follows:

1. Polyacrylated and polymethylacrylate
2. Polymethylmethacrylate
3. Polyacrylonitrile.

We shall discuss these one by one.

1. Polyacrylate and Polmethylacrylate

It is produced by polymerisation of methyl acrylate which is produced by the methanolysis of acrylonitrile or ethylene cyanohydrin.

Crawford developed cyanohydrin process for acrylic and methacrylic acid as follows:

$$CH_2{=}CH_2 + HOCl \dashrightarrow OHCH_2\,CH_2Cl \xrightarrow{KCN}$$

Ethylene chlorohydrin

$$HOCH_2{-}CH_2CN \xrightarrow[175°C]{H_2SO_4} H_2C = CH{-}COOH$$

Ethylene cyanohydrin Acrylic acid

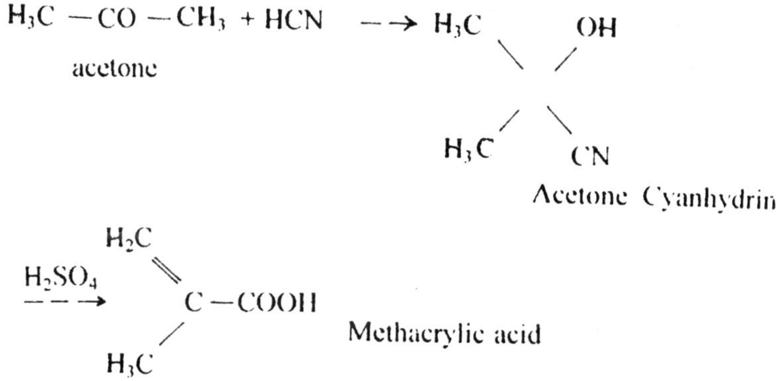

$$H_3C-CO-CH_3 + HCN \longrightarrow (H_3C)_2C(OH)CN$$

acetone Acetone Cyanhydrin

$$\xrightarrow{H_2SO_4} H_2C{=}C(H_3C){-}COOH$$

Methacrylic acid

Ethylene cyanohydrin can be obtained by the hydrocyanation of ethylene oxide

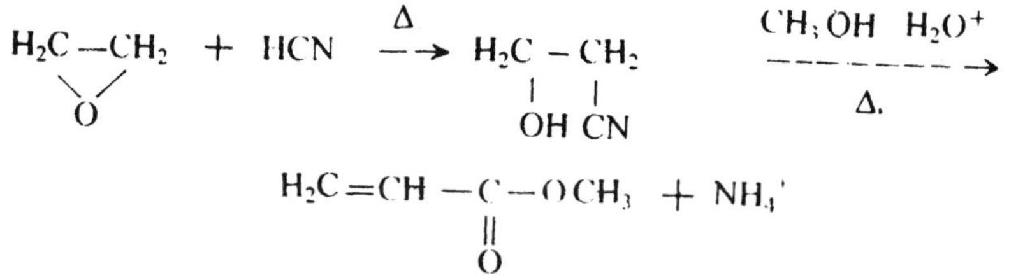

$$\underset{\text{(epoxide, O)}}{H_2C-CH_2} + HCN \xrightarrow{\Delta} \underset{OH\quad CN}{H_2C - CH_2} \xrightarrow[\Delta.]{CH_3OH\ \ H_3O^+}$$

$$H_2C{=}CH-\underset{\underset{O}{\|}}{C}-OCH_3 + NH_4^+$$

Both acrylic acid and methacrylic acid polymerise to give water soluble hard resins. The viscous solutions so formed have been used as emulsifying agents, adhesives and as thinckening agents for inks and dyes. Polymers of esters of these acids are of greater commercial importance. Esters can be prepared from cyanhydrins by reaction with an alcohol:

$$HOCH_2{-}CH_2CN + CH_3OH \xrightarrow[H_2SO_4]{} CH_2{=}CHCOOCH_3$$

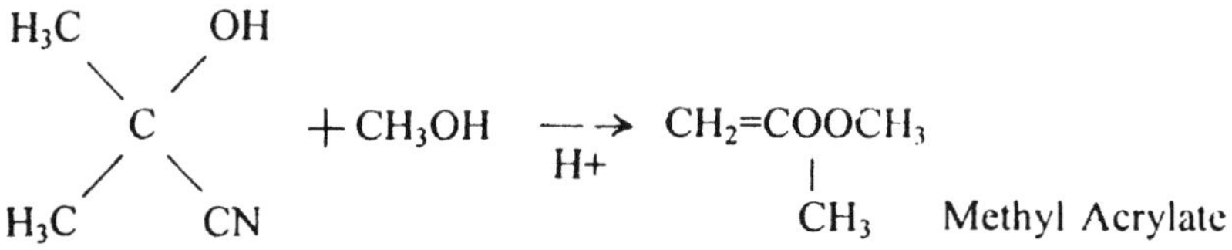

$$(H_3C)_2C(OH)CN + CH_3OH \xrightarrow[H+]{} CH_2{=}C(CH_3)COOCH_3 \quad \text{Methyl Acrylate}$$

Several new methods of synthesis have been established.

$$CH{\equiv}CH + CO + CH_3OH \xrightarrow{Ni(CO)} CH_2{=}CHCOOCH_3$$

$$CH_2 = CH - CH_3 \xrightarrow{HCl_{30°c}} CH_2{=} CH - CHO \rightarrow$$

$$\rightarrow CH_2{=}CH{-}COOH$$

In this process cobalt and molybdenum catalyst are used:

$$CH_3{-}CH(OH){-}COOH \rightarrow CH_2 = CHCOOH$$

Lactic Acid

Organic peroxide are used to polymerise the esters by solvent or emulsion polymerisation. They form tough and pliable film. They are also used as plasticizing agents for vinyl polymers. The polymer is soluble in benzene, toluene, chloro form, ethylene dichloride, ethyl acetate etc.

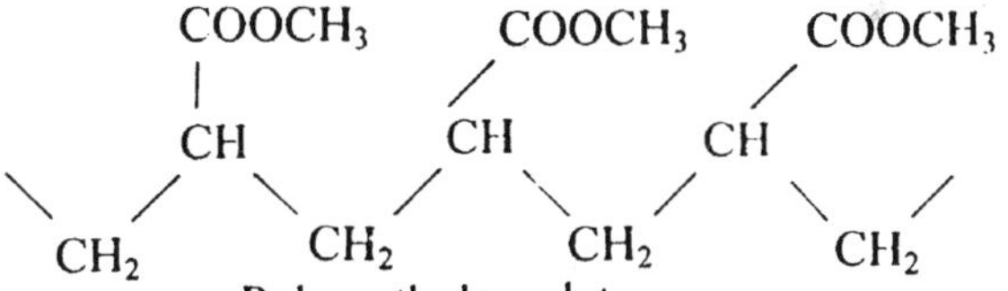

$$\sim CH_2{-}CH(COOCH_3){-}CH_2{-}CH(COOCH_3){-}CH_2{-}CH(COOCH_3){-}CH_2 \sim$$

Polymethylacrylate

Methyl acrylate is also produced by the carboxylation of acetylene in aqueous methanol, by the dehydration of methyl lactate and by the methanolysis of p-propiolactone or acrylic acid.

$$HC{\equiv}CH + CO + H_3COH \xrightarrow{Ni(CO)_4} H_2C{=} CH{-}C(O){-}OCH_3$$

$$H_3C{-}CH(O){-}C(=O){-}OCH_3 \xrightarrow[\Delta]{H_3O^+} H_2C{=} CH{-}C(=O){-}OCH_3$$

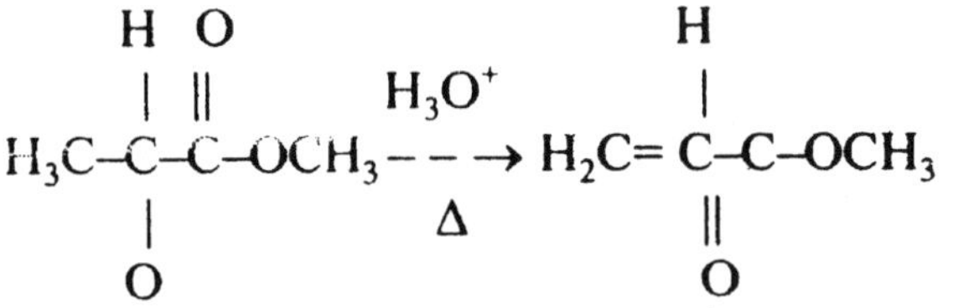

$$\begin{array}{c} H_2C-C=O \\ | \quad | \\ H_2C-O \end{array} \xrightarrow[H_3O^+]{H_3COH} HO=\begin{array}{c} H \quad H \\ | \quad | \\ C-C-C-OCH_3 \\ | \quad | \quad | \\ H \quad H \quad O \end{array} \rightarrow$$

$$H_2C=CH-\underset{\underset{O}{\|}}{C}CH_3$$

Polymerisation of methyl acrylate is catalysed by heat, light or peroxide.

2. Polymethylmethacrylate

This polymer has the following structure:

$$-\left[-CH_2-\underset{\underset{COOCH_3}{|}}{\overset{\overset{CH_3}{|}}{C}}----\right]_{-n}-$$

$$\begin{array}{c} CH_3 \\ CH_2 \end{array}\!\!>C<\begin{array}{c} COOCH_3 \\ CH_2 \end{array}\!\!-\!\!\begin{array}{c} H_3C \\ \end{array}\!\!>C<\begin{array}{c} COOCH_3 \\ CH_2 \end{array}\!\!-\!\!\begin{array}{c} H_3C \\ \end{array}\!\!>C<\begin{array}{c} COOCH_3 \\ CH_2 \end{array}\!\!-$$

Polymethyl methacrylate

The monomer methyl methacrylate can be obtained from acetone as under:

$$CH_3COCH_3 \xrightarrow{+HCN} \begin{array}{c} CH_3 \\ | \\ C\begin{array}{l} -OH \\ -CN \end{array} \\ | \\ CH_3 \end{array} \xrightarrow[\text{conc. } H_2SO_4]{150\ ^0C} \begin{array}{c} CH_3 \\ | \\ C\begin{array}{l} -OSO_2OH \\ \ \ CONH_2 \end{array} \\ | \\ CH_3 \end{array}$$

$$\xrightarrow{+CH_3OH \mid H^{\oplus}} \left[SO_4\right]^{\ominus\ominus} + \left[NH_4\right]^{\oplus} + CH_2=C\begin{array}{l} CH_3 \\ COOCH_3 \end{array}$$

The polymerisation of methyl methacrylate can also be carried out by all the techniques. The sheet is made by bulk polymerisation and casting. The monomer is heated to 95°C with benzoyl peroxide at 90°C to form a syrup. Which is put in a casting cell that is held by spring located clamps and maintained at 40°C for about 15 hours. About 20% shrinkage in volume takes place on polymerisation.

Commercially, polymethylmethacrylate is made by either suspension or bulk polymerisation using a peroxide free-radical initiator:

$$n\,CH_2{=}\underset{\displaystyle CH_3}{\overset{|}{C}}.COOCH_3 \longrightarrow -CH_2-\overset{\displaystyle COOCH_3}{\underset{\displaystyle CH_3}{\overset{|}{\underset{|}{C}}}}-CH_2-\overset{\displaystyle COOCH_3}{\underset{\displaystyle CH_3}{\overset{|}{\underset{|}{C}}}}-CH_2-\overset{\displaystyle COOCH_3}{\underset{\displaystyle CH_3}{\overset{|}{\underset{|}{C}}}}-$$

polymethyl methacrylate

Both isotactic and syndiotactic polymethyl methacrylate have been prepared. However, the commercial polymer is generally atactic because of the random arrangement of the bulky side groups.

The commercially produced polymethyl methacrylate is a transparent thermoplastic that softens at 120°C and becomes pliable at 160°C.

It has an excellent outdoor life period and good strength. It is amorphous because of the presence of bulky side groups in the molecules. It is resistant to many chemicals but soluble in organic solvents like ketones, chlorinated hydrocarbons and esters. It can be thermally depolymerised to give back the entire quantity of monomer.

The main feature of this plastic is its optical clarity. It is an excellent substitute for glass. It has good mechanical properties also. But as compared to glass, it has poor scratch resistance. PMMA is used to make attractive signboards and durable lenses for automobile lighting. It is also used in buildings for decorative purposes.

It is used in windows, windscreens, sun visions and in air crafts. It is also used in optical fibres, contact lens, lighting fixtures, biological specimen preservation, dentures, presentation articles, automobile surface coatings etc.

Long chain alcohol esters have lower softening temperature and lower shrinkage on moulding. Poly-cyclohexyl methacrylate has refractor index similar to crown glass. The esters are prepared by alcoholysis of methyl ester.

3. Polyacrylonitrile (PAN)

It is also known as polyvinyl cyanide. Its structure is as follows:

$$\left\{ \begin{array}{c} CH_2 - CH \\ \quad\quad | \\ \quad\quad CN \end{array} \right\}$$

On a large scale it is prepared from acrylonitrile by the radical polymerisation technique using peroxide initiators.

The monomer acrylonitrile can be prepared by the following methods:

(i) Acetylene and hydrogen cyanide are passed through hydrochloric acid containing cuprous chloride at 80.90°C under pressure. Acrylonitrile vapour is dissolved in water, steam distilled and purified by distillation.

$$CH_3CHO + HCN \longrightarrow CH_3 - \underset{\underset{OH}{|}}{CH} - CN$$

$$\longrightarrow H_2O + CH_2 = CH - CH$$

(ii) Ethylene is converted to ethylene oxide which is converted to ethylenecyanhydrin. Its reaction over alumina at 300°C gives acrylonitrile

$$C_2{=}CH_2 \xrightarrow{O_2} \underset{\diagdown_{O}\diagup}{CH_2 - CH_2} \xrightarrow{HCN} CH_2OH - CH_2CN$$

$$\xrightarrow[\text{or } 200^\circ\ MgCO_3]{300^\circ C\ Al_2O_3} CH_2{=}CHCN \qquad \text{Acrylonitrile}$$

(iii) Ammoxidation of propylene is the most important commercial method today. Propylene, ammonia and air are passed in a fluidised bed reactor containing bismuth, molybdenum or uranium compounds as catalysts at 400-500°C and 1-3 atom pressure. The gases are scrubbed with water to dissolve acrylonitrile. It is purified by distillation. The conversion in this process is not very good but because of low raw material costs this process has become important.

$$CH_2 = CH - CH_3 + NH_3 + \frac{3}{2} O_2 \xrightarrow[\text{Catalyst}]{500^\circ C} 2CH_2 = CH - CN + 3H_2O$$

Polyacrylonitrile is prepared by solution or suspension polymerisation. The solution process is used for obtaining fibres. Suspension or emulsion polymerisation is carried out in an aqueous redox system containing copper or iron salts as catalysts. Hydrogen bonding between nitrogen and hydrogen on adjacent chains leads to close packing of polymer chains giving a polymer with high melting point ($\simeq 300°C$). The polymer dissolves in solvents such as dimethyl formamide, dimethylsulphoxide etc. It is difficult to work with it as on heating it turns yellow and red due to linking of nitrile groups:

```
        CN          CN          CN          CN
        |           |           |           |
        CH          CH          CH          CH
   \   /  \       /   \      /   \       /     \  Polyacrylonitrile
    CH2     CH2          CH2          CH2
                          |
                  heating ↓

   C = N  - C = N - C  = N  - C = N  - C =
   |        |       |         |
   CH       CH      CH        C
  /  \     /  \    /   \     / \    /
CH2    CH2      CH2      CH2     CH2
       Yellow  |
               ↓ Further heating

      N       N        N       N
   \ //  \  //  \   //  \   //  \
    C      C       C       C         1500°C-3000°C
    |      |       |       |         ------------>
    C      C       C       C     /   Carbour fibre
  / \\ CH / \\ CH / \\ CH   \ CH
```

Polypyridine Polymer

The major use of polyacrylonitrile is in acrylic fibres.

Properties

It is a colourless liquid with *b.p* 77°C. It (PAN) is soluble in dimethyl formamide, dimethyl sulphoxide, adipo nitrile, etc. It has a remarkable resistance to heat upto around 200°C and exhibits very good mechanical properties.

Uses

Polyacylonitrile is used to produce what are known as PAN fibres (**Acrilan, orlon etc.**)

The copolymer of acrylonitrile with butadiene (nitrile rubber) is a material or great industrial importance.

4. Acrylic Copolymers

The acrylic monomers readily get copolymerised with several other monomers to yield important plastics such as ABS. Thermosetting acrylic copolymers for thermosetting finishes have also been developed. These terpolymer are made from 3 monomers. One monomer confers hardness, other one flexibility while the third one allows crosslinking. Solution polymerisation is done and after reaching appropriate degree of polymerisation the viscous solution is treated with initiators and final thermosetting done at temperatures of 100-140°C. These thermosetting finishes are used on metals and have required degree of flexibility of hardness as per requirements.

9.4. Polyesters

These polymers contain ester groups in the main chain of the polymer itself. The polymers are of various types like linear saturated, linear unsaturated, network polyesters, polyallylesters and polycarbonates. These fibres will be discussed separately.

1. Linear Polyesters

Linear saturated polyesters of low molecular weight (less than 10,000) can be obtained by condensing a diol with a diacid. Molecular weight can be controlled by adding a monohydric alcohol or monocarboxylic acid. These polymers are used as plasticisers or as polyester diols for making polyurethanes.

Linear unsaturated polyesters contain aliphatic unsaturation that can result in subsequent crosslinking. The first such product was prepared in 1946 by the condensation of maleic anhydride with diethyleneglycol. This polymer can be crosslinked with styrene at low temperature and pressure as under:

$$OH\text{–}CH_2\text{–}CH_2\text{–}O\text{–}CH_2\text{–}CH_2\text{–}OH + HO\text{–}OC\text{–}CH{=}CH\text{–}CO\text{–}OH$$

$$\longrightarrow [\text{–}O\text{–}CH_2\text{–}CH_2\text{–}O\text{–}CH_2\text{–}CH_2\text{–}O\text{–}CO\text{–}CH{=}CH\text{–}CO\text{–}O\text{–}]_n$$

The product obtained is a hard thermosetting resin. By using a mixture of saturated and unsaturated acids a more flexible polymer can be obtained because of less sites for crosslinking.

The polyester resin can be obtained by heating maleic anhydride and glycol at 150-200°C for 6-16 hours in inert atmosphere of N_2 or CO_2. Water formed during the reaction is continuously removed. Peroxide initiator and accelerators like methyl ethyl ketone peroxide are used for crosslinking with vinyl monomers.

The use of these resins is made for obtaining glass fibre reinforced plastics which have high impact and tensile strength similar to steel. They are good insulators and resistant to chemicals. They can be used upto 150°C—200°C.

They are resistant to solvents except chlorinated hydrocarbons, esters and ketones. They are used in boat hulls, lorry cabs, car bodies, caravans, roofing panels, tea trays, insulation plates, motor cycle, side cars, chairs, fishing rods, sinks, suit cases and in chemical plant.

Glass fibre reinforced polyester melamine-formaldehyde laminate with copper foil attached is used for printing circuits.

The melting point of polyesters can be increased by introducing aromatic rings into the polymer chain. It can be illustrated by taking the example given below:

$$\left[O\text{–}(CH_2)_2\text{–}O\text{–}\underset{\|}{\overset{}{C}}\text{–}(CH_2)_6\text{–}\underset{\underset{O}{\|}}{C} \right]_n \quad \text{(meting point } 65^0\text{C)}$$

(O printed below the first C)

$$\left[O\text{–}(CH_2)_2\text{–}O\text{–}\underset{\underset{O}{\|}}{C}\text{–}C_6H_4\text{–}\underset{\underset{O}{\|}}{C} \right]_n \quad \text{(melting } 265^0\text{C)}$$

The remarkable increase in the melting point of polyesters formed by the incorporation of the aromatic ring is because of the stiffening of the polymer backbone. Thus, a polyester like polyethylene terephthalate (PETP) has a high melting point due to the presence of the aromatic ring and is commercially the most popular polymers marketed under the trade name of Terylene or Terene.

The starting materials for PETP are ethylene glycol and terephthalic acid. In commercial practice, however, dimethyl terephthalate (DMT) is taken instead of terephthalic acid. DMT is melt condensed with ethylene glycol to yield PETP.

The reaction taking place can be shown as under:

$$n\ H_3C-O-\underset{\underset{O}{\|}}{C}-C_6H_4-\underset{\underset{O}{\|}}{C}-O-CH_3 + n\ OH-CH_2-CH-OH$$

$$\downarrow \begin{array}{l} -CH_3OH \\ \text{Condensation} \end{array}$$

$$n\ H_3-C-O-\underset{O}{C}-C_6H_4-\underset{O}{C}-O-CH_2-CH_2-OH$$

$$\updownarrow \begin{array}{l} \text{Trans-Esterification} \\ HOCH_2CH_2OH \end{array}$$

$$HO-CH_2-CH_2-O\left[\underset{\underset{O}{\|}}{C}-C_6H_4-\underset{\underset{O}{\|}}{C}-O-CH_2-CH_2-O\right]_n H$$

The first stage of the reaction (*condensation*) is done at the reflux temperature of ethylene glycol with a low vacuum, when methanol is recovered.

The second stage (*trans-esterification*) is done at high temperatures about 200-250°C, under a very high vacuum so that the ethylene glycol can be effectively removed so as to push the reaction in the forward direction.

Polyethylene terephthalate melts at round 265°C and is resistant to heat and moisture. It remains virtually unattached by most of the chemicals. It is extensively used to make textile fibres. It has good mechanical strength up to 175°C. Garments prepared from its fibres resist the formation of wrinkless. PETP can also be made into films, that which are used in the manufacture of magnetic recording tapes, aluminised sheets source of these are as under for certain special applications.

Because flexible films have excellent strength and good electrical and thermal properties, have low permeability to gases, water, oil, odour and could be sterilised so they are used in capacitors, cables, magnetic recording tapes, typewriter ribbons, food packaging, tracing film, book covers etc. The film is coated with aluminium by vacuum disposition and used for decoration and in textiles.

The unsaturated polyester-styrene combination, is used as the resin matrix, in fibre-reinforced plastics (FRP) structures. These resins also find use as decorative coatings.

2. Net Work Polyesters

Smith condensed phthalic anhydride with glycerol in 1901 to prepare a glassy brittle material. Kienle in 1924 started investigating these resins and called them alkyds, He modified the polyesters with drying oils. By this modification they become soluble, on exposure to oxygen rapid crosslinking occurs and crosslinked films are flexible and durable. This led to rapid acceptance of oil modified polyesters for surface coatings. Unmodified alkyds are used for bonding mica, asbestos etc.

The most important alkyds are obtained from glycerol and phthalic anhydride. A linear polymer can be obtained by heating them at 180°C in ratio 2 : 3 (*i.e.*, excess of acid). This polymer when heated forms crosslinks and yields a thermosetting insoluble resin.

$$C_6H_4\begin{matrix} CO \\ CO \end{matrix}\!\!>O \;+\; \begin{matrix} CH_2OH \\ | \\ CHOH \\ | \\ CH_2OH \end{matrix} \longrightarrow$$

$$\left[-O-\overset{O}{\overset{\|}{C}}-C_6H_4-\overset{O}{\overset{\|}{C}}OCH_2-\underset{OH}{\underset{|}{CH}}-CH_2- \right]_n$$

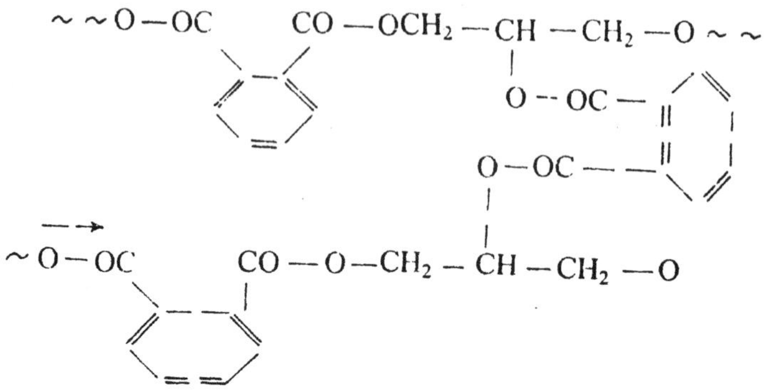

~ ~O—OC   CO—OCH_2—CH—CH_2—O ~ ~

O--OC—

O—OC—

→

~O—OC   CO—O—CH_2—CH—CH_2—O

Instead of phthalic anhydride, isophthalic acid can be used as is done in USA. Its use is in increasing. Polyhydric alcohols like pentaerythritol, trimethylolpropane and sorbitol are also used. Sometimes small quantities of diols like ethylene glycol and propylene glycol are used to reduce crosslinking.

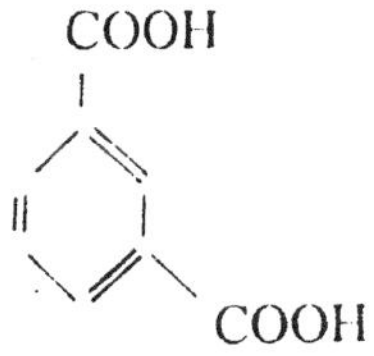

COOH

COOH

Isophalic Acid

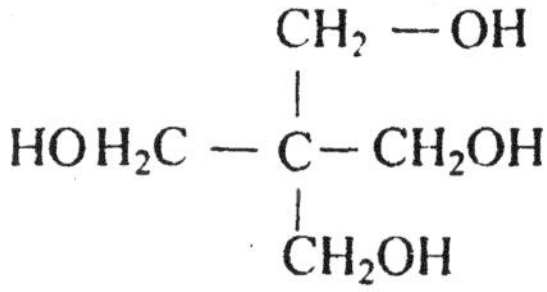

CH_2—OH

HOH_2C — C— CH_2OH

CH_2OH

Pentaerythritol

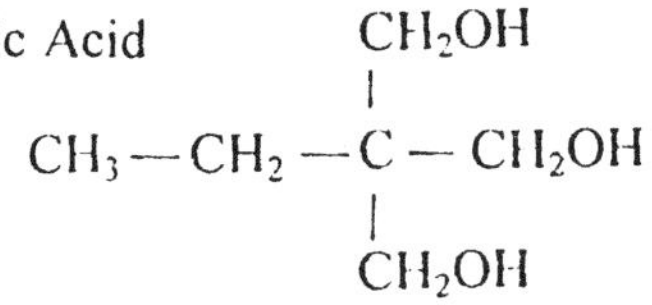

CH_2OH

CH_3—CH_2—C—CH_2OH

CH_2OH

Trimethylol Propane

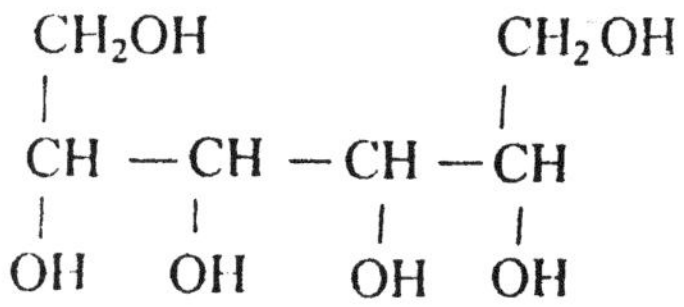

CH_2OH CH_2OH

CH—CH—CH—CH

OH OH OH OH

Sorbital

Drying, semidrying and non-drying oils are used for modifying alkyd resins. Alkyd resins can also classified according to the percentage of oil present in them. Short oil resins contain less than 40% oil. Medium oil resins contain 50–70% oil, while long oil resins contain more than 70% oil.

Resins

Drying oil resins contain linseed, perilla or tung oil. Short oil resins dissolve in aromatic solvents and yield very hard glossy finishes when cured at higher temperature used for appliances, toys, sign boards etc. Medium drying oil resins dissolve in aliphatic or aliphatic aromatic mixed solvents which may air dried or stoved to give durable glossy finished used for farm implements, hardware and metal furniture. Long oil resins are soluble in aliphatic solvents, like naptha. They dry rapidly in air and are used in household paints.

Properties

Semidrying oil resins are similar to drying oil resins but the films do not yellow appreciably on ageing. Oils like dehydrated castor oil, sunflower oil and soybean oil are used for high gloss white finishes.

Non-drying oil resins are soluble only in aromatic hydrocarbons. They are used with amino resins for stoving finishes for appliances. Medium resins are used as plasticisers for cellulose nitrate. Along with natural oils several natural occurring and synthetic acide like resin (abiotic acid) pelargonic acid and isooctanoic acid are added to modify alkye resins. The alkyd resins are obtained by two processes, *i.e.*, (1) **Fatty acid process** and (2) **Alcoholysis process**.

(i) Fatty Acid Process

In Fatty acid process the oil is hydrolysed to yield a fatty acid. A mixture of fatty acid, ''dibasic acid and polyalcohol'' are heated at 200-240°C in inert atmosphere. Sometimes small quantities of xylene is used as solvent.

(ii) Alcoholysis Process

In alcoholysis process the oil is heated with polyol at 240°C in presence of a base like calcium hydroxide. The principal product formed is a monoglyceride. The dibasic acid is then added and further esterification done:

```
                               R
                               |
  /\\   — CO —                 CO
 ||  |                         |
 ||  |  — CO                   O
  \//     |                    |
          O — CH2 — CH — CH2 — O — CO —    /\\
                                          ||
          O — CH2 — CH — CH2 — O — CO —   || \//
  /\\     |           |
 ||  | — CO           O
 ||  |                |
  \//\                CO
       CO             |
       |              R
       O —
```

Commercial alkyd resins contain catalysts which speed up crosslinking reaction. These catalysts is called **Driers**.

Primary driers like cobalt, lead and manganese napthenates or linoleates establish redox systems that lead to rapid auto-oxidation and cross-linking of resin.

Several other resins can be blended with alkyd resins to introduce desired improvement in properties *e.g.* cellulose nitrate, chlorinated rubber, phenolics, amino resins or silicons oils. Vinyl monomers like styrene can be added to alkyd resins along with initiator to get a tougher resin with shorter drying times and lighter colour.

3. Poly Allyl Esters

Diallyl phthalate can be obtained by condensing phthalic anhydride with allyl alcohol as under:

$$C_6H_4(CO)_2O + 2\ CH_2{=}CH{-}CH_2OH \longrightarrow C_6H_4(-CO-O-CH_2-CH{=}CH_2)_2$$

Diallyl phthalate when heated to 100°C with an initiator like *t*-butyl per benzoate yields a linear polymer linked through one allyl unit per monomer. This polymer having mol. wt. 10000–25000 is then compounded with fillers to get thermosetting moulding powder. Mouldings have thermal stability upto 180° for a long time. Polymer is used only under severe thermal operating conditions. Similar polymer from diallyl isophthalate are capable of withstanding temperatures upto 220°C and organic solvents.

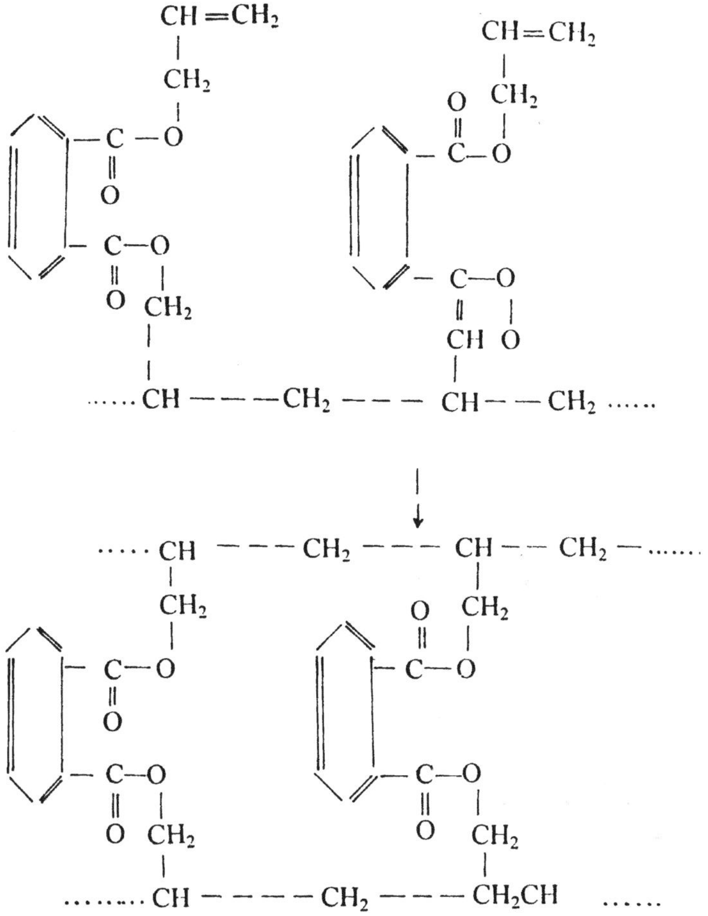

$CH{=}CH_2$ — CH_2 — O — $C(=O)$ — [benzene ring] — $C(=O)$ — O — CH_2 — CH ...... / $CH{=}CH_2$ — CH_2 — O — $C(=O)$ — [benzene ring] — $C(=O)$ — O — CH

$$\cdots CH - - - CH_2 - - - CH - - CH_2 \cdots$$

↓

$$\cdots CH - - - CH_2 - - - CH - - CH_2 - \cdots$$

CH_2 — O — $C(=O)$ — [benzene ring] — $C(=O)$ — O — CH_2 / CH_2 — O — $C(=O)$ — [benzene ring] — $C(=O)$ — O — CH_2

$$\cdots CH - - - - CH_2 - - - CH_2CH \cdots$$

4. Polycarbonates

Polycarbonates are polyesters of phenols and carbonic acid. Polycarbonates are polymer containing –O–CO–O– groups. They were prepared accidentally in 1989 by Einhorn by the action of phosgene with hydroquinone. He also prepared a resin by reacting phosgene with resorcinol. Bischoff and Hedenstrom in 1902 reacted dihydricphenols with diphenylcarbonate to get insoluble materials. Carothers and others prepared aliphatic polycarbonates in 1930 but they were of no commercial importance.

In 1953 Farbenfabrican Bayer made linear polycarbonates. By 1957, they were commercially produced by Bayer and General Electric. These polymers are based on 4,4'-dihydroxy diphenylpropane (bisphenol A) anu phosgene.

Bisphenol A of high purity is needed for preparing polycarbonates.

Preparation

Bisphenol—A with diphenylcarbonate given a polycarbonate as under:

$$n\ HO-C_6H_4-C(CH_3)_2-C_6H_4-OH + n\ O{=}C(OC_6H_5)_2 \xrightarrow{-C_6H_5OH} H\!\left[O-C_6H_4-C(CH_3)_2-C_6H_4-O-\underset{\underset{O}{\|}}{C}\right]_n\!-O-C_6H_5$$

Like PETP, the polycarbonate reaction is also carried out in two stages, *i.e.*, the first stage at 200°C with low vacuum when the oligomers are obtained and the second stage at 300°C under high vacuum when the solid polymer is produced.

Interfacial condensation of sodium salt of bisphenol A and phosgene also yields the same polymer:

$$n\ NaO-C_6H_4-C(CH_3)_2-C_6H_4-CNa + n\ O{=}C\genfrac{}{}{0pt}{}{Cl}{C} \xrightarrow{NaCl} Na\!\left[O-C_6H_4-C(CH_3)_2-C_6H_4-O-\underset{O}{C}\right]_n\!-Cl$$

Phosgene needed is made by reaction of chlorine and carbon-monoxide at 200°C in presence of charcoal. It is a poisonous gas having unpleasant odour.

Diphenylcarbonate is prepared by passing phosgene in solution of phenol in sodium hydroxide in presence of methylchloride. The organic phase yields disphenyl carbonate as white crystalline solid:

$$2\ C_6H_5-ONa + COCl_2 \rightarrow C_6H_5-O-CO-O-C_6H_5 + 2NaCl$$

Two methods which are used for preparation are: (i) *direct phosgenation* (2) *ester interchange.* In direct phosgenation method phosgene is passed into a solution of Bisphenol A in chloroform and pyridine at 25–35°. The solution is washed with dilute hydrochloric acid or remove pyridine. The solvent is removed by evaporation.

$$n\ HO-C_6H_4-C(CH_3)_2-C_6H_4-OH + n\,COCl_2 \xrightarrow{\text{base}}$$

$$\left[-O-C_6H_4-C(CH_3)_2-C_6H_4-O-CO-\right]_n$$

Direct phosgenation can also be carried out in aqueous sodium hydroxide rapidly stirred with methylene chloride containing a tertiary amine as catalyst.

In the ester interchange method a mixture of bisphenol A and diphenylcarbonate along with a basic catalyst like lithium hybride is melted and stirred at 150°C under nitrogen. Later on when the temperature is raised to 210° at 20 mm mercury pressure most of the phenol formed gets distilled off. In the last stage the temperature is raised to 300°C at 1 mm Hg.

$$n\ HO-C_6H_4-C(CH_3)_2-C_6H_4-OH + n\ C_6H_5-O-CO-O-C_6H_5 \longrightarrow$$

$$\left[-O-C_6H_4-C(CH_3)_2-C_6H_4-O-C-\right]_n + 2n\ C_6H_5OH$$

This method is preferred because it yields a purer polymer in a molten stage.

Properties

These polymers are transparent and white. They have outstanding rigidity and tougness at low and high temperatures and also possess outstanding dimensional stability in humid atmosphere. They tend to crack under strain or on ageing. It yellows due to impurities. It dissolves in organic solvents and alkalis.

Uses

The electrical and electronics industries are the largest users. It is used also in blender housings, kitchen utensils and babies milk bottles. The physical properties are improved by stretching to form crystalline polymers like "**Lexan**".

Many useful articles such as safety goggles, safety shields, telephone parts and machinery housings can be made from this plastic.

9.5. Poly (Vinyl Chloride): PVC

Introduction

The vinyl polymers are important plastics. Polyvinyl chloride is a very widely used plastic in India.

Poly (vinyl chloride) prepared by the polymerization of vinyl chloride. Vinyl chloride is a gas which is prepared by the reaction of ethylene with HCl and oxygen to yield 1,2—dichloroethane. The compound is then pyrolysed to yield vinyl chloride and HCl.

$$CH_2{=}CH_2{+}Cl_2 \longrightarrow CH_2Cl{-}CL_2Cl \longrightarrow \underset{\text{Vinyl Chloride}}{CH_2{=}CHCl} + HCl$$

$$CH_2{=}CHCl \xrightarrow{\text{Polymer}} \underset{\text{Polyvinyl Chloride}}{\ldots{-}CH_2{-}\underset{\displaystyle Cl}{\underset{|}{CH}}{-}CH_2{-}\underset{\displaystyle Cl}{\underset{|}{CH}}{-}\ldots}$$

Vinyl chloride is also obtained from acetylene.

$$CH{\equiv}CH + HCl \xrightarrow[130°C;\ 5\ atm]{Hg_2Cl_2} CH_2{=}CHCl$$

Vinyl chloride (b.p. —14°C) is purified by distillation at low temperature.

Preparation

Poly (vinyl chloride) is produced from vinyl chloride by three methods: (i) suspension polymerisation, (ii) bulk polymerisation, and (iii) emulsion polymerisation. Generally poly (vinyl chloride) is produced by suspension polymerisation.

Industrial polymerisation of vinyl chloride is carried out either in suspension or emulsion. Limited quantities of PVC are also made by bulk polymerisation.

Suspension Polymerisation

A mixture of vinyl chloride (100 parts by weight), water, poly (vinyl alcohol) (0.04 parts by weight), trichloromethylene (0.2 parts by weight) and lauryl peroxide (0.2 parts by weight) are transferred to a reactor which is fitted with a brusting disc and also gets connected to a vacuum line.

Now the reactants are heated to about 50°C and the pressure in the reactor increased to about 100 lb/-in^2. This temperature is maintained for nearly 15 hours after which time the pressure starts to drop as the last of the monomer gets consumed. The slurry is now discharged to a centrifuge where the polymer gets separated and washed. Finally the polymer gets dried in hot air at about 100°C to obtain the product.

Emulsion polymer isolation gives polymers in the shape of tiny hollow spheres called **cenospheres**. The pure polymers are rarely used. They are generally compounded with a variety of additives such as fillers, plasticisers, lubricants, pigments and stabilisers to provide a variety of materials with differing physical, chemical and electric properties.

Structure

Structurally PVC molecule is partially syndioactic and so it has low crystallinity. Poly (vinyl chloride) is, having a head-to-tail structure:

```
        Cl          Cl          Cl          Cl
        |           |           |           |
        CH          CH          CH          CH
 \     /  \        /  \        /  \        /  \       /
   CH2      CH2        CH2        CH2          CH2
```

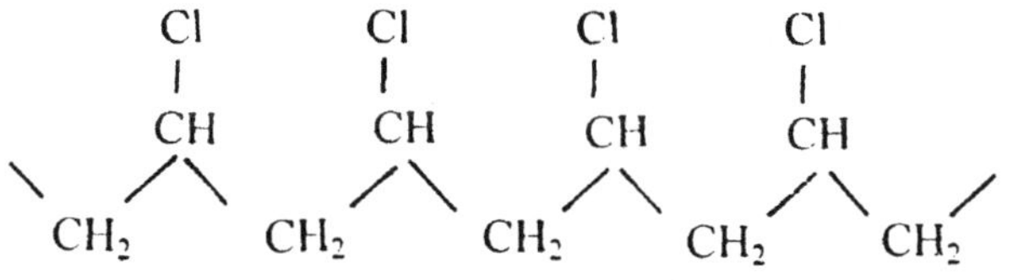

When poly (vinyl chloride) in dioxane solution is made to treat with zinc dust the resulting polymer gets saturated and is having a small chlorine content. This reveals that it is having a head-to-tail structure.

Properties

Poly (vinyl chloride) occurs as a colourless rigid material. It is having a high density and low softenting point. It is also having a higher dielectric constant and power factor. The high chlorine content of poly (vinyl chloride) makes it flame retardant polymers.

In dissolves in dioxan, tetrahydrofuran, cyclohexamine, methyl isobutyl ketone and nitrobenzene. It does not get affected by acids, alkalies and aqueous solution. Chemically it is an inert material.

Application

During the second world war there occurred a great demand for poly (vinyl chloride) for cable insulation. After the war the civilian use of poly (vinyl chloride) got expanded rapidly.

It has been one of the three most important plastics which are currently in use. It is mainly used in cable insulation, chemical plant, leather cloth, packaging and toys.

It aslo finds use in the manufacturing of film, sheet, and floor covering. The film and sheeting are used for making rainwear, handbag, shower curtains, food covers etc.

Modifications

Chlorination of PVC at low temperature gives a polymer which finds use in industrial plumbing for hot effluents.

Chlorinated PVC obtained above 100°C is largely used in adhesive and protective coatings and to make filter cloth for corrosive liquids.

9.6. Polyformaldehyde

Polyformaldehyde is made up of – CH_2– O – repeat units and can be produced by the *cationic polymerisation* of either formaldehyde or trioxane.

$$n\ HCHO \xrightarrow[\text{(Moisture)}]{BF_3} HO\text{-}(CH - O)_n\text{-}H$$

$$n\ \underset{\text{Trioxane}}{O\langle\begin{smallmatrix} CH_2 - O \\ CH_2 - O \end{smallmatrix}\rangle CH_2} \xrightarrow{HClO_4} HO\text{-}(CH - O)_n\text{-}H$$

The polymer produced can not be stored because it is prone to degradation by unzipping on storage so it is usually stabilised by esterifying the hydroxyl end groups with acetic acid and pyridine. Commercial grade polyformaldehyde is a fibrous polymer and has 50-80% crystallinity. Its melting point is around 185°C. It has a high degree of hardness and rigidity and is useful as an engineering plastic.

9.7. Polyparaphenylene

This polymer is made up entirely of aromatic rings which form the chain backbone:

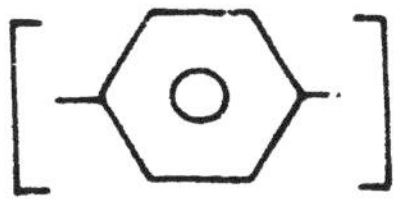

$$\left[-\langle \bigcirc \rangle - \right]$$

Preparation

For preparing this polymer the starting material used is paradibromobenzene, paradibromobenzene on treatment with activated copper powder at high temperatures, forms parapoly-phenylene as under:

$$n\ Br-\langle \bigcirc \rangle-Br \xrightarrow{\text{Copper}} \left[\langle \bigcirc \rangle \right]_n + n\ CuBr_2$$

This polymer can also be obtained by the cationic polymerisation of benzene with aluminium trichloride and cupric chloride.

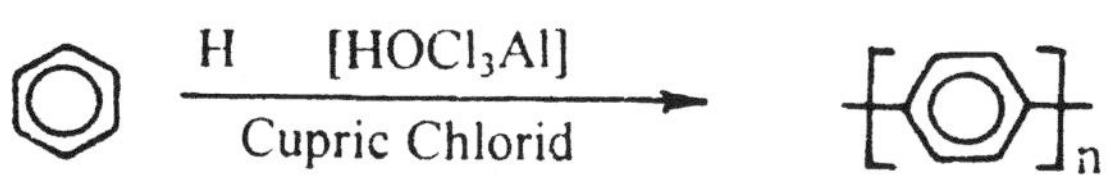

$$\langle \bigcirc \rangle \xrightarrow[\text{Cupric Chlorid}]{H \quad [HOCl_3Al]} \left[\langle \bigcirc \rangle \right]_n$$

Properties

It is quite brittle and is insoluble, because of the right nature of the polymer chain that is caused by the presence of aromatic rings linked together through parapositions. It can withstand temperatures even up to 560°C. This property is attributed to the presence of resonance-stabilised conjugated double bonds in the aromatic rings.

Preparation

Polysulphone is obtained by condensation of bisphenol—A and

dichlorodiphenyl sulphone as under:

$$n\ HO-C_6H_4-C(CH_3)_2-C_6H_4-OH + n\ Cl-C_6H_4-SO_2-C_6H_4-Cl \xrightarrow{NaOH} H\!-\!\left[O-C_6H_4-C(CH_3)_2-C_6H_4-O-C_6H_4-SO_2-C_6H_4\right]_n\!-\!Cl + NaCl$$

Properties

It is a thermoplastic having a high thermal stability and good mechanical strength over a wide range of temperature (from – 180°C to + 140°C) The extraordinary thermal stability is attributed to the resonance stabilisation because of presence of aromatic rings near –SO_2- groups. Polysulphone is used for making injection-moulded items and films.

9.9. Polyimides

Preparation

These are produced by polycondensation of pyromellitic anhydride and p,p'-diamino diphenyl ether. The reaction, is carried out in stages. To start with, the reaction is carried out in suitable solvents, like DMF, at around 50°C, when a polyaddition reaction occurs with the formation of polyamic acid as under:

$$n\ O(CO)_2C_6H_2(CO)_2O + n\ H_2N-C_6H_4-O-C_6H_4-NH_2 \longrightarrow \left[OC(HOOC)C_6H_2(COOH)CO\ NH-C_6H_4-O-C_6H_4-NH\right]_n$$

In the second stage, the polyamic acid is cast as a film and the solvent is evaporated and the film is baked at 300°C in a nitrogen atmosphere, when the condensation reaction occurs as under:

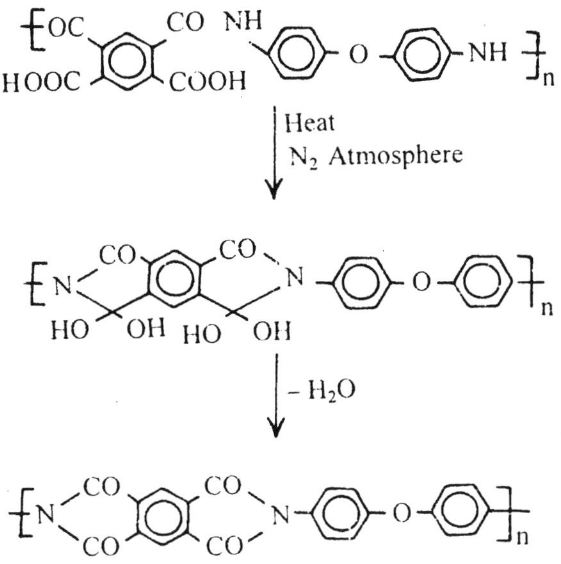

$$\left[\text{OC}(\text{HOOC})\text{-C}_6\text{H}_2\text{-(COOH)CO-NH-C}_6\text{H}_4\text{-O-C}_6\text{H}_4\text{-NH} \right]_n$$

$$\downarrow \text{Heat, } N_2 \text{ Atmosphere}$$

$$\left[\text{N}\langle\text{CO, C(HO)(OH)}\rangle\text{C}_6\text{H}_2\langle\text{CO, C(HO)(OH)}\rangle\text{N-C}_6\text{H}_4\text{-O-C}_6\text{H}_4 \right]_n$$

$$\downarrow -H_2O$$

$$\left[\text{N}\langle\text{CO, CO}\rangle\text{C}_6\text{H}_2\langle\text{CO, CO}\rangle\text{N-C}_6\text{H}_4\text{-O-C}_6\text{H}_4 \right]_n$$

The product produced in the second stage from polyamic acid is found to be insoluble and infusible and has to be produced in the final required shape and form.

Uses

Polyimide is extensively used in electrical industry for insulation coating of electromagnetic wirings. Actually the wire is coated with polyamic acid and baked at 300°C under an inert atmosphers when the polyimide gets deposite on the surface of the wire.

It is also used as surface coating in supersonic aircraft because it can withstand temperatures as high as 420°C for short time exposures.

9.10. Polyureas

These are polyamides of carbonic acid.

Their structure resembles that of polyurethanes and polyamides, because they contain the group.

$$\text{—NH—}\underset{\underset{\text{O}}{\|}}{\text{C}}\text{—NH—}$$

They are obtained by stepwise polymerization of diisocyanates with diamines:

$$nOCN-R-NCO+nH_2N-R'-NH_2 \rightarrow$$
$$\rightarrow \left[-\underset{\underset{O}{\|}}{C}-NH-R-NH-\underset{\underset{O}{\|}}{C}-NH-R'-NH- \right]$$

Polyureas can also be prepared by other methods, e.g. by polycondensation of diamines with diurethanes, carbonic acid esters with diamines, etc.

Polyureas have higher melting points than the corresponding polyamides and polyurethanes, *e.g.*;

	Melting point,°C
$\left[-NH\underset{\underset{O}{\|}}{C}NH(CH_2)_5- \right]_n$ Polyureas	270
$\left[-NH\underset{\underset{O}{\|}}{C}NH(CH_2)_5- \right]_n$ Polyurethane	150
$\left[-NH\underset{\underset{O}{\|}}{C}NH(CH_2)_5- \right]_n$ Polyamide	215

Polyureas are inferior in heat resistance to polyurethanes and polyamides. Polyureas are used in the plastics industry and for dressing textiles: in Japan they are used for spinning fibre.

9.11. Polytriazoles

Their general structure is as follows:

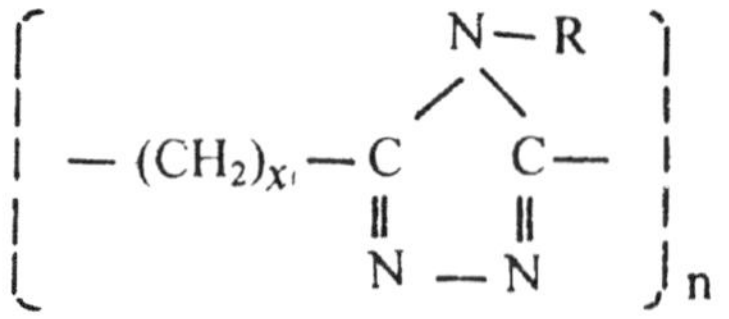

$$\left[-(CH_2)_x-\underset{\underset{N}{\|}}{C}\overset{\overset{N-R}{\diagup\diagdown}}{\quad}\underset{\underset{N}{\|}}{C}- \right]_n \quad (N-N \text{ bond in ring})$$

where R=H or NH_2

Preparation

These are obtained by polycondensation of dihydrazides of dicarboxylic acids (in an excess of hydrazine hydrate);

$$nH_2NHNOC{-}(CH_2)_8{-}CONHNH_2 \rightarrow$$

$$\rightarrow \left[-(CH_2)_8 - \underset{\overset{\|}{N}}{C} \overset{N-NH_2}{\diagup\diagdown} \underset{\overset{\|}{-N}}{C} - \right]_n + 2nH_2O$$

They can also be prepared by the action of diiminoesters or dinitriles of dicarboxylic acids with hydrate:

$$\underset{OC_2H_5}{NH=\overset{|}{C}}-(CH_2)_8-\underset{OC_2H_5}{\overset{|}{C}}=NH \xrightarrow{NH_2NH_2 \cdot H_2O} \left[-OC_2H_5 - \underset{\overset{\|}{N}}{C} \overset{NH_2}{\diagup\diagdown} \underset{\overset{\|}{-N}}{C} - \right]_n$$

$$N{\equiv}C-(CH_2)_8-N{\equiv}C \xrightarrow{NH_2NH_2 \cdot H_2O} \left[-OC_2H_5 - \underset{\overset{\|}{N}}{C} \overset{N-NH_2}{\diagup\diagdown} \underset{\overset{\|}{-N}}{C} - \right]_n$$

They can also be obtained by the action of diamides with dihydrazides of dicarboxylic acids as under:

$$\underset{H_2N}{\overset{O}{\gg}}C-R-C\underset{NH_2}{\overset{O}{\ll}} + \underset{O}{\overset{H_2NHN}{\gg}}C-R'-C\underset{C}{\overset{NHNH_2}{\ll}} \rightarrow$$

$$\rightarrow \left[-R - \underset{\overset{\|}{N}}{C} \overset{NH}{\diagup\diagdown} \underset{\overset{\|}{-N}}{C} - R' \right]_n + 2nH_2O$$

When sebacic dihydrazide reacts with hydrazine hydrate the product formed is polyoctamethylene-4-amino-1,2,4-triazole, that resembles poly (hexamethylene adipamide) in properties.

Properties

The melting points of some polyaminotriazoles prepared from acid dihydrazides are listed here: from azelaic dihydrazide –230°C, from suberic dihydrazide–294°C, and from adipic dihydrazide–280°C.

Of various heterochain polymers polyaminotriazoles are found to be the most stable to hydrolysis; they are soluble in cyanohydrins and polyhydric alcohols.

Polyaminotriazoles are used for making fibres and films.

9.12. Polyurethanes

Polyurethane polymers are characterised by the presence of urethane –NHCOO– linkages in their repeat units:

$$\left\{ \underset{\underset{O}{\parallel}}{C} - \underset{\underset{H}{|}}{N} - \{R\} - \underset{\underset{H}{|}}{N} - \underset{\underset{O}{\parallel}}{C} - O - \{R'\} - O \right\}_n$$

The structure of these resemble with the structure of polyamides because both of them contain –CONH– groups. However the principal linkage in polyurethane, is –NHCOO–. The presence of additional oxygen in the chain increases its flexibility, because of this the melting point of polyurethane is much lower than that of the corresponding polyamide as illustrated below:

$$\left[O - \overset{\overset{O}{\parallel}}{C} - (CH_2)_5 - NH \right]_n \qquad \text{Polyurethane m.p=150°C}$$

$$\left[\underset{\underset{O}{\parallel}}{C} - (CH_2)_5 - NH \right]_n \qquad \text{Nylon 6 (Polymide m.p=250°C}$$

Preparation

Some other groups such as ester, ether, amide, or urea are present in the polymer chain of commercial polymers. In 1937 O. Bayer found that reaction of diisocyanates with glycols fields polyurethanes which are useful as plastics, fibres, adhesives, rigid foams and surface coatings.

For commercial preparation of these polymers polyaddition reaction between a diisocyanate with a diol or triol is very frequently used:

$$(n+1)HO-R-OH + n\ OCN-R-NCO$$

$$\downarrow$$

$$H\left[O-R-O-\underset{\substack{\|\\ O}}{C}-\underset{\substack{|\\ H}}{N}-R'-\underset{\substack{|\\ H}}{N}-\underset{\substack{\|\\ O}}{C}\right]_n O-R-OH$$

Isocyanates can be prepared by reaction of an amine with phosgene:

$$R-NH_2+COCl_2 \longrightarrow R-N=C=O + 2HCl$$

Raw Materials

The raw materials needed are:

(i) Toluene diisocyanate
(ii) Diphenylmethane, 4,4' –isocyanate
(iii) Napthylene 1,5–diisocyanate and
(iv) Hexamethylene diisocyanate.

The most important raw material is toluene diisocyanate. Nitration of toluene produces O-nitrotoluene (60%) and p-nitrotoluene (40%) which on further nitration gives 2,4-dinitrotoluene (80%) and 2,6-dinitrotoluene (20%). Nitration of pure 2-nitrotoluene yields 65% 2, 4 and 35% 2,6-dinitrotoluenes while nitration of 4-nitrotoluene gives 100% 2,4-dinitrotoluene. Reduction of dinitrotoluene with iron and acid yields diaminotoluene which on subsequent phosgenation forms diisocyanates of toluene.

Depending on starting diamines these are three toluene diisocyanates i.e. 80, 20; 65, 35 or 100% 2,4 toluene diisocyanate. All these are commercially available. They are respiratory irritants and should be handled carefully 80, 20 toluene diisocyanate is used widely for production of flexible foams.

Many diols and polyols like 1,4-butanediol and hydroxy-terminated polyesters or polyethers or polyesteramides are used for reaction with diisocyanates commercially.

Properties

Isocyanates are quite reactive and they react with compounds which contain active hydrogen such as hydroxy, amine, carboxylic acids, amide, urea etc. They can also undergo addition reaction:

$$R-N{=}C{=}O + H_2N-R' \longrightarrow R-NH-\overset{\overset{\large O}{\|}}{C}-NH-R'$$

$$R-N{=}C{=}O + H_2O \longrightarrow R-NH-COOH$$

$$R-N{=}C{=}O + R'-NH-CO-R' \longrightarrow R-NH-CO-\underset{}{\overset{\overset{\large R}{|}}{N}}-CO-R'$$

$$2R-N{=}C{=}O \longrightarrow R-N\genfrac{}{}{0pt}{}{\diagup CO \diagdown}{\diagdown CO \diagup}N-R$$

$$3R-N{=}C{=}O \longrightarrow \text{cyclic trimer: } R-N(CO)_3\text{ ring with } N-R,\ N-R$$

For the reaction of isocyanate with active hydrogen compounds, tertiary amines or tin compounds act on catalysts. The addition reactions are slow but catalysed by tertiary amines.

Polyurethanes are used to make foams, coatings, adhesives and elastomers.

To get polyurethane foams the polymer is formed along with gas evolution. When these two processes take place simultaneously the gas bubbles are trapped in polymer matrix yielding a cellular product. Slightly cross-linked products are flexible while highly cross-linked products are rigid. Both flexible and rigid foams are of commercial importance.

The foams can be obtained by the action of a diiscyanate on a polyol and water. The reaction with water forms carbondioxide and the reaction with polyol forms a urethane polymer. Catalysts play a crucial role in the process. Tin octeate and dibutyl tin dilaurate are preferred catalysts along with tertiary amines.

Uses

Flexible polyetherpolyurethane foams are used in cushions, upholstery etc. While polyester based foams are used in textile laminates, coat shoulder pads etc. Polyester foams can withstand drycleaning.

Rigid polyurethane foams are mostly based on polyether alcohol and are highly crosslinked. Rigid foams are many times blown by halogenated alkanes like trichlorofluoromethane. These foams have closed cell structures and are used for thermal insulation. Semirigid foams are used in car crashpads and packaging.

Solid polyurethane elastomers are of three types, cast, mallable and thermoplastic. They are used in types, bearings, shoe heels etc.

Polyurethane fibres (Perlon U) were prepared by reacting hexamethylene diisocyanate with 1,4 butanediol in 1942. Because they were difficult to dye and had harsh feel so went out of the market. But these resins are used for small mechanical components like gears and bearings because of dimensional stability and retention of electrical resistance in humid conditions.

Attempts were made later on to prepare synthetic stretch fibres based on polyurethane. At present many such fibres are marketed as **"Spandex"**, **"Lycra"** etc.

These fibres contain long chains of polyglycols or polyesters between polyurethane blocks. Urethane gets copolymerised with suitable polyol or polyester and then melt spun as monofilament or polyfilament yarn. The urethane blocks are in a randomly disordered fashion in the yarn. When stretched they uncoil and straighten out.

When the stress is released they again go back to original shape. The urethane blocks act like rubber but can withstand washings and action of chemicals. They are difficult to dye. These **"Spandex"** fibres are replacing rubber and stretchable crimp nylon from foundation garments, belts and surgical stockings. Swimming suits are also manufactured from these fibres. They withstand oil, perspiration and mild bleaching. Synthetic leather like **Corfoam** is made by impregnating fibre material in polyurethane.

Polyurethane surface coatings are used in wood varnishes and quick drying enamels. Electrical wires coated with polyurethane can be soldered without removing the polymer coating.

Polyurethane adhesives are used for bonding wood, plastics, metals and leather. They can bond rubber to rubber, metal, glass or synthetic fibres.

9.13. Polysilicones

These polymers are not organic but contain SiO bond. The organic groups are attached to the silicon atom.

Like carbon, silicon is able to form covalent compounds. Unlike carbon, silicon is not able to form double or triple bonds. Hence silicon is able to form compound by condensation reaction.

The commercial interest in silicon polymers started in 1930's when searches for heat-resistant electrical insulating materials were made. Silicon polymers were used to some extent during World war II. However, the commercial production of silicon polymers got started by the Dow Corning Crop (U.S.A.) in 1943 and by the General Electric Co. (U.S.A) in 1946.

Polymerization

The monomers for silicone polymer have been silanols which are very unstable compounds and are produced by the hydrolysis of chlorosilanes. Then, they condense intermolecularly to yield silicon polymers.

(i) *Linear silicones.* Hydrolysis of dichlorosilanes, give rise to linear silicones. However, a small amount of cyclic polymer is also obtained.

$$\mathrm{Cl-\underset{\underset{R}{|}}{\overset{\overset{R}{|}}{Si}}-Cl \xrightarrow{H_2O} HO-\underset{\underset{R'}{|}}{\overset{\overset{R}{|}}{Si}}-OH}$$

dichlorosilane

$$\mathrm{HO-\underset{\underset{R'}{|}}{\overset{\overset{R}{|}}{Si}}-OH + HO-\underset{\underset{R'}{|}}{\overset{\overset{R}{|}}{Si}}-\ddot{O}\,H + HO-\underset{\underset{R'}{|}}{\overset{\overset{R}{|}}{Si}}-O\;H}$$

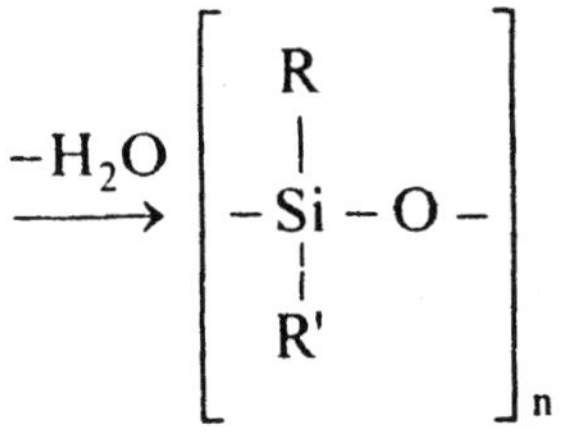

$$\xrightarrow{-H_2O} \left[\begin{array}{c} R \\ | \\ -Si-O- \\ | \\ R' \end{array} \right]_n$$

Linear silicones

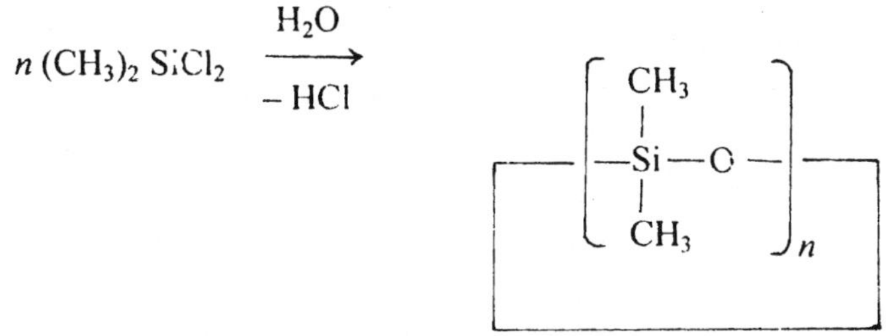

$$n\,(CH_3)_2\,SiCl_2 \xrightarrow[-HCl]{H_2O} \left[\begin{array}{c} CH_3 \\ | \\ -Si-O- \\ | \\ CH_3 \end{array} \right]_n \text{ (cyclic)}$$

The formation of the linear cyclic polymers is dependent upon the reaction conditions. Hydrolysis with water alone gives rise to 50.80% linear polydimethyl **siloxane**—α, ω-**diols** and 50-20% polydimethyl-cyclosiloxanes.

Hydrolysis with 50-85% sulphuric acid yields mostly high molecular weight linear polymers having only small amount of cyclosiloxanes.

(ii) **Branched and cross-linked silicones**. On hydrolysis, trichlorosilance gives rise to branched as well as cross-linked silicones

$$\begin{array}{c} R \\ | \\ Cl-Si-Cl \\ | \\ Cl \end{array} \xrightarrow{H_2O} \begin{array}{c} R \\ | \\ HO-Si-OH \\ | \\ OH \end{array} \xrightarrow{H_2O}$$

$$\begin{array}{ccc} & & | \\ R & & O \\ | & & | \\ -Si & -O- & Si-O- \\ | & & | \\ O & & R \\ | & & | \\ -Si & -O- & Si-O \\ | & & | \\ R & & O \end{array}$$

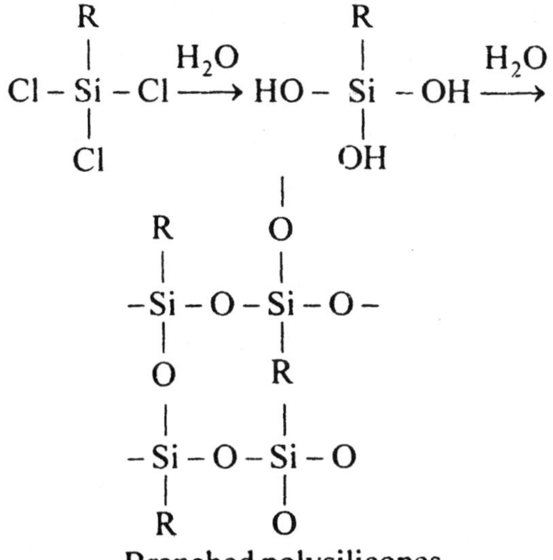

Branched polysilicones

In the presence of inert solvents, there occurs the promotion of intramolecular condensation and the formation of ring compounds.

Types of Silicone Polymers

There are three groups of silicon polymers such as (a) Fluids (b) Elastomers (c) Resins.

1. Fluids

The silicone fluids are linear polymers which are having very low molecular weight within the range 4000–25,000. These are mainly obtained from dimethyldichlorosilane which is continuously hydrolysed by mixing with dilute hydrochloric acid (20%).

The product gets separated as an oil. Now the mixture is made to equilibrate by heating with a small amount of aqueous sulphuric acid or sodium hydroxide at 150°C for many hours. The fluid so obtained is filtered, dried and heated under reduced pressure to get rid of the volatile material.

These compounds find use as cooling and dielectric fluids, in polishes and waxes as release and antifoam agents, and for paper and textile treatment.

Silicone fluids are used as lubricants under vacuum and at high temperature. They are used in vacuum pumps, jet turbines, and guided missiles as lubricants. These fluids can be used for long periods without replacement as they do not degrade to given deposits of carbon like other organic lubricants. They are used in heating baths for industrial heating and as damping fluid in instruments and electrical switchgears. They are used as antifoaming agents in several industries such as dyes, textiles, brewing, paint, glue paper, food processing and sewage disposal.

2. Silicone Rubbers or Elastomers

These are having high molecular weight than the fluids. Generally they are polydimethyl siloxanes. General purpose silicone elastomers are made from polydimethyl siloxanes.

Dimethyl Silicone Elastomers

They are having linear polymers of very high molecular weight within the range of 30,000–700,000. In order to prepare these polymers, dimethyl dichlorosilane is made to dissolve in ether and the solution is then mixed with excess of water.

Cyclic polymers form nearly 50% of the product while the tetramer, octamethylcyclotera siloxane (I) constitutes the main cyclic compound. Now the polymerisation of the tetramer is done by heating at 150–200°C with a trace of sodium hydroxide and a very small amount of nonfunctional material so as to control the molecular weight. The products obtained is a highly viscous gum having no elastic properties.

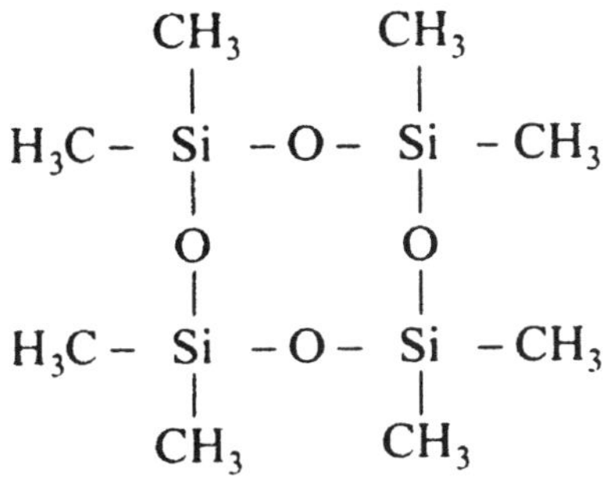

$$
\begin{array}{ccccccc}
 & CH_3 & & CH_3 & \\
 & | & & | & \\
H_3C- & Si & -O- & Si & -CH_3 \\
 & | & & | & \\
 & O & & O & \\
 & | & & | & \\
H_3C- & Si & -O- & Si & -CH_3 \\
 & | & & | & \\
 & CH_3 & & CH_3 &
\end{array}
$$

(I)

Some other silicon elastomers include vinyl silicone elastomers, phenyl silicon elastomers, nitrile silicone elastomers, and fluorosilicone elastomers.

The silicone elastomers combine the properties of rubber and silicones. They retain flexibility over –90° to 250°C temperature.

They can withstand exposure to ozone, ultraviolet light and weathering. They have excellent dielectric properties and are not affected by oils. They get more easily torn than rubber and dissolve in aromatic solvent.

Curing of silicone rubbers can be done with high energy radiation such as gamma rays.

Silicone elastomers find use as gaskets and seals, wire and cable insulation and hot gas and liquid conduits. They also find use in surgical and prosthetic devices. The RTV elastomers are used for sealing and encapsulating.

Cold cure silicone rubbers and available as pastes. These pastes are mixed with an organometallic catalyst and silicate and cured at room temperature. These are used as adhesives, and as encapsulating materials for electronic components. They are also used for textile coating and in moulds.

3. Silicone Resins

As silicone resin are highly branched polymers, they are obtained by the hydrolysis of trichorosilanes. But the product obtained by the hydrolysis of only trichlorosilanes are highly cross-linked and unsuitable for normal applications. Therefore, a blend of tri- and dichlorosilanes are hydrolysed to obtain the desired product.

A blend of chlorosilane is made to dissolve in a solvent like toluene, or xylene and then stirred with water. The reaction is extremely exothermic when the blend is having mainly methyl chloroilanes and cooling becomes necessary to avoid gelation.

In the presence of phenychlorosilanes, it often becomes necessary to raise the temperature to 70 --75°C for ensuring complete hydrolysis. At the end of the reaction, the mixture is allowed to separate into two layers.

The organic layer is repeatedly washed free from hydrochloric acid and then some of the solvent gets distilled off to leave the solution with a solid content of about 80%. At this stage, the resin is having silanolterminated linear, branched and cross-linked polymers and cyclic polymers of low molecular weight.

Now the resin solution is **'bodied'** by heating at about 150°C in the presence of a catalyst like zinc octate. During this process, the molecular weight of the resin gets increased by involving the condensation of some of the remaining silanol groups.

Silicone resins find use as insulating varnishes, impregnating and encapsulating agents and in industrial paints.

Silicon resins given a non-stick surface to frying pans, baking vessels, etc. Silicon laminates with glass-fibre or asbestos making them fire proof, damp proof with good insulation properties. It is used for making fire barriers in aircrafts etc. Silicon resin emulsions are sprayed on bricks, concrete mortar or stone to give water repellant finish.

9.14. Organotin Polymers

These are also known as **polystannoxanes**. They are having following structure:

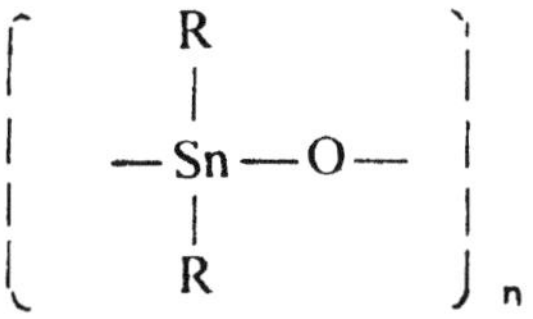

$$\left[\begin{array}{c} R \\ | \\ -Sn-O- \\ | \\ R \end{array} \right]_n$$

They are prepared by the hydrolysis of dialkyl-or diaryl-chlorostannanes, followed by polycondensation:

$$n \; R_2SnCl_2 \longrightarrow n\,HO-\underset{\displaystyle R}{\overset{\displaystyle R}{\underset{|}{\overset{|}{Sn}}}}-OH \xrightarrow{H_2O} \left[\begin{array}{c} R \\ | \\ -Sn-O- \\ | \\ R \end{array} \right]_n$$

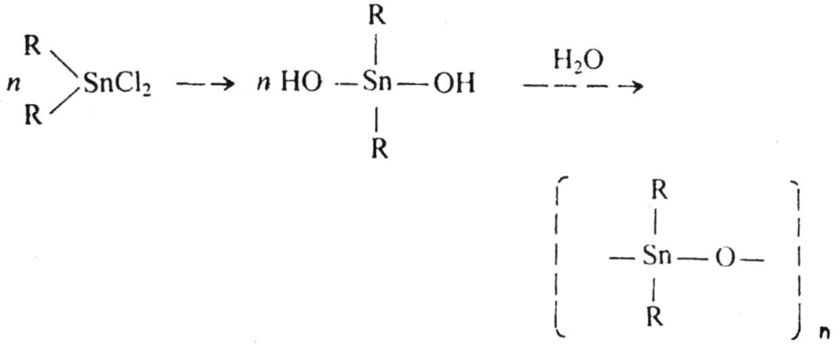

Organotin polymers are brittle substances. They generally melt at 300–400°C.

9.15. Organotitanium Polymers

These polymers are known as **polytitanoxanes**. These are having the following structure:

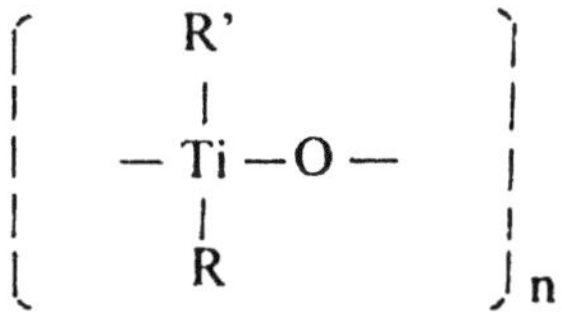

$$\left[\begin{array}{c} R' \\ | \\ -Ti-O- \\ | \\ R \end{array} \right]_n$$

These are obtained by partial hydrolysis of organic o-titanates followed by polycondensation *e.g.*, *o*-butyl titanate:

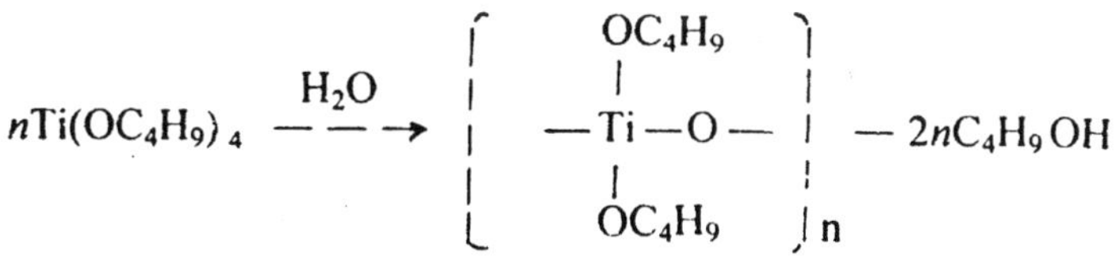

$$n\,Ti(OC_4H_9)_4 \xrightarrow{H_2O} \left[\begin{array}{c} OC_4H_9 \\ | \\ -Ti-O- \\ | \\ OC_4H_9 \end{array} \right]_n - 2n\,C_4H_9OH$$

Hydrolysis under rigorous condition results in branched and three dimensional polymers.

Polytitanoxanes are transparent, heat-resistant and hydrophobic. They can be used in potties and as surfactants resins.

9.16. Organoaluminium Polymers

These are also known as **polyalumoxanes**. They are having the following structure:

$$\left[\begin{array}{c} S \\ | \\ R \\ | \\ -Al-O- \\ | \\ R \end{array} \right]_n$$

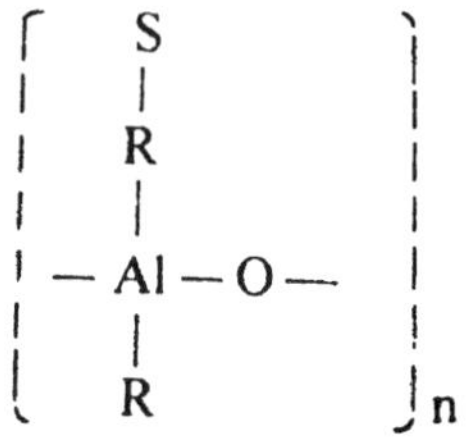

(where R is an alkoxy group or a fatty or β-hydroxyacid residue) are obtained by polycondensation of aluminium alcoholates with simultaneous or subsequent saponification of the polymeric alcoholates.

Uses

Polyalumoxanes include aluminium soaps obtained for polymeric aluminium alcoholates and fatty acids. The polymer is regarded to have the following structure:

$$RO-\left[\begin{array}{cc} -\underset{|}{Al}-O- & \underset{|}{Al}-O- \\ OR & OCOR \end{array} \right]_n -OR$$

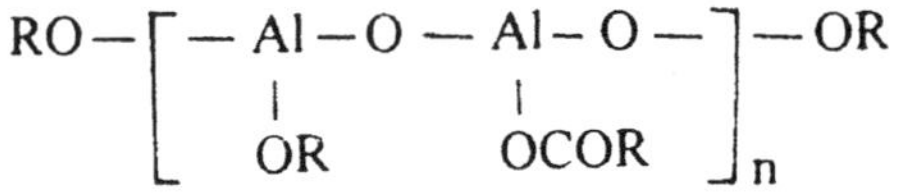

Poly (acetyl acetonate alumoxane) results from they hydrolysis of dialkoxyaluminium acetyl acetonate:

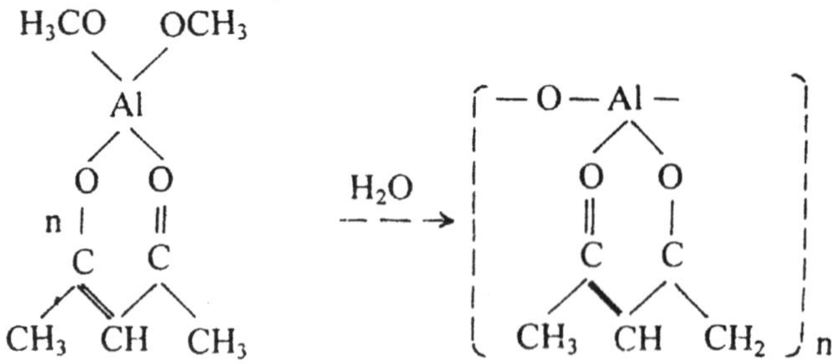

$$\begin{array}{c} H_3CO \quad OCH_3 \\ \diagdown \diagup \\ Al \\ \diagup \diagdown \\ O \quad O \\ n\,| \quad \| \\ C \quad C \\ \diagup \diagdown \diagup \diagdown \\ CH_3 \;\; CH \;\; CH_3 \end{array} \xrightarrow{H_2O} \left[\begin{array}{c} -O-Al- \\ \diagup \diagdown \\ O \quad O \\ \| \quad | \\ C \quad C \\ \diagup \diagdown \diagup \diagdown \\ CH_3 \;\; CH \;\; CH_2 \end{array} \right]_n$$

Uses

The polymer has been stable to hydrolysis and does not alter its properties when heated to 500°C. Polyalumoxanes find use as binders in the preparation of heat-resistant paints and for special treatment of textiles.

9.17. Polyamides

Introduction

Polyamides can be obtained by the melt polycondensation between dicarboxylic acids and diamines. They have the general structure as follows:

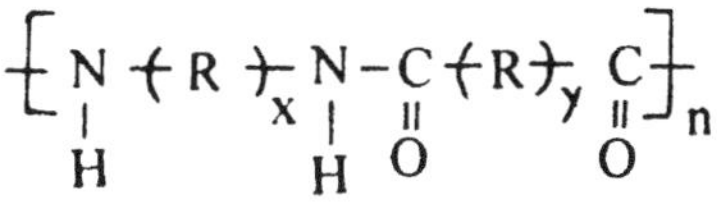

$$\left[\underset{\substack{|\\H}}{N} -(R)_x- \underset{\substack{|\\H}}{N} - \underset{\substack{\|\\O}}{C} -(R)_y- \underset{\substack{\|\\O}}{C} \right]_n$$

Where R =— CH_2 – OH –⟨benzene ring⟩

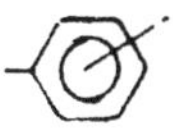

Polymide polymers contain amide group NH–CO in the main chain. The most important are linear aliphatic polyamides. These are known as nylons. Polyamides are also used as adhesives, coatings and in engineering applications.

The first synthetic polyamide was made by Carrothers at du Pont in 1935. Commercial production began in 1938. Commercial mouldings were made in 1941. It became popular only after 1950.

Raw Materials

Principal raw materials are adipic acid, sebacic acid, hexalmethylene diamine, caprolactam. 10-aminoundecanoic acid and dodecyl lactam.

Adipic Acid

It can be obtained on commercial scale from benzene. It is white solid having m.p. 152°C.

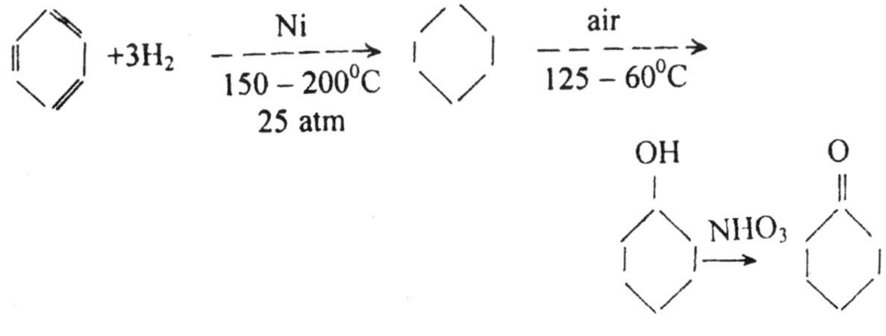

$$\text{Benzene} + 3H_2 \xrightarrow[\substack{150-200^{\circ}C\\25\text{ atm}}]{Ni} \text{Cyclohexane} \xrightarrow[125-60^{\circ}C]{air} \text{Cyclohexanol (OH)} \xrightarrow{NHO_3} \text{Cyclohexanone (=O)}$$

$$\xrightarrow[\substack{NHO_3 \\ NH_4\text{-vanadate} \\ + \text{ copper}}]{\substack{50\text{—}150°C \\ 4\text{—}18 \text{ atm}}} HOOC\text{–}CH_2\text{–}CH_2\text{–}CH_2\text{–}CH_2\text{–}COOH$$

Sebacic Acid

It can be prepared by heating castor oil with NaOH at 250°C. During the reaction ricinoleic acid is produced which forms octane-2-ol and sebacic acid:

$$\underset{\text{ricinoleic acid}}{CH_3(CH_2)_5\text{–}\overset{\overset{\large OH}{|}}{C}H\text{–}CH_2\text{–}CH{=}CH\text{–}(CH_2)_7\text{–}COOH}$$

$$\longrightarrow \underset{\text{Octanol}}{CH_3(CH_2)_3\text{–}CHOH\text{–}CH_3}\text{–}\underset{\text{Sebacic Acid}}{HOOC\text{–}(CH_2)_8\text{–}COOH}$$

Hexamethylene Diamine

Adipic acids gets converted to adiponitrile on treatment with NH_3 at 350°–450°C in presence of boron phosphate catalyst. Hydrogenation of adiponitrile can be done in presence of NH_3 to suppress cyclisation. Cobalt catalyst is used at 100-135°C and 600 atm pressure.

$$HOOC\text{–}(CH_2)_4\text{–}COOH \longrightarrow NC\text{–}(CH_2)_4\text{–}CN$$

$$\xrightarrow{H_2} H_2N\text{–}(CH_2)_6\text{—}NH_2$$

It can also be synthesised starting from butadiene as under:

$$CH_2{=}CH\,CH{=}CH_2 + Cl_2 \longrightarrow CH_2Cl\text{–}CH{=}CH\text{–}CH_2Cl$$

$$\xrightarrow{NaCN} CN\text{–}CH_2\text{–}CH\text{–}CH\text{–}CH_2\,CN$$

$$\xrightarrow{\text{Hydrogenation}} Pd/300°C$$

$$NC\text{–}(CH_2)_4\text{–}CN \xrightarrow{H_2} H_2N\text{–}(CH_2)_6\text{–}NH_2$$

Caprolactam

It can be obtained from cyclohexane. Cyclohexane is air oxidised to yield a mixture of cyclohexanol and cyclohexanone. Cyclohexanol is dehydrogenated to cyclohexanone over copper catalyst. Cyclohexanone when treated with hydroxylamine sulphate at 20–95°C gives an oxime. The oxime when treated with concentrated sulphuric acid undergoes Beckmann rearrangement to yield caprolactam.

In another method a photochemical reaction between cyclohexane and nitrosyl chloride directly yields cyclohexane oxime hydrochloride. Caprolactam is made by Gujarat State Fertilizer Corporation.

Cyclohexane $\xrightarrow{O_2}$ Cyclohexanol (OH) + Cyclohexanone (=O)

Cyclohexanol (OH) $\xrightarrow{NHO_3}$ Cyclohexanone (=O) $\longrightarrow$ Cyclohexanone oxime (=NOH) $\xrightarrow{H^+}$

Cyclohexane + NOCl $\xrightarrow[-10 \text{ TO } 16^{\circ}C]{h\nu}$ Cyclohexanone oxime (=NOH) $\rightarrow$

$$\begin{array}{c} CH_2 - - \rightarrow CH_2 \\ CH_2 \qquad\qquad CO \\ \qquad\qquad\quad NH \\ CH_2 - CH_2 \end{array}$$

Caprolactin

Dodecyl Lactam

Butadiene when trimerised over Ziegler-Natta catalyst to yields 1,5,9-cyclododecatriene. Hydrogenation form cyclododecane which yields dodecyl lactam.

$$3CH_2{=}CH - CH{=}CH_2 \longrightarrow \begin{array}{c} CH{=}CH - CH_2 - CH_2 \\ H_2C \qquad\qquad\qquad CH \\ | \qquad\qquad\qquad\quad \| \\ H_2C \qquad\qquad\qquad CH \\ CH{=}CH - CH_2 - CH_2 \end{array} \xrightarrow{H_2}$$

1,5,9 cyclododecathiene

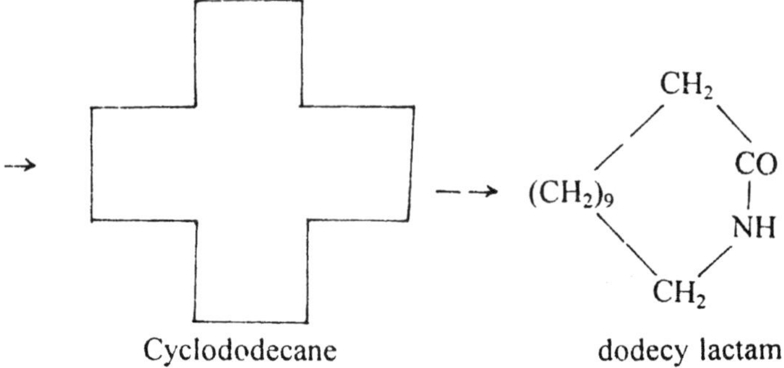

Cyclododecane dodecy lactam

ω-Aminodesanoic Acid

It can be obtained from castor oil. Methanolysis of castor oil yields methylester of ricinoleic acid. Pyrolysis at 500°C forms *n*-heptaldehyde and methyl-undecylenate. Hydrolysis of latter yields undecylenic acid. Reaction with HBr gives ω-bromoundecanoic acid, which on ammonolysis forms ω-amino undecanoic acid which as a white solid having m.p. 189°C.

$$CH_3\text{–}(CH_2)_5\text{–}\underset{}{\overset{OH}{\overset{|}{C}}}H\text{–}CH_2\text{–}CH{=}CH\text{–}(CH_2)_7\text{–}COOCH_3$$

$$\xrightarrow{H\Delta} CH_3-(CH_3)_5\text{–}CHO+CH_2{=}CH\text{–}(CH_2)_8\text{–}COOCH_3$$

$$\xrightarrow[\text{(2) BHr}]{\text{(1)Hydrolysis}} BrCH_2\text{–}(CH_2)_9\text{–}COOH \longrightarrow$$

$H_2N\text{–}(CH_2)_{10}\text{–}COOH$ ω-aminoundecanoic acid

Preparation

The polyamides can be made by three methods:

(i) Reaction of a diamine with dicarboxylic acid *e.g.*, hexamethylene diamine + adipic acid

$$nH_2N\text{–}(CH_2)_6\text{–}NH_2+nHOOC\text{–}(CH_2)_4\text{–}COOH$$

$$\longrightarrow [\text{–}HN\text{–}(CH_2)_6\text{–}NHCO\text{–}(CH_2)_4\text{–}CO\text{–}]_n+2nH_2O]$$

Nylon 6, 6

The product obtained is known as nylon 6, 6 as there are six carbon atoms in diamine and 6 carbon atoms in diacid. Nylon 6, 10 is obtained by a similar method from hexamethylene diamine and sebacic acid:

$$nN_2N\text{–}(CH_2)_6\text{–}NH_2 + nHOOC\text{–}(CH_2)_8\text{–}COOH \dashrightarrow$$
$$[\text{–}HN\text{–}(CH_2)_6\text{–}NH\text{–}CO\text{–}(CH_2)_8\text{–}CO\text{–}]_n + 2nH_2O$$

Nylon 6, 10

(ii) Self-condensation of an ω-aminoacid *e.g.* ω-amino-undecanoic acid. The product is known as nylon 11 as there are eleven carbon atoms in the acid:

$$nH_2N\text{–}(CH_2)_{10}\text{–}COOH \dashrightarrow [\text{–}HN\text{–}(CH_2)_{10}\text{–}CO\text{–}]_n + nH_2O$$

(iii) Ring-opening polymerisation yields a polyamide e.g. caprolactam and dodecyl lactam:

```
              CH2
            /     \
      CH2             CH2
  n   |               |          ——→   [(CH2)5 —CO — NH —]n
      CH2             CO                    Nylon 6
        \            /
          CH3    NH

           CO
          / |            ——→  [(CH2)11  — CO — NH — ]n
n [(CH2)11  |                      Nylon 12
          \ |
           NH
```

The commercially importan' nylons for fibre are nylon 6, 6, and nylon 6.

Experimental Details

For the first method exact e ui 'alence of reactants are used to obtain high molecular weight polymer. In the first step an aqueous solution of adipic acid is neutralised with hexamethylene diamine to form nylon salt.

$$H_2N\text{–}(CH_2)_6\text{–}\overset{+}{N}H_3\,\overset{-}{O}OC\text{–}(CH_2)_4\text{–}COOH \text{ Nylon salt.}$$

Nylon salt can also be obtained by boiling methanol. 65.70% nylon salt solution in water along with a small quantity of acetic acid is heated in an autoclave at 220°C when 20 atm pressure developes. After 1-2 hours the temperature is raised to 270-80° and steam gradually permitted to escape at 20 atm pressure and then pressure is gradually lowered to atmospheric pressure by permitting steam to escape slowly.

Nitrogen is pumped in and molten polymer is extruded and cooled on a drum to form ribbons. Nylon 6, 10 is formed in similar way at a lower temperature. Nylon 6, 6 can also be prepared in a continuous process in three separate tubes wherein reaction is started, steam removed and polymerisation completed. The product is directly melt spun into fibres.

Nylon 11 is prepared by a continuous process wherein ω-aminoundecanoic acid is heated 200-220°C and steam formed is continuously removed. Later on pressure is reduced to produce high molecular weight polymer.

Nylon 6 can be prepared both by batch as also by continuous process. In **batch process** caprolactam, water (5 to 10%) and acetic acid (0.1%) are fed to a reactor under nitrogen and heated to 250°C at 15 atm for 12 hours. Steam formed is vented. The product contains about 10% low molecular weight products which are removed either by leaching with water at 85°C or by heating at 180°C at 0.005 atm. pressure.

In **continuous process a** mix ure of molten Caprolactam, water and acetic acid is pumped continuously at 260° through a reactor. Steam is removed at intervals. The residuence time in reactor is 18-20 hours. The product is directly spun into fibres:

$$(CH_2)_5 \begin{matrix} CO \\ | \\ NH \end{matrix} \; H_2O \longrightarrow H_2N{-}(CH_2)_5{-}COOH$$

$$\rightleftharpoons H_2N{-}(CH_2)_5{-}CO{-}NH{-}(CH_2)_5{-}COOH$$

Caprolactam can also be prepared by **bulk polymerisation** process using anionic catalysts like strong bases and metal hybrides.

Nylon 12 is prepared by heating dodecyl lactam at 300° in presence of phoshoric acid to obtain a high molecular weight polymer.

Nylon copolymers can be prepared by either heating the blend of different nylons together or by polymerising mixed monomers.

Properties

The polyamides have high impact strength, toughness, flexibility and abrasion resistance. Their mechanical properties depend on degree of crystallization, temperature and humidity. Since amide group, leads to intermolecular hydrogen bonding so they have higher melting points and tensile strength.

With increase in length of aliphatic chain there occurs a decrease in melting point and ease of processing. Copolymers have reduced ability to crystallize and have lower melting points and tensile strengths. Nylons are soluble in a few solvents such as acetic acid, Phenol etc. at room temperature. They swell in alcohol. They are resistant to oils and fuels. Concentrated mineral acids attack nylon rapidly. Nylon get affected by alkalis and oxidising agents.

Uses

Nylon is used in gears, cams, brushes and other parts. Nylon moving parts do not need lubrication. Nylon can be sterilised with steam so it is sued in hypodermic syringes and surgical accessories.

Curtain runners, sinks, zips, combs and switches are manufactured from nylon. Extruded nylon is used in covering wire ropes, in packaging film for pharmaceuticals, bottles, tubing etc. Nylon laminates are used for heavy duty driving belts. Monofilaments are used in brushes, sports equipment, surgical sutures etc. Monofilaments are prepared from nylon 6, 10 or nylon 11.

Methylmethoxy Nylons

Preparation

Methylmethoxy nylons are prepared by treating nylon with methanol and formaldehyde in presence of phosphoric acid:

$$\sim\sim CO-NH\sim\sim + CH_2O + CH_3OH \;\; \sim\sim CO-\underset{}{\overset{\displaystyle (CH_2O)_n\,CH_3}{\overset{|}{N}}}\sim\sim$$

Properties and Uses

Such polymers are soluble in a alcohols. They are commercially available and used for producing films, coatings and adhesives resistant to chemicals and abrasion. Crosslinking can be done with weak organic acids such as citric acid to made them strong and insoluble.

Classified nylon is now available which can be vacuum

metallised and chromium plated. It is used in electric drill housing telephone relay bobbins, car door handles etc.

9.18. Environmental Pollution

The synthetic polymers are used for making so many items and it is impossible to imagine modern civilisation without them. However the main concern is that once they are made, they do not decay and tend to remain for all times. In other words, they cannot be disposed off unlike other waste products, which are degradable. This problem has been solved to a certain extent by reuse or recycling of some of the polymers.

The normal method of disposing of the waste products is by burning them but this can not be used for polymers, since they evolve poisonous gases like hydrogen cyanide (from polyacrylonitrile). The increasing use of synthetic polymers for almost all purposes requires to be looked into. Attempts are being made to have biodegradable polymers in the near future.

10

Synthetic Fibres

10.1. Introduction

Fibres

Fibres are linear polymers having relatively high tensile strength but low elasticity in the direction of the fibre axis. For a material to be classified as a fibre, its length must be at least 100 times its diameter. To be useful as a textile fibre, a polymer must have *adequate tensile strength* over a wide temperature range, a *high softening point* to permit ironing, *solubility* to permit spinning and *good textile properties* like dye-ability, stability, crease resistance, comfort, etc.

In the fibres, generally the linear polymer chains are arranged in orderly three-dimensional structure having strong interchain forces.

Fibres played an important role in the development of polymer chemistry. H. Staudinger's work on cellulose and rubber led to the realization of the fact that these are long-chain molecules. The only fibres for making textiles were naturally occurring proteins like wool, silk, and cellulosic polymers like cotton till synthetic fibre forming polyamides and polyesters were developed. Fibres have been divided conventionally into three classes (i) natural; (ii) regenerated and (iii) synthetic.

10.2. Natural Fibres

These are either of animal origin such as wool and silk, or of vegetable origin such as cotton, jute, etc. Since wool fibres are not homogeneous, so structure determination becomes difficult. However, the mechanical properties of wool are due to the fact that individual chains are held by H-bonding and covalently bound disulphide links.

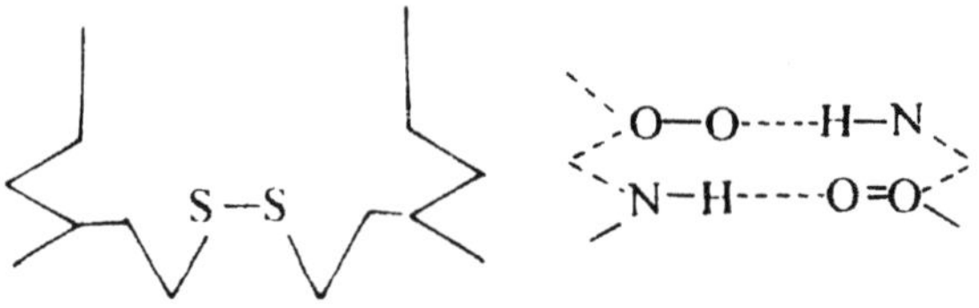

S–S

O—O----H—N
N–H----O=O

Because of the cross-linking, wool is insoluble in many solvents which do not disturb or affect these linkages However, by suitable chemical treatment, it is possible to improve its crease resistance, ability to be dyed, etc.

Silk

Silks are these fibrous proteins which are produced by spiders, moths and butterflies. The cocoon spun by the larva of the moth *Bombyx mori* to protect the pupa, is the main source of commercial silk. The silkworm is thought to produce a single continuous filament about one kilometre long. This is unwound from the cocoon and spun to form a yarn or twisted bundle of filaments. The smoothness of the fibre is due to the long filaments from which it is made up.

Cotton

These are vegetable fibres which are either derived from the stem (jute) or seed hairs (cotton) of plants. Cotton is a pure form of cellulose, containing thousands of β-glucose units as monomers. Cotton readily absorbs water and is chemically inert towards aqueous alkali, if the alkali concentration is <5M. Concentrated alkalis *mercerise* the cotton and improve its properties such as strength, lustre and dyeability. Cross-linking reaction of cotton with ureas, derivatives like *bis*-(N-hydroxmethyl) ureas, i.e. $HOCH_2NHCONHCH_2OH$ to form compounds of the type $ROCH_2NHCONHCH_2OR$, where ROH stands for cellulose, are mainly used to provide crease retention properties of cotton.

Some important varieties of cotton fabric are calico, cambric, canton flannel, chiffon, chintz, crepe, denim, drill, duck, gabardine, Jeans, madras, muslin, oilcloth, poplin, print cloth, sateen, tapestry, terry cloth, twill, typewriter ribbon cloth, voile, etc. These can be distinguished by some of the properties associated with each of these.

Calico

A low-count or medium-count cotton print cloth with special designs called calico designs. Many early American designs are in use.

Cambric

Originally linen, now cotton print cloth or lightweight sheeting construction. Given special sizing and calendered finish.

Canton Flannel

Warp-faced twill flannel with a heavy, soft filling yarn and lighter warp than filling yarn. Has a long nap on one side only. Comes unbleached or bleached. For men's work gloves, infants' wear, linings.

Chiffon

A sheer voile with a dull finish; of fine, hard-twisted yarns. Often dyed or printed.

Chintz

A print cloth or high-count fine sheeting with bright attractive floral or geometric designs, both large and small. Often given a permanent or semi-permanent glaze; then known as glazed chintz. For draperies, slip covers, frocks.

Crepe

Has a pebbly or crinkled surface produced by use of special crepe yarns. Can be either crepe, granite, or plain weave. Generally, mixed-twist crepe yarns used in both warp and filling, occasionally crepe yarns used only in the warp or the filling. Mostly woven on a box loom.

Denim

A durable fabric of the twill family. Usually of single hard-twisted yarns with coloured warp and natural or white filling.

Drill

A durable fabric of medium weight. Usually, three-harness warp-faced twills made of carded sheeting yarns. Comes in various weights and thread counts ranging from 60×36 to 80×48. When dyed, known as khaki, tickings, silesia, herringbones. One thread goes over two filling yarns, then under one.

Duck

The name ''duck'' covers a wide range of fabrics. It is one of the most durable fabrics made. A closely woven, heavy material. The most important fabrics in this group are known as *number duck, army duck,* and *flat* or *ounce duck.* Number and army ducks are always of plain weave with medium or heavy ply yarns; army ducks are the lighter. Ounce ducks always have single warp yarns woven in pairs and single or ply filling yarns. Other names for variations of these fabrics are sail duck, belt duck, hose duck, tire duck (such as breaker, cord, chafer), wide and narrow duck, biscuit duck, harvester duck, oil press duck, wagon duck, enameling duck, boot duck, canvas, and so on. Generally of ply yarns in warp and yarns of various sizes and weights in filling. Thread counts range from 54×40 to 72×40.

Gabardine

A distinctive steep-diagonal warp-faced twill of carded or combed yarns. Twill is to the left if made with all single yarns, and to the right when ply warp and single filling are used. Thread counts range from 110×76 to 130×80.

Jeans

Three-harness warp-faced twills of lightweight sheeting yarns,. One warp thread goes over two or more filling threads, than under, moving one pick versions.

Madras

Usually woven in stripes, cords, dobby, and Jacquard. Mostly all white, but in a number of cases done with a patterned warp and while filling, such as white and colour warp, or colour-and-colour warp and white filling, or vice versa. Often a stripe is woven in solid colours. A number of variations of the above.

Muslin

Coarse type of cotton fabric made of carded yarns; of various thread counts up to 140. Gray print cloths and lightweight sheetings known as unbleached muslins.

Oilcloth

Sheetings or print cloths that are printed, bleached or dyed and given a special linseed oil and pigment preparation.

Poplin

Has a heavier rib effect than broadcloth. Heavier filling than warp yarns, and more threads in the warp than filling. Prial cloth yarns mostly. Combed yarns in the better cloths. Many mixtures made. Poplin comes in many variations, such as slubs, nubs, and yarn-dyed checks and stripes. Mainly dyed solid colours or printed. Often given-special finishes, such as water-repellent, fire-retardant, mildew-proof.

Print Cloth

Carded cloth made with same yarns as cheesecloth, but with more warp and filling threads to the inch. Most print clothes are made in narrow widths up to 40 inches. Given a range of finishes, thus producing cambric, muslin, lawn, loincloth, and printed percales.

Sateen

A satin weave. Usually woven so that the surface is smooth and the finish lustrous, resembling satin. Can be either a strong warp-faced sateen or a softer filling-faced sateen. Often, though not always, filling sateens have a soften finish than warp sateens. Either carded or combed yarns.

Tapestry

A closely woven yarn-dyed figured material with two or more sets of warp and filling, often plied. Woven on a jacquard loom comes in both fine and coarse versions. The reverse side is smooth.

Terry Cloth

A fabric having uncut loops on both sides Woven on a dobby loom with terry arrangement. Various sizes and numbers of yarns used in picks and ends, forming different versions of terry cloth. Can be woven on Jacquard loom to form designs. Can be yarn-dyed in different colours to form patterns. Bleached, piece-dyed, even printed.

Twill

Cotton fabric in twill weave. Right-hand or left-hand twills. Known as drills, jeans, gabardines, denims, serge, canton flannel, cottonade, tickings and others, depending on finish and thread count.

Typewriter Ribbon Cloth

Fine fabric of plain weave. Often balloon cloth. Made with finest regular yarns, then cut into strips and specially finished.

10.3. Regenerated Fibres

These are the **Synthetic fibres derived from natural fibres (e.g. cellulose or protein base).** Examples of these are cellulose nitrate, cellulose acetate, etc.

Resins

Cotton yarns are obtained from short cotton fibres, and the filaments in the yarn are imperfectly aligned. These protruding ends are responsible for the rough feel of the cottons. For obtaining a single continuous filament like silk, the artificial silk industry was started. To produce a continuous filament, it is essential that the starting materials be in a liquid state, so that it can be forced through a die or *spinneret* (a die containing a large number of holes <0.1 mm dia.) Cellulose is dissolved and regenerated as a filament. Either cotton or wood pulp can be used. The regenerated cellulose is known as *rayon*. Viscose rayon is produced by converting cellulose into a soluble xanthate

$$ROH + S{=}C{=}S + NaOH = R{-}O{-}\underset{\underset{\displaystyle S}{\|}}{C}{-}S{-}Na$$

and subsequently regenerating cellulose by the forcing the viscous solution through *spinnerets* in an acid bath. The filaments are washed, dried and wound. Coprammonium or Bemberg rayon is produced by dissolving cellulose in tetrammine copper (II) hydroxide (*Schweizer reagent*), $[Cu(NH_2)_4](OH)_2$, and regenerating the cellulose by spinning into an acid bath. Natural cotton fibre is made up of countless tiny *fibrils* bound together, but rayon fibre consists of a solid filament. It is possible to have rayon spun in the form of a fibrillar structure. However, unlike rayons, these fibrillar rayons known as **polynosic fibres** has high wet strength, Rayon is used as a textile where high tensile strength is needed, but is unsuitable for use as ropes or fishing nets. Viscous rayon can also be extruded as thin sheets to give transparent plastic films known as *cellophane*, mostly used as packaging material.

Cellulose Acetates

Each monomer unit in cellulose contains three —OH groups that can be substituted. The triacetate formed by acetylating cellulose, using acetic anhydride, is hydrophobic and is difficult to dye. The diacetate, formed by the partial hydrolysis of the triacetate, is not so hydrophilic and can be dyed. These acetates are largely used as dress materials.

Synthetic Fibres Derived from Cellulose

Previously, the term '*rayon*' was used to describe the synthetic fibre obtained from a solution of cellulose. However, this term has now been limited only to regenerated cellulose type fibres.

Cellulose is converted into synthetic fibres either in the form of *regenerated cellulose* or *cellulose derivatives*. The characteristic properties of the regenerated cellulose type of fibres are similar to that of the original cotton while the characteristic properties of the cellulose derivatives are those of the plastic from which they are made.

The synthetic fibres are obtained from cellulose by the following four methods:

1. Chardonnet (Nitrocellulose) Process

In this process, cellulose is first of all nitrated with nitric acid and sulphuric acid under such a condition that only cellulose dinitrate is obtained. The cellulose dinitrate is commonly known as *pyroxylin*. Pyroxylin is now dissolved in a mixture of alcohol and ether to get *collodion*. This is forced through *spinnert* into a warm air chamber when the solvent evaporates leaving being threads of cellulose dinitrate. But these fibres are highly inflammable. Therefore, they are denitrated by treatment with sodium or ammonium hydrosulphides to give the regenerated cellulose.

$$\left[C_6H_7O_2(OH)_3\right]_n \xrightarrow[H_2SO_4]{HNO_3} \left[C_6H_7O_2\,OH(ONO_2)_2\right]_n$$

cellulose dinitrate (pyroxylin)

$$\left[C_6H_7O_2OH(ONO_2)_2\right]_n + 3nNH_4HS \rightarrow$$

$$\left[C_6H_7O_2OH(OH)_2\right]_n + 3NH_4NO_2 + nS$$

This process is costly.

2. The Cuprammonium Process

The cuprammonium process of making rayon was first developed by L. H. Despaisses in France in 1890. However, the process was abandoned a few years later as it could not compete commercially with the more economical viscose process. It was subsequently successfully revived in Germany in 1919 by the J.P. Bemberg, A.G. company as a result of the invention and application by Dr. E. Thiele and Dr. E. Elsaesser of a *stretch spinning* technique, that produced very fin denier fibres with a soft hand and high flexibility. In 1926 the American Bemberg Corporation began production of cuprammonium rayon in Elizabethton, Tennessee.

The cuprammonium process makes use of purified cellulose obtained from cotton linters or wood pulp. The pulp is mixed with copper sulphate and ammonia at low temperatures. The name *cuprammonium* is derived from these two substances. Caustic soda and stabilizing agents are added to maintain the solution's equilibrium and control its cellulose content. The solution is filtered for removal of impurities and cleared, before extrusion through the *spinneret* for further treatment, to form pure regenerated cellulose filaments and yarn.

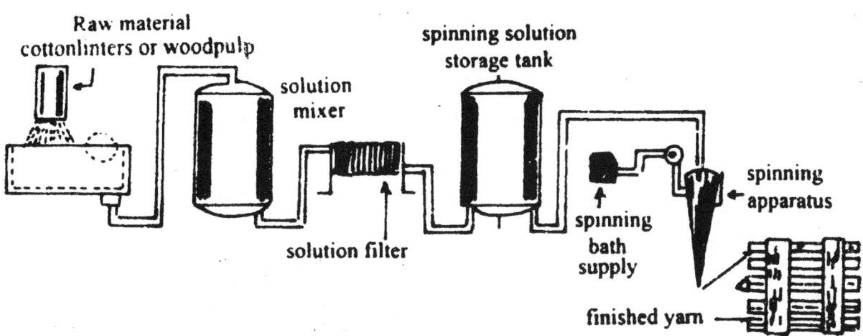

Fig 10.1: Flow chart of the manufacturing process of cuprammonium rayon yarn.

Upon extrusion from the spinneret, the cuprammonium solution is processed into multifilament yarns.

Processing

The two methods of spinning cuprammonium rayon are:
(i) Conventional reel spinning and (ii) continuous spinning.

(i) Conventional Reel Spinning

As the solution is pumped through large-sized holes of the spinneret, it passes into a glass funnel where most of the ammonia and about one-third of the copper sulphate is removed, that causes the solution to coagulate into filaments. When it passes through the funnel, the filaments are uniformly stretched to finer diameters. When the filaments leave the funnel, they enter a bath of mild sulphuric acid that hardens them as its reaction removes the remaining ammonia and copper salt, leaving filaments of pure cellulose. The yarn is then wound on reels and subsequently washed, rinsed, lubricated and dried.

The stretch-spinning feature of the cuprammonium process gives fine yarn that can be given a great amount of twist. For its weight, it has greater elasticity than viscous rayon filaments. The fine threads can be woven into the sheerest of fabrics. The high twist contributes to drapability.

(ii) Continuous Spinning

The technique for continuous spinning was introduced in 1944 with the development of continuous machine by Dr. Hugo Hofmann.

The cuprammonium solution is pumped through 120 spinning funnels that from two banks of 60 parallel filaments. These filaments are passed through a solution of hot sulphuric acid in a pretreatment pan in which they contract to approximately one-third their original diameter and get ultimately converted into cellulosic fibre. The banks of filaments are passed from the pretreatment pan through an acid trough for further purification and then into water to wash off the acid. The filaments are then lubricated or sized, dried, oiled wound on spools or skeins, and conditioned in controlled humidity for a few days. This method is more economical as it requires no handling of the filaments from the spinneret through the winding machine.

Cuprammonium rayon is produced in multifilament form only. These yarns may be twisted or untwisted because, unlike viscose rayon, cuprammonium filaments are not separated; they adhere to each other except when subjected to relatively high friction. Cuprammonium yarn generally ranges from 40 to 900 denier having from 30 to over 700 filaments each. The most popular deniers for fabrics are 50 to 100.

Variations in certain characteristics of cuprammonium rayon can be done. Dyes may be added to the solution to produce *solution-dyed* yarns. A cuprammonium rayon yarn that is solution-dyed at varied intervals along its length has been produced under the trade name of *Parfe*. Techniques have been developed to produce novelty filament yarns having smooth, slub, and nubby effects. The finer deniers are suitable for a wide range of apparel, such as dresses, blouses and sportwear. The heavier deniers are used for interesting textural effects for upholstery and related fabrics.

Trade Names of Cuprammonium Rayon

Cuprammonium rayon is marketed under any one of several trade names listed in Table 10.1. The fibres are all chemically the same–that is they, are all made from cellulose and have the same chemical properties. Variations in certain physical properties, like solution-dyed effects and slub effects, can given the fabrics a unique appearance. These variation can be identified by the respective trademarks.

Table 10.1: Cuprammonium Rayon Trade Names

Trade Name	*Type*	*Special Characteristics*
Bemberg, Regular	Multifilament	Reel spun
Bemberg, 44 HH	Multifilament	Spool spun
Cupioni	Multifilament	Entangled slub
Cuprel	Multifilament	Texturized
Cupracolor	Multifilament	Solution-dyed
Dream Slub	Multifilament	Large slub
Flaikona	Multifilament	Flak slub
Multi-Cupioni	Multifilament	Short-entangled slub
Nublite	Multifilament	Short-entangled slub
Ondelette	Multifilament	Flake slub
Parfe	Multifilament	Space-dyed
Strate	Multifilament	Thick and thin
Stratella	Multifilament	Thick and thin

Because of several limitations viscose and cuprammonium rayon could not be used as substitutes for cotton. The most significant limitation is their wet strength: both of these rayons lose up to 70 per cent of their strength while wet. The **tensile strength** of a fibre is known by its *tensile modulus i.e.* The relative amount of pulling force that it can endure before breaking.

In 1960, commercial production was initiated for a rayon that was nearly same as the natural strength of cotton and retained most of that strength, even when wet. It was identified as a *high wet-modulus rayon*. Later on it also became known as a *modified rayon*.

The three general varieties of this type of rayon in commercial production are *Zantrel, Lirelle,* and *Avril.* While all these are cellulosic and have basic characteristics similar to high-grade cotton, they differ from each other in certain respects Table 10.2. Commercial production of a fourth variety, *Nupron,* has recently begun; *Nupron* and *Xena* were also introduced in 1966.

Table 10.2: Comparison of Properties–The High Wet-Modulus Rayon Fibre

	Zantrel	*Lirelle*	*Avril*
Strength	Fair; similar to Upland cotton	Very good similar to Pima cotton	Good; approxi-mates Pima cotton
Abrasion resistance	Fair	Good	Fair
Elasticity	Little	Little	Little
Absorbency	Good	Very good	Superior
Ironing temperature	At cotton setting	At rayon setting	At rayon setting
Shrinkage	Shrinks unless preshrunk	Shrinks unless preshrunk	Shrinks unless preshrunk
Reaction to alkalies	Good resistance	Good resistance	Good resistance
Reaction to acids	Poor resistance	Poor resistance	Poor resistance
Resistance to perspiration	Fair	Fair	Fair
Colourfastness	Good	Good	Good

3. Viscose Process

It is very useful method for producing regenerated type of rayon. It this process, cellulose is socked in 15-20% caustic soda solution for about 3 hours when cellulose is converted into *soda* or *alkali cellulose.*

This is then treated with carbon disulphide to form the alkali soluble cellulose xanthate. This is allowed to stand for sometime during which there occurs the depolymerisation of the desired amount and a viscous yellow liquid called *viscose* is obtained.

Viscose solution is carefully filtered and forced through a spinneret into a bath of dilute sulphuric acid which hardens the gum-like threads into rayon fibres (Fig. 10.2.)

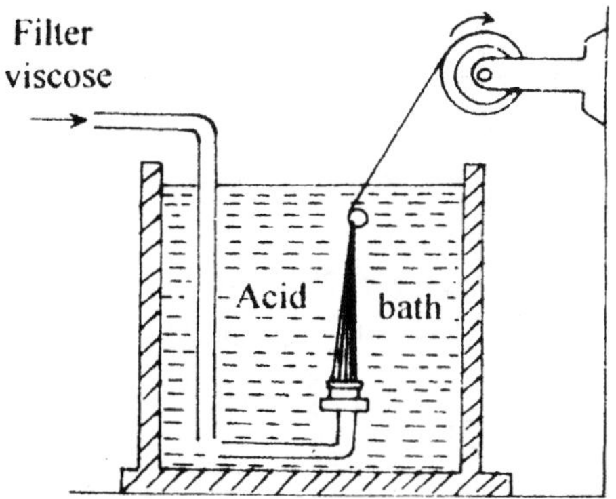

Fig. 10.2: Spinning of viscose rayon yarn.

It is the cheapest process and commonly used. This process is used these days for producing a large amount of rayon.

The various chemical reactions taking place in the viscose process are as follows:

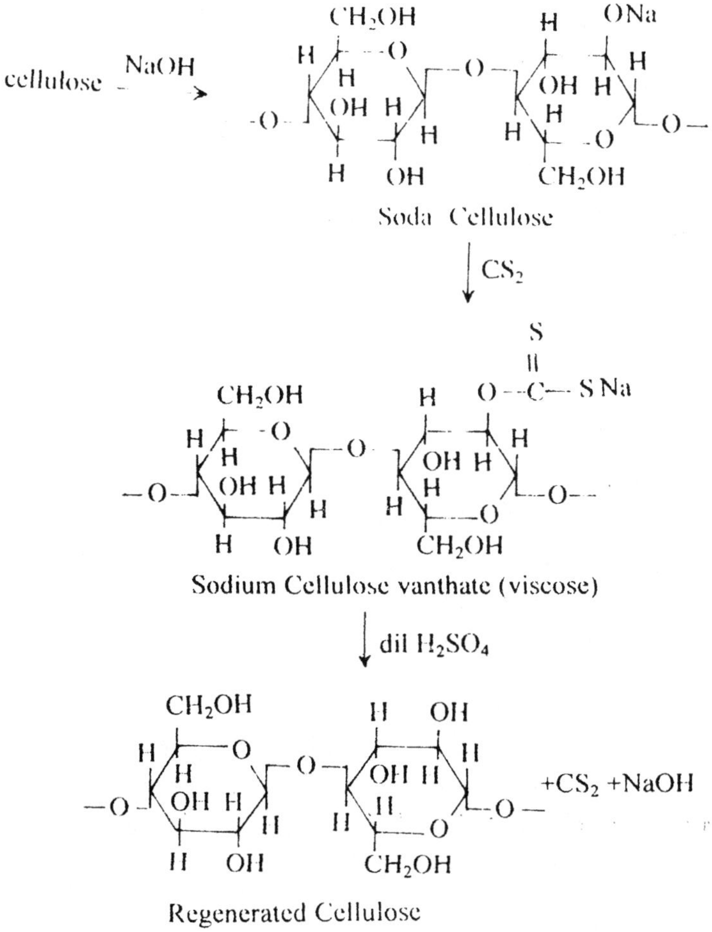

cellulose $\xrightarrow{NaOH}$

Soda Cellulose

$\downarrow CS_2$

Sodium Cellulose vanthate (viscose)

$\downarrow$ dil H_2SO_4

$+CS_2$ $+NaOH$

Regenerated Cellulose

When the viscose solution is allowed to pass through a narrow slit instead of a spinneret, a thin transparent sheet called cellophane is obtained. The cellophane is used as protecting and wrapping material.

The viscose process is much cheaper than the Chardonnet process and cuprammonium process because the former (viscose) process employs the cheap wood cellulose whereas the latter two methods require a costly pure starting material generally cotton linters.

4. Acetate Rayon Process

This process involves the acetylation of cellulose with a mixture of acetic anhydride and glacial acetic acid in the presence of a small amount of sulphuric acid as a catalyst when cellulose triacetate is obtained. This is decomposed with water to get the diacetate which is first precipitated in water, then filtered, and dried and finally dissolved in acetone mixed with other organic solvents. Now the solution is forced to flow through a spinneret Fig. 10.3 into a warm air stream when the solvent evaporates leaving behind threads of cellulose diacetate. The product is termed *celanese yarna*.

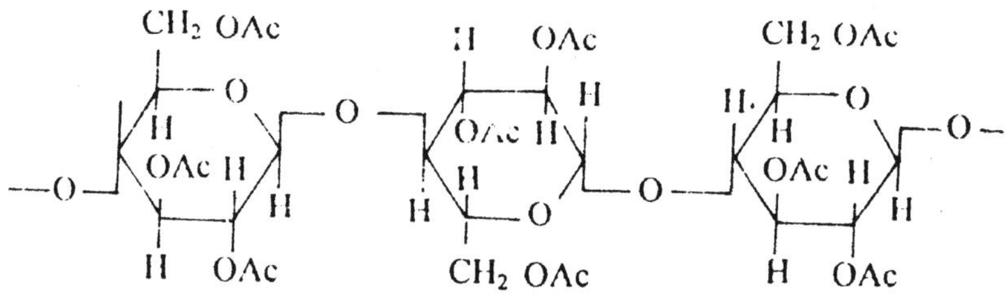

Cellulose triacetate

The advantages of cellulose diacetate are as follows:

(i) It burns with difficulty.

(ii) It is more resistant to moisture than cotton and regenerated-cellulose rayons.

(iii) It is less susceptible to wrinking.

(iv) Regenerated cellulose rayon is insoluble in organic solvents and it can, therefore be dyed with such dyes that could not be applied on the cellulose type fibre.

Some synthetic fibres are also obtained from simple esters (e.g., propionates, butyrates, stearates, etc.) and some mixed esters of cellulose.

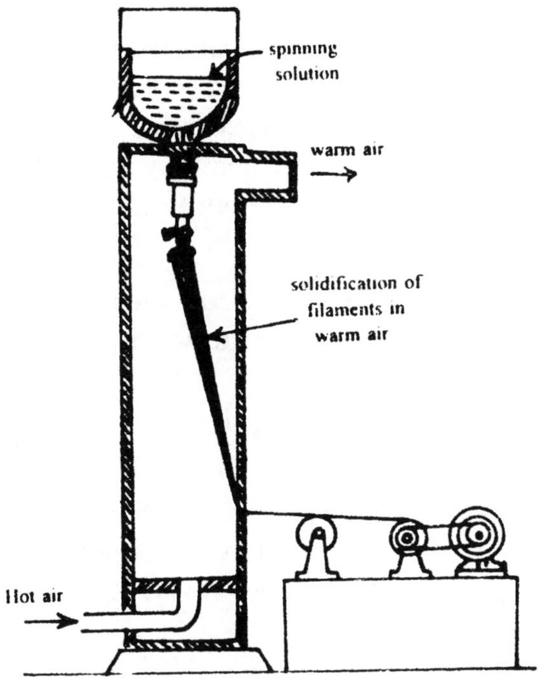

Fig. 10.3: Spinning of cellulose acetate (celanese yarn.

The flow chart of the manufacturing process of acetate is shown in Fig. 10.4.

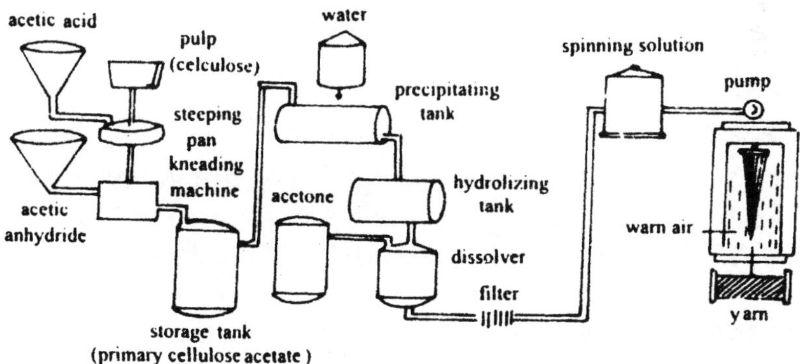

Fig.10.4 : Flow chart of the manufacturing process of acetate.

Synthetic Fibres Derived from Proteins (Azlons)

Synthetic fibres are produced from natural proteins as casein, zein, peanut and soyabean.

When casein obtained form milk is farced through a die, it is extruded as a filament (*lanital*). The filament so obtained is hardened by treating it with formaldehyde which on reacting with the free amino groups of the proteins bring about crosslinking of the polymer chains.

Similarly, maize (zein) and peanut proteins are also extruded as filaments called *vicara* and *ardil* respectively.

As protein fibres resemble wool, they are blended with wool because they themselves have a low tensile strength when wetted.

10.4. Synthetic Fibres

Synthetic fibres derived are generally prepared from compounds like ethylene, acetylene, cyclohexane, butadiene, vinyl derivatives, etc.

Introduction

It was established by X-ary diffraction studies, that natural fibres like cotton and wool are long-chain molecules. Wallace Carothers, while investigating the synthesis of such molecules, chose polyamides and polyesters. Among the many fibre-forming polymers; polyamides, polyesters and polyolefins have achieved great commercial and technological success.

Synthetic Fibres from Simple Molecules

These fibres are produced from simple compounds which are mainly derived from petrochemical industry. These fibres are cheap.

We shall now describe the various synthetic fibres from simple molecules.

1. Nylons

Introduction

The word "*nylon*" is now accepted as a generic term for the **synthetic polyamides** which are characterised by a repeating amide linkage–NHCO– Nylon is not a trade mark–it is textile just like cotton, linen, wool, silk, rayon etc.

Nylon was the first direct product of the technological breakthrough achieved by W.H. Carothers and his co-workers in 1928. Initially they produced polymer with molecular weights ranging from about 2500–5000 but later work produced "superpolymers" having molecular weights as high as 25,000.

History of Nylon

Credit for the discovery or invention of nylon goes to E.I. du Pont de Nemours and Company. Specially, to Dr. Wallace H. Carothers *et al.* Dr. Carothers was interested in obtaining a better understanding of polymerization and wanted to know how and why certain molecules join to from "giant" molecules, like those that occur in cotton, silk, and rubber. After long research work Dr. Carother discovered that one polymer, which looked like clear, heavy molasses when molten, could be drawn out into a long fibre. On cooling, the fibre could be drawn out farther to several times its original length. This stand was strong, lustrous, and silky.

Carother *et al* were working with a type of compound called *Polyesters*. Many were synthesized. But each was deficient in one or more vital textile properties. To get a better fibre, they mixed polyesters and polyamides, which are related to polyesters but were not successful and therefore decided to discontinue the mixtures and concentrate on the polyamides. One of the chemists filled a container, fitted with the tip of a hypodermic needle, with a hot viscous polyamide solution. He squired a stream of it into the air, and the stream cooled into a fine filament After the lustrous fibre was drawn and tested, it was found to be strong and pliable.

For the next several years, researches continued with a series of polyamides. In February, 1935, one of these was made from hexamethylene diamine and adipic acid. This polymer was called "6, 6" because each of these component chemical compounds contains six carbon atoms per molecule. This "6, 6" polymer proved to be best at that time.

The name *nylon* was given to this material. Since nylon is protein like in composition, the fibre was said to be similar in certain of its properties to silk, wool, and hair. Moreover, this new fibre had many wonderful properties of its own Nylon began to be used for a wide variety of garments.

Other experimentations with polyamides eventually led to the creation of other forms of nylon. In 1939. Schlak and Kleine of the I.G. Farben combine produced a German counterpart, nylon 6, from a polyamide called *caprolactum*, which contains 6 carbon atoms. Since then, other forms have been developed: examples are nylon 7 produced in the U.S.S.R, nylon 11 produced in Italy, and nylon 4 and nylon 6, 10 produced in the United States. These nylons are all polyamides created from various raw materials in different ways, as found most economical by the producer. Chemically, they are similar but differ in molecular structure, which sometimes results in subtle differences in properties. Today, the major production is nylon 6,6 with less production of nylon 6, used primarily for carpeting and industrial fabrics.

Nomenclature

The nylons have been named on the basis of the number of carbon atoms in the monomer chain. Polymers based on amino acids are designated by a single number such as nylon 6 for polycaprolactam. Polymers based on diamines and dibasic acids are designated by two numbers, the first giving the number of carbon atoms in the diamine monomer and the second the number of carbon atoms in the acid monomer. Thus, nylon 66 is made from hexamethylenediamine and adipic acid. Various commercially important nylons are 6, 66, 610, and 11. Among which nylon 66 is the most important.

(i) Nylon 6

It is prepared from the single monomer called caprolactam which is prepared from ε-aminocaproic acid in the following manner:

$$H_2N{-}(CH_2)_5{-}COOH \rightarrow -H_2O + \begin{array}{l} CH_2{-}CH_2{-}CO \\ | \qquad\qquad\quad | \\ CH_2 \qquad\quad\ \ | \\ | \qquad\qquad\quad | \\ CH_2{-}CH_2{-}NH \end{array}$$

ε-Aminocaproic acid

ε-caprolactam is also manufactured from cyclohexanol in the following manner:

$$\text{Cyclohexanol (OH)} \xrightarrow{[O]} \text{Cyclohexanone (O)} \xrightarrow{NH_2OH} \text{Cyclohexanone oxime (NOH)} \xrightarrow[\text{or } PCl_5/\text{Toluene (Beckmann reaction)}]{80\%\ H_2SO_4}$$

Cyclohexanol Cyclohexanone Cyclohexa – no oximene

$$\begin{array}{l} CH_2{-}CH_2{-}CO \\ | \qquad\qquad\quad | \\ CH_2 \qquad\quad\ \ | \\ | \qquad\qquad\quad | \\ CH_2{-}CH_2{-}NH \end{array}$$

ε-Caprolactam

Mechanism

Although ε-caprolactam can undergo a ring-opening polymerisation to yield 6-nylon by involving an ionic chain mechanism. However, in practice it is polymerised by practical conversion to ε-aminocaproic acid which reacts directly with caprolactam *via* a ring opening and with itself to eliminate water. In this way, the chain is built up.

Initiation

$$\begin{array}{l} CH_2{-}CH_2{-}CO \\ | \qquad\qquad | \\ CH_2 \qquad\quad | + H_2O \dashrightarrow \overset{+}{N}H_3\,(CH_2)_5{-}CO\overset{-}{O} \\ | \qquad\qquad | \\ CH_2{-}CH_2{-}NH \end{array}$$

ε-Caprolactam

$$\overset{+}{N}H_2\,(CH_2)_5\,CO\overset{-}{O} + \begin{array}{l} CH_2{-}CH_2{-}CO \\ | \qquad\qquad | \\ CH_2 \qquad\quad | \\ | \qquad\qquad | \\ CH_2{-}CH_2{-}NH \end{array} \rightleftharpoons$$

$$\overset{+}{N}H_3{-}(CH_2)_5{-}CO{-}NH{-}(CH_2)_5\,COO^-$$

Propagation

$$NH_3^+{-}(CH_2)_5{-}CO{-}NH{-}(CH_2)_5\,COO^- + \begin{array}{l} CH_2{-}CH_2{-}CO \\ | \qquad\qquad | \\ CH_2 \qquad\quad | \\ | \qquad\qquad | \\ CH_2{-}CH_2{-}NH \end{array}$$

$$\rightleftharpoons NH_{3+}{-}(CH_2)_5{-}CO{-}NH{-}(CH_2)_5{-}CO{-}NH{-}(CH_2)_5{-}COO^-$$

or

$$\cdots\cdots{-}\left[{-}(CH_2)_5\;\overset{\displaystyle O}{\overset{\|}{C}}{-}\overset{\displaystyle H}{\overset{|}{N}}{-} \right]_n\cdots\cdots$$

During the polymerisation process, there also occurs a side reaction which gives rise to the formation of cyclo oligomers. Therefore, for getting high molecular weight polymers, the process requires specific conditions such as time, temperature and the catalyst. Acids such as acetic, benzoic or adipic acid find use as stabiliser for controlling molecular weight and accelerating the reaction. In order to control molecular weight of the polymers, compounds such as diamines, tertiary amines and metal complexes of bis-hydroxy phenyl sulphones are employed.

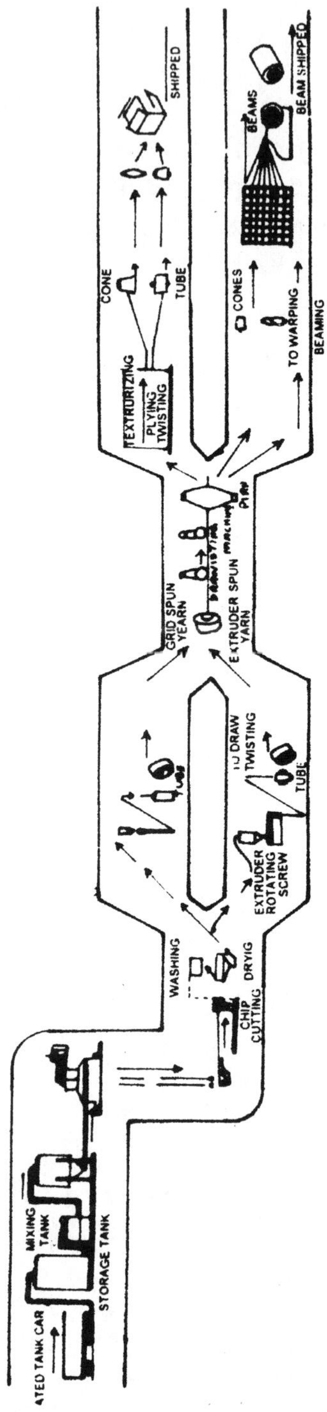

Fig. 10.6: Schematic of Nylon 6 Production

Preparation. In the commercial production of nylon 6, a mixture of caprolactam, water (5–10% by weight) and acetic (about 0.1%) is made to introduce into a reactor in the presence of nitrogen. Now the mixture is made to heat at 250°C for nearly 12 hours and at about 15 atmospheric pressure.

After polymerisation, the unreacted monomer and the low molecular weight polymer are removed by washing with water before nylon 6 is melt spun. During the spinning more monomer is formed and again removed from the fibre.

The polymer obtained by this method consists of 90% of the high molecular weight polymer and 10% low molecular weight compounds.

Properties. Nylon 6 is polycaprolactam which is a polycondensation polymer having a number average molecular weight ranging from 14,000–20,000 for fibre production. It gets softened around 210°C and its crystalline melting point is 220–225°C. The specific gravity of unoriented polymer is 113.

Fibre Properties

Tenacity. A staple nylon 6 fibre is having a tenacity of 4–5.5 g/den and high tenacity continuous filament would break at 5.6–9 g/den.

Elongation. Elongation of staple is somewhat quite high i.e., 35–50%. Its elastic recovery is almost 100% upto about 2–5% strain.

Water absorbency. It is 2.5–5% and the fibre loses a small amount of strength when wet.

Abrasion resistance. Its abrasion resistance is quite excellent.

Electrical resistivity. The electrical resistivity of a dry nylon fibre has been of the order of $6 \times 10^{14} \Omega$ cm and decreases to $10^{8} \Omega$ cm under high humidity conditions.

Action of chemicals. Nylon 6 fibres are soluble in mineral acids such as HCl and H_2SO_4. They are also soluble in organic acid as acetic acid and formic acid. These fibres are also soluble in m-creasol, trichloroethanol and methanol saturated in carbon tetrachloride.

Nylon-6 gets hydrolysed by dilute acids and not by concentrated acids. It is little attacked by basic compounds. Reducing agent fail to degrade polycaprolactam.

Action of radiation. When nylon 6 fibre is exposed to high energy source such as radiation or U.V. light, it gets degraded by involving chain scission and loss of desirable tensile property. The degradation process is dependent on the presence or absence of TiO_2 in the fibre. In the absence of TiO_2, the polymer would absorb the energy directly producing free radicals. In the presence of Tio_2 which acts as a photosensitiser, the TiO_2 absorbs the energy and forms the metastable TiO_3 and loosing atomic oxygen by this process, thereby the molecular weight and tensile property of the fibre gets decreased.

Use

It is used in the manufacture of types and non-woven fabrics. It is also used in the molding process.

(ii) Nylon 6, 6

It is also known as poly (hexamethylene adipamide). The term nylon 6, 6 is derived from the fact that the polymer is produced from a six-carbon aliphatic diamine, hexamethylene diamine and a six-carbon aliphatic dicarboxylic acid, adipic acid.

Raw-materials

The raw materials for the manufacture of nylon 66 are adipic acid and hexametylene diamine. These may by obtained in several ways:

(i) From Phenol and Cyclohexane

Adipic acid is obtained from cyclohexanol in the following way:

$$\text{Phenol} \xrightarrow[160^{0}C]{H_2/Ni} \text{Cyclohexanol} \xrightarrow{HNO_3}$$

$$\text{Cyclohexanone} \xrightarrow{[O]} \underset{\text{Adipic Acid}}{HOOC(CH_2)_4\,CH_2}$$

Adipic acid is also obtained from cyclohexane in the following manner:

$$\text{Cyclohexane} \xrightarrow{HNO_3 \text{ or } O_2} \text{Cyclohexanol (OH)} + \text{Cyclohexanone (O)} \xrightarrow[\text{Copper acetate (Catalyst)}]{HNO_3}$$

$$HOOC(CH_2)_4COOH$$
Adipic Acid

Adipic acid obtained by the above methods can be converted into hexamethylene diamine through the catalytic dehydration of ammonium adipate followed by catalytic hydrogenation:

$$HO.\overset{O}{\overset{\|}{C}}-(CH_2)_4-\overset{O}{\overset{\|}{C}}-OH \xrightarrow{NH_4OH}$$
Adipic acid

$$H_4NO-\overset{O}{\overset{\|}{C}}-(CH_2)-\overset{O}{\overset{\|}{C}}-ONH_4$$
Ammonium adipate

$$\xrightarrow[-2H_2O]{heat} H_2N-\overset{O}{\overset{\|}{C}}-(CH_2)_4-\overset{O}{\overset{\|}{C}}-NH_2 \xrightarrow[-H_2O]{P_2O_5}$$
Adipamide

$$N{\equiv}C-(CH_2)_4-C{\equiv}CN$$
Adiponitrile

$$\xrightarrow{H_2Ni} H_2N(CH_2)_6\,NH_2$$
Hexamethylene diamine

(ii) *From furfural.* Adipic acid as well as hexamethylene diamine may be synthesised from furfural in the following manner:

$$\text{Furfural (CHO)} \xrightarrow[\text{Zn and Mn dicomites}]{\text{Steam. }400^0C} \text{Furan} \xrightarrow[\text{100 atmospheres } 150^0C]{H_2-Ni} \text{Tetrahydrofuran}$$

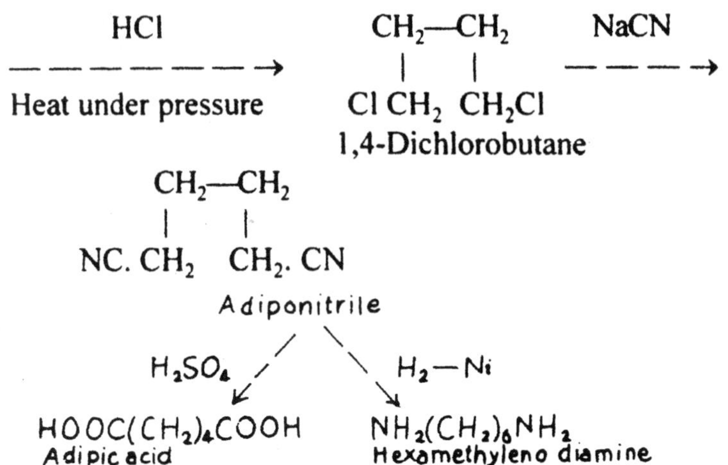

It is also possible to prepare 1, 4-dichlorobutane from butadiene which is a cheaper compound.

$$\underset{\text{Butadiene}}{CH_2{=}CH{-}CH{=}CH_2} \xrightarrow[\text{(ii) } H_2]{\text{(i) } Cl_2} \underset{\text{1,4-Dichlorobutane}}{ClCH_2CH_2CH_2CH_2Cl}$$

Preparation. Hexamethylene diamine and adipic acid are polymerised in 1:1 molar ratio according to the equation:

(i) *Salt Separation*

$$\underset{\text{Adipic acid}}{HO{-}\overset{\overset{O}{\|}}{C}{-}(CH_2)_4{-}\overset{\overset{O}{\|}}{C}{-}OH} + H_2N{-}(CH_2)_6{-}NH_2 \longrightarrow$$

$$\left[\overset{-}{O}{-}\overset{\overset{O}{\|}}{C}{-}(CH_2)_4{-}\overset{\overset{O}{\|}}{C}{-}\overset{-}{O}\right]\left[\overset{+}{N}H_3-(CH_2)_6-\overset{\oplus}{N}H_3\right]$$

(ii) *Poly Condensation*

(i) Dissolve in water in an autoclave.

(ii) 220-230 psi (under steam pressure)

n 66 salt ——————————→

(iii) 270–280°C

$$\cdots-\left[-\overset{\overset{O}{\|}}{C}-(CH_2)_4-\overset{\overset{O}{\|}}{C}-NH-(CH_2)_6-NH-\right]_n-\cdots+H_2O$$

Polyamide (nylon 66)

The reaction takes place through the intermediate salt formation. The polycondensation reactions continue till a high molecular weight nylon 66 is obtained.

Procedure. In a 50-60% aqueous solution having 0.5% acetic acid and nylon salt is heated in an autoclave. Heating at 270-280°C is continued until a pressure of 200-230 psi is obtained and then released. An inert atmosphere must be maintained above the melt at all times to prevent oxidation and discoloration. When the viscosity (molecular weight) of the nylon has reached the desired value, the molten polymer is extruded as a ribbon and cut into chips [see Fig. 10.6].

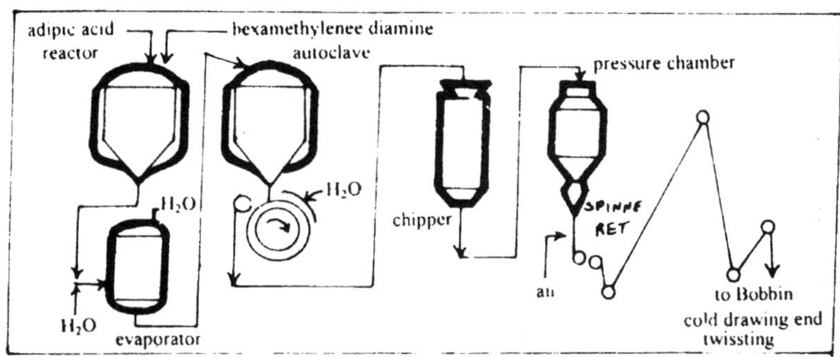

Fig. 10.6: Flow chart of the manufacturing process of nylon 6, 6.

Properties. This polymer is a linear polymer having molecular weight above 10,000. Its crystal structure exists in 1200 forms: α and β. Little information is available on the β-form. Its glass transition temperature is nearly 50°C and its melting point is 250–252°C.

Fibre Properties. After drying, the nylon clips are melt spun into filaments which then need drawing. The drawing is able to improve orientation and allows the development of crystallinity especially if heat applied to the fibre before or during drawing. The resulting fibre is having a high tenacity which ranges from 3-0 g/den to 10 g/den. When wetted fibre strength decreases. Elongation of commercial fibres has been found to range between 16 and 65% and they are having excellent recovery properties upto about 5% extension. The wide variation in tenacity and elongation is attributed principally to variables in molecular weight, polymer purity and the spinning and drawing conditions.

The fibre properties remain unaffected by water at normal temperatures. However, the fibres in water at 150°C under pressure get hydrolysed completely. In the presence of alkali and dilute mineral acids, the depolymerisation of the fibre takes place. Oxidising agents like hydrogen peroxide, hypochlorite and potassium permaganate are able to degrade the fibre.

Uses

Nylon 66 fibres are mainly used in the production of tyre cord. Other important uses of nylon 66 fibres include the manufacture of carpets, rope, thread, fibre cloth, etc. where high strength and good resistance to abrasion and chemical attack are needed.

(iii) Nylon 610

It is manufactured by the polymerisation of hexamethylene diamine and sebacic acid.

$$\underset{\text{Hexamethylene diamine}}{H_2N{-}(CH_2)_6{-}NH_2} + \underset{\text{Sebacic acid}}{HO{-}\overset{\overset{\displaystyle O}{\|}}{C}{-}(CH_2)_8{-}\overset{\overset{\displaystyle O}{\|}}{C}{-}OH} \dashrightarrow$$

$$\left[\overset{+}{H_3N}-(CH_2)_6-\overset{+}{NH_3}\right]\left[\overset{-}{O}-\overset{\overset{\displaystyle O}{\|}}{C}-(CH_2)_8-\overset{\overset{\displaystyle O}{\|}}{C}-\overset{-}{O}\right] \xrightarrow{\text{Heat}}$$

$$\cdots-\left[-\overset{\overset{\displaystyle O}{\|}}{C}-(CH_2)_8-\overset{\overset{\displaystyle O}{\|}}{C}-NH-(CH_2)_6-NH-\right]_n-\cdots$$

Nylon 610

Sebacic acid may be produced by heating castor oil with strong caustic to obtain sebacate and 2-octanol.

Nylon 610 is not used as a textile fibre but is mainly used in the production of monofilaments for brushes, bristles, sporting equipment, etc.

(iv) Nylon 11

It is more, expensive than other nylons. It is manufactured by heating w-aminodecanoic acid at 200-220°C with continuous removal of water. The latter part of the reaction is performed under reduced pressure for producing polymers of high molecular weight. ω-aminodecanoic acid is produced from castor oil.

$$\text{Castor oil} \xrightarrow[\text{in vacuum}]{300°C} \underset{\text{undeclynic acid}}{CH_4{=}CH{-}(CH_2)_8{-}COOH} + (CH_3)_2\,(CH_2)_5\,CHO$$

$$CH_2{=}CH{-}(CH_2)_8{-}COOH \xrightarrow{HBr}$$

$$Br{-}CH_2{-}(CH_2)_9{-}COOH \xrightarrow{NH_3} H_2N{-}(CH_2)_{10}{-}COOH$$

$$\xrightarrow{\text{Polymerisation}} \ldots{-}\left[{-}(CH_2)_{10}{-}\overset{O}{\overset{\|}{N}}{-}\overset{H}{\overset{|}{N}}{-}\right]_n{-}\ldots$$

Nylon 11

Nylon 11 is less water sensitive than other nylons because of the lengths hydrocarbon chain in the monomer. Due to its hydrophobic nature, it finds little commercial use. It has very good electrical insulating properties and melts at 185°C.

Nylon 11 finds use mainly as a textile fibre.

(v) Nylon 7: Poly (ω-enanthamide)

It is produced from 7-aminoheptanoic acid. Its melting point is 223°C and it is having a lower water absorption.

(vi) Nylon 9: Poly (ω-perlargon amide)

It is produced by melt condensation of ω-amino pelargonic acid at temperature above 230°C. This polymer could not be evaluated extensively as a fibre due to relatively high cost.

(vii) Nylon 12

It is produced by heating dodecyl lactam at about 300°C in the presence of aqueous phosphoric acid.

Properties and Applications

The main differences between the various types of nylon lies in the length of the aliphatic chain segments which are separating adjacent amide groups.

Properties

The amide groups give to intermolecular hydrogen bonding so

that crystallization readily take place to produce materials of high melting point and tensile strength. As the length of the aliphatic chain gets increased the melting point gets decreased. There occurs the reduction of the properties like tensile strength, heat distortion temperature and water absorption and an increase in elongation and impact strength.

Uses

The use of a particular type of nylon for a given application is dependent upon the relative importance of mechanical properties, water resistance and ease of processing.

Textile fibres are usually obtained from nylon 6, 6 and nylon 6. Nylon 6, 10 and 11 find use for the manufacture of brushes, sports equipments and surgical equipments. The nylons are having good electrical insulation properties at low frequencies and in dry conditions.

Nylons are only affected by concentrated mineral acids but not by dilute acids. They do not get affected by the alkali at room temperature. Nylon gets oxidised by chlorine, hydrogen peroxide and on exposure to ultraviolet light.

2. Terylene (Dacron, Teron, Cronar, Nylar)

It is a polyster fibre. Polyster fibres are long chain polymers produced from coal, air, water and petroleum. These fibres are composed of at least 85% by weight of an ester of dihydric alcohol and terephthalic acid.

Polyethylene terephalate (terylene) is the most important lenier polymer and is described afterwards in this chapter.

As already discussed the ground work for the development of polyester fibres was laid by Dr. W.H. Carothers and he had published much research information on polyesters. After studying his works, British research chemists decided to examine the possibilities of polyesters. During the period of 1939-1941, investigations were conducted in the laboratories of the Calico Printers Association, Ltd., by J.R. Whinfield, J.T. Dickson, W.K. Birtwhistle, and C.G. Ritchie. The work resulted in the development of a polyester fibre called in England as *Terylene*.

In U.S.A. Du Pont conducted intensive development work. By 1951, the company was producing commercially this fibre which is known as *Dacron* (pronounced daykron).

3. Polyethylene Terephthalate

It is also known as terylene. It is a polyester fibre. It was first of all prepared by Whinfield and Dickion of the Calico Printers Association Ltd. (U.K.). The manufacture of this fibre started in 1953 by E.J. du Pont de Nemours and Co. and Imperial Chemical Industries Ltd (ICI, U.K).

Polyethylene terephthalate was first of all manufacture by the transesterification reaction between dimethyl terephthalate (DMT) and ethylene glycol (EG) which is followed by polycondensation reaction.

Transesterification

$$\underset{\text{DMT}}{CH_3OC(=O)-C_6H_4-C(=O)-OCH_3} + \underset{\text{EG}}{HOCH_2CH_2OH} \xrightarrow{\text{catalyst}}$$

$$\underset{\text{Bis (2 – hydroxyethyl) terephthalate (Bis HET)}}{HO\,CH_2\,CH_2\,OC(=O)-C_6H_4-C(=O)-OCH_2CH_2OH} + CH_3OH$$

Polycondensation

$$HO\,CH_2\,CH_2\,COC(=O)-C_6H_4-C(=O)-OCH_2CR_2OH \xrightarrow[\text{280 – 290°C, vacuum}]{\text{Catalyst}}$$

$$HOCH_2CH_2\,[O-C(=O)-C_6H_4-C(=O)-O-CH_2-CH_2]\,OH + HOCH_2CH_2OH$$

The catalysts employed in the transesterification reaction include the salts of zinc, manganese, calcium and lead salts. For polycondensation reaction antimony compounds find use as the catalyst.

It the free acid finds use as the monomer, a catalyst is not used but the reaction is performed at high temperature and pressure. The by-product obtained is water when the terephthalic acid is employed as the monomer. Polyethylene terephthalate was first of all produced in a batch process, but continuous polymerization processes using dimethyl terephthalate or terephthalic acid have been developed which are known to exert both technical and economic advantages.

Ethylene glycol is a byproduct which gets eliminated continuously from the system and hence their occurs a progressive increase of the molecular weight. The polymerization is performed in presence of a catalyst.

The catalysts employed are usually carbonates, oxides, alkanoate or alkoxides of metals like antimony, barium, calcium, cobalt, lead, magnesium, manganese, titanium and zinc. The process of polymerization gets continued until an average molecular weight of about 20,000 gets reached and then molten polymer gets extruded from the reaction and disintegrated. The polymer is melt spun into fibre of extruded as film.

Terephthalic acid required for the manufacture of terylene is produced by the catalytic air oxidation of p-xylene.

$$CH_3-C_6H_4-CH_3 \xrightarrow[V_2O_5\ 360^0C]{O_2} HOOC-C_6H_4-COOH$$

Fibre Properties

Polyethylene terephthalate fibres are obtained by a melt spinning process. The ash-spun fibre has been amorphous. So as to develop useful fibre properties, the ash-spun yarn gets drawn at temperature ranging from 180 to 200°C. The drawn fibre gets highly oriented and gets characterised by a well-developed crystalline structure. When wet fibre tenacity and the tensile modulus gets reduced only slightly. The fibre is able to maintain its strength during proloned periods of heating. Fibre shrinkage will take place and gets linearly related to temperature. If the yarn is held at a fixed length during a heat treatment, the load-extension properties does not get affected very much even though thermal annealing decreases shrinkage from 12–15% to 2% or less.

Properties

Polyethylene terephthalate occurs as a colourless rigid substance. The polymer readily gets crystallised due to its structural regularity. In the production of the fibre, the molten polymer gets extruded through spinners into air at normal temperature.

Therefore, the filaments are rapidly cooled and are largely amorphous and are weak. The filaments are heated at temperatures above 80°C so that the molecular orientation and crystallinity gets reduced. The filaments are made to heat at about 200°C under tension to yield a dimensionally–stable material of maximum crystallinity.

The completely amorphous material is transparent and is having good impact strength. The poly (ethylene terephthalate) used for producing fibres and films is not suitable for injection moulding and extrusion since the mouldings become partially crystalline/ partially amorphous and are very brittle.

The main application of poly (ethylene terephthalate) film is in electrical insulation.

At elevated temperatures, the polymer is soluble in various organic liquids, like anisol, aromatic ketones, dibutylphthalate and dimethyl sulphone. It is having good resistance to water and dilute mineral acids but gets degraded by concentrated nitric and sulphuric acids. The polymer gets attacked by basic reagents like ionic bases and aqueous sodium hydroxide.

Another polyester fibre was introduced by Eastman Kodak called Kodel. It is made from DMT and 1,4-cyclohexanedimethanol.

10.5. Processing of Fibres

The essential requirements for a polymer to act as a fibre are: (i) high tensile strength; (ii) flexibility; (iii) resistance to abrasion; (iv) chemical stability (resistance to hydrolysis under washing conditions and to degradation by light); and (v) dyeability. A fibre is an oriented polymer, that is stronger along the longitudinal axis of the fibre, than across it. If the polymer chains are short, they are not entangled and are free to move. However, as the chain length increasses, intertwining becomes possible and the fibre strength improves. The optimum molecular weight should be between 10000 and 50000. It must also have a value of T_m>200°C so as to allow ironing, but not greater than 300°C to enable spinning from the melt. Melt spinning is used when the polymer can be readily melted without any decomposition. The molten polymer is then forced through a spinneret and the threads that solidify are drawn, so as to strengthen the fibre and orient the chains.

It is the amorphous region of a fibre which is affected during the stretching operation. The randomly oriented chains in the amorphous regions arrange themselves in line with the longitudinal axis of the fibre. This has been likened to the self alignment of randomly arranged logs of wood floating on water, when water has to pass through narrow strips of land. This Thinning down of the polymer is referred to as *'necking-down'* and it is the necked-down region that is elongated by further drawing. The enhanced intermolecular attraction promotes crystallinity and avoids the reversal of the drawn fibre, back to the old form. The tensil strength of a fibre increases with crystallinity. For proper alignment of the chains, the polymer should be linear because cross-linking and branching of chains tends to interfere with the alignment. Low density polyethylene, a highly branched polymer gives only a weak fibre compared to the practically unbranched high-density polyethylene. Molecular alignment is more important than crystallinity. The cohesive force between polymer chains may be van der Waals forces in the case of polyolefins; but in polyamides like nylon, it is the hydrogen bonding that is responsible for the attraction. The individual chains in nylon are aligned in such a manner that the $-\overset{|}{C}O$ group of one chain is near the $-\overset{|}{N}H$ group of another. This favours H-bond formation. This increased cohesive force accounts for a higher T_m (265°C) for nylon 6,6 compared to polyethylene (T_m = 160°C).

Fibre Spinning

Fibre spinning is a process by which fibres are made from polymers. The following three principal methods are used for fibre spinning:

(1) Melt Spinning
(2) Dry Spinning
(3) Wet Spinning

In the melt spinning method, the polymer in the molten state is used while in both dry spinning and wet spinning methods the solution of polymer in the appropriate solvent is used. However, in all the three cases the polymer either in the molten or in the solution form is allowed to stream through a spinneret which is a special type of plate having extremely small holes for the fibres to emerge.

It is possible to melt a polymer under reasonable conditions, the production of a fibre by *melt spinning* is preferred over solution processes. It melt spinning cannot be performed, a distinction as to type of process used is made depending upon whether the solvent can be removed by evaporation (*dry spinning*) or by leaching out into another liquid which is miscible with the spinning solvent but is not itself a solvent for the polymer (*wet spinning*). Dry spinning owes some advantages over wet spinning. The three processes are having many features in common.

The conversion of the spun polymer melt or solution to a solid fibre has been involving cooling solvent evaporation, or coagulation, depending on the type of spinning used. The rates of these processes have been found to decrease in the order listed. Cooling of a fine filament is generally very rapid and can be controlled within relatively narrow limits.

Solvent evaporation has been involving simultaneous outward mass transfer and inward heat transfer, the rate-controlling step invariably being out ward diffusion of solvent. Coagulation has been involving two-way mass transfer, the coagulating agent (e.g., acid) diffusing inward, and the products of coagulation (e.g., salts, H_2S) diffusing out.

Because of these facts, it is very easy to get a melt-spun fibre that is having uniform properties throughout its cross section, but almost impossible to do so with solvent-spun or coagulated fibres. In addition, the rapid cooling of melt-spun fibres tends to produce an almost circular cross section, while with solvent-spun fibres the cross sections are generally elliptical, and with coagulated fibres (because of the absence of strong surface-tension effects) the cross section have been ordinarily highly convoluted.

It is reported, however, that convoluted fibres are having desirable aesthetic properties, especially hand and luster, so that an effort is made for producing the effect at will. This can be carried out in melt spinning by the use of noncircular orifices in the spinneret or in several types of spinning by extruding two different polymers through the same orifice to produce a single *conjugate* fibres (Hicks 1967). The production of such bicomponent and biconstituent fibres is getting increasingly important especially because they provide a means for producing the textured products.

1. Melt Spinning

Melt spinning can be applied to polyamides, polyesters and polyolifines.

The process of melt spinning is very simple. In this process molten polymer is allowed to pump at a constant rate under high pressure through a plate called a *spinneret* which is having a large number of small holes. The liquid polymer streams gets emerged downward from the face of the spinneret, generally into air. They solidify and are brought together to yield a thread and wound up on bobbins. A subsequent drawing step becomes necessary to orient the fibres.

The polymer is made to melt by containing a hot grid in the form of steel tubing, which gets heated by electric current or by hot vapours. It is generally necessary to protect the polymer melt from oxygen by blanketing it with steam or an inert gas like carbon dioxide or nitrogen. If the viscosity of the molten polymer has been low, it may pass directly to the metering (constant-rate) pump. For melts of higher viscosity a booster pump is generally used.

Other methods of melting include the use of an extrusion-type screw, one section of which can serve as its own metering pump. Methods in which the metering pump has been replaced, for example, by source of gas pressure or other device, do not appear to offer sufficiently precise control for holding the denier of fine yarns constant.

The spinneret is having a 2- to 3-in. diameter steel disk about one-quarter in. thick with 50–60 countersunk holes 0.010 in. or less in diameter. It is possible to determine the denier of the filament not by the diameter of the hole but by the rate at which polymer gets pumped through the spinneret and the rate at which the filaments get wound up.

The filaments emerge from the spinneret face into air and start cooling. An air blast can be used to speed up the cooling process. After the filaments have travelled far enough to become solid (about 2 ft) they have been brought together and wound up. Speeds of nearly

2500 ft/min are usually used.

The filaments as spun are almost completely unoriented. Most of the stretching taking place between the spinneret and wind up does so while the filament is still molten, and there is sufficient time for the molecular orientation to relax before the fibre cools and gets crystallized.

Consequently a separate drawing step becomes necessary for producing the orientation of the crystallites necessary for optimum physical properties. In the drawing step somewhat lower speeds were traditionally needed than can be achieved in spinning, so the two steps were generally carried out separately; now they are generally carried out in a single continuous operation.

The drawing step used two sets of rolls (Fig. 10.7), one to feed the undrawn yarn from a supply package at velocity v_1 and the other, moving about four times as fast, for effecting the drawn yarn at velocity v_2. The filaments may be passing over a metal pin between the two sets of rolls; drawing gets localized in the neighbourhood of the pin. The yarn is then collected on a strong metal bobbin. Freshly drawn yarn is having a tendency to contract somewhat in length. Heat gets evolved during the drawing process, because work is performed on the polymer.

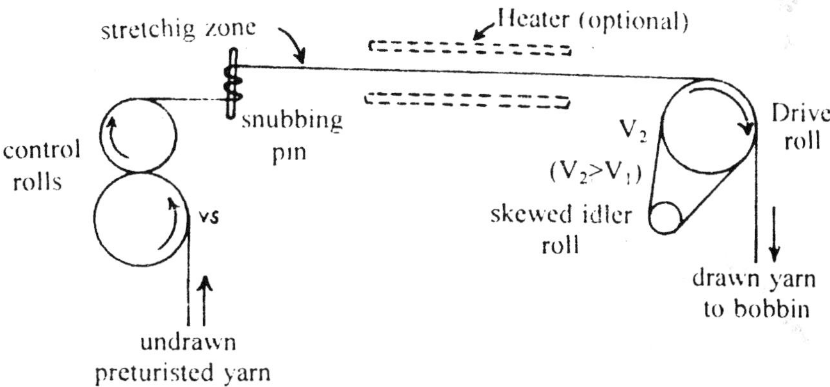

Fig. 10.7: Schematic diagram of a typical drawing process.

Melt spinning has been used for polyester and nylon fibres.

An important point to remember in the melt-spinning process is that the diameter of the filament is dependent as much on the rate which the molten polymer streams through the holes as on the rate at which the filament gets pulled and wound up as on the hole diameter.

2. Dry Spinning

In this process a concentrated solution of the polymer is used for spinning and the solvent in evaporated from the thread. Thus in this process, the filament is obtained by the evaporation of solvent from the polymer solution into air or an inert gas atmosphere. The solvent must be volatile. Dry spinning is being used for many years for spinning cellulose acetate from acetone solution, and is now also used for spinning polyacrylonitrile and the acrylamide fibres from solutions in dimethyl formamide or dimethyl acetamide.

So far no process is employed in which the polymer is made in the same solvent from which it is dry spun. Hence the first step in dry spinning is essentially the preparation of the polymer solution. The solvent used for spinning is based on consideration of solvent power, boiling point, heat of evaporation, stability, toxicity, ease of recovery, and so no. Nonpolar solvents are preferred due to their convenient boiling points and nonhygroscopic nature, but may bring about hazardous buildup of static charge. Low-boiling solvents having high heats of evaporation may bring about polymer precipitation or coagulation on the surface of the fibre, in part because of condensation of moisture as the fibre cools when solvent gets evaporated. This gives rise to loss of luster and strength. High solution concentration are desired, and elevated temperatures are generally used to keep viscosities from becoming too high. Cellulose acetate is spun from 20–45% solutions with 400–1000 poise viscosity at 40°C. Almost 22% of the acrylic fibres gets produced by spinning polyacrylonitrile from solution in dimethyl formamide.

Due to the higher boiling point of this solvent, up to 50% residual solvent is left in the fibre after it gets spun; this must be washed out in a subsequent step. As in other types of spinning, polymer solutions are generally filtered just before dry spinning.

Spinning is carried out in a vertical tubular cell jacketed for temperature control, in which air, steam, or inert gas may get passed through either concurrently or counter currently as needed.

Downward spinning has been preferred for small-denier fibres and upward spinning for high deniers, for controlling draw better by eliminating the influence of gravity. Dry spinning is performed at rates as high as 2500–3000 ft/min, about the same as melt spinning.

3. Wet Spinning

Wet spinning, like dry spinning, uses a fairly concentrated polymer solution. The polymer may be made to precipitate as threads by forming the polymer solution into a solvent in which the polymer is insoluble, the increase in polymer solution viscosity is controlled by spinning the solution at a higher temperature. The details of wet spinning process are shown in Fig. 10.8. Wet spinning also changes a viscous polymer solution into fine jets through a spinneret. These jets, are led into a coagulation bath which contains a large volume of a non-solvent that can precipitate the polymer from its solution. When continuous jets of the polymer solution come in contact with a non-solvent, they get precipitated in the form of fine filaments. The filaments so produced are gathered on a spindle after undergoing washing, drying, etc. An important point to be borne in mind while carrying out wet spinning is that sometimes droplets are formed instead of continuous filaments. This happens because of the breaking of continuity in the jet as it emerges from the spinneret hole and is caused by the forces of interfacial tension acting on the newly formed jet. This should be avoided by increasing the viscosity of the polymer solution.

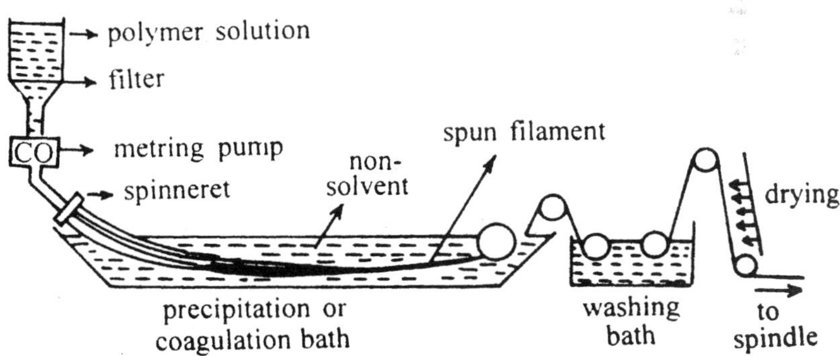

Fig. 10.8: Schematic diagram of wet spinning process.

Coagulation which is an essential step in wet spinning is a rather slow process. Hence as compared to the other two processes, the rate of wet spinning may be as slow as 50 m/min. The wet spinning process is used commercially to obtain fibres from cellulose, viscose rayon, polyacrylonitrile, etc.

10.6. Comparative Characteristics

To make a satisfactory textile choice for a specific use, it is essential to understand not only the separate and combined properties of the various textile fibres but also their comparative merits.

The commercial value of any fibre depends largely on the extent to which it processes certain characteristics, like tensile strength, elasticity, fineness, cohesiveness, and sufficient length, all these facilitate the process of spinning into yarn that will produced satisfactory fabrics for the consumer. The effectiveness of finishes and care of fabrics depend upon such properties as the effect of heat, reaction to alkalies and acids, and affinity for dyes.

All such factors have to be considered before making choice of a suitable fabrics for particular use.

The choice, is thus a matter of selecting the fabric of the fibre or combination of fibres which offers the most desirable properties and satisfies the particular purpose for which the fabric or garment is intended to be used. This kind of determination has become increasingly difficult with the greater number of fibre groups and the variations within each group. The wide variations of properties existing within each group of fibres such as polyester, spandex etc. makes precise statements difficult.

The expression "the longer the fibre, the stronger the yarn" is true only when fibres within a single groups are to be compared; for example, two or more wool fibres or two or more cotton fibres. Comparing wool with cotton reveals that wool fibre, though longer than cotton fibre, is not stronger.

Long-staple yarns within each group of fibres produce stronger, smoother, more serviceable fabrics which cost more than the less durable fabrics made from short-staple yarns. In wool, for example, the short-staple fibre is used in the manufacture of woollens, and the longer staple is reserved for worsteds. Worsteds are considered more serviceable. A similar distinction is found in every textile fibre.

Flexibility

Flexibility of the polymer chain is very important because a highly flexible chain being capable of reverting to random coiled contaminations is less desirable in a fibre. Polyethylene oxide is highly flexible because of the free rotation around the $-CH_2-O-CH_2-$ linkage. It is not fibre forming.

However, by stiffening the chain by introducing phenylene groups as in polyethylene terephthalate, a fibre-forming polymer results.

Resistance to Degradation

The resistance of a fibre, to degradation is related to its chemical structure. Esters such as terylene, and amides like nylon are capable of undergoing hydrolysis. However, because of their high orientation, they are not so readily attacked. Polypropylene and polyethylene are quite stable to chemical attack, but the former is liable to suffer photochemical degradation because of the greater ease with which a hydrogen atom can be abstracted from a tertiary carbon than a secondary carbon.

Dyeability of Fibre

The ability of a fibre to be coloured permanently by a dye is the **dyeability of a fibre.** For the dye to diffuse into the fibre, the chains must not be rigid at the temperature of dyeing. The dye molecules should also be covalently or ionically bound to the chain. These conditions are fulfilled in the case of nylon, whose glass-transition temperature is <50°C and the amino groups at the chain ends provides sites for salt formation by acidic dyes. In the case of terylene (T_2>100°C), dyeing is done at higher temperatures or in the presence of carriers such as diphenyl, methyl salicylate, etc. that act as plasticizers. The dyeability can also be improved by making terpolymers, containing, besides the major monomer, one to provide suitable sites for dyeing and one to act as plasticizer.

Absorbancy

Generally the synthetic fibres, tend to absorb only very little moisture and are rather uncomfortable to wear in comparison to cotton, wool, etc. The moisture absorbing tendency also helps to reduced the buildup of static electricity. The ability to absorb moisture depends on the ability of the fibre to form hydrogen bonds with water. Thus, nylon absorbs moisture and terylene does not.

10.7. Polyacrylic Fibres

These are made from acrylonitrile and other monomers. Polymers which contain 85% or more acrylonitrile e.g. Orlon, Acrilan etc.

Polymers which contain between 35.65% of acrylonitrile are known as modacylics *e.g. Verel, Dynel* etc. du Pont made first acrylic fibre in 1945 and called it *Orlon*. These are copolymers in which monomers like vinyl acetate, vinyl chloride, styrene or acrylamide are co-polymerised with acrylonitrile. The copolymers are insoluble and heat sensitive so they cannot be melt spun. The insolubility is because of hydrogen bonding with nitrogen atom of nitrile group. After many trials dimethylformamide was found to rupture these hydorgen bonds, Polyacrylonitrile dissolves in concentrated aqueous solutions of certain salts like zinc chloride and sodium thiocyanate.

$$
\begin{array}{ccc}
\diagdown & & \diagdown \\
H-C-C\equiv N & \cdots\cdots\cdots & H-C-C\equiv N \\
\diagup & & \diagup \\
CH_2 & & CH_2 \\
\diagdown & & \diagdown \\
H-C-C\equiv N & \cdots\cdots\cdots & H-C-C\equiv N \\
\diagup & & \diagup \\
CH_2 & & CH_2 \\
\diagdown & & \diagdown
\end{array}
$$

Initially acrylic fibres prepared had buff tint because of the presence of CN groups but suitable copolymers are white.

Acrylonitrile gets polymerised when added to warm water containing catalyst like ammonium persulphate and activator like sodium bisulphite, small quantities of monomers like vinylpyridine are stirred in. The copolymer gets precipitated, filtered and washed. A solution of polymer in dimethylformamide is prepared. Both dry and wet spinning can be used. In dry method the filaments are dried

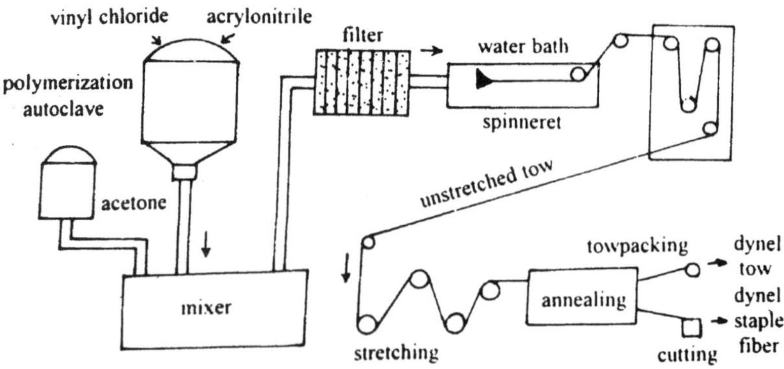

Fig. 10.9: Flow chart of the manufacturing process of Dynel.

in a hot blast of air. In wet spinning coagulation is done in a bath of hot glycerol or propylene glycol. The filaments are stretched. The fibres shrink in water. By blending stretched and unstretched fibres and treating with hot air or water bulky high lofted fibres can be obtained. These are used in sweater and pullovers. They have a soft warm feel and are quick drying.

Acrylic staple fibres can be dyed in presence of cuprous ions which get attached to nitrile groups. Acrylic fibres containing about 10% acrylamide monomers can be dyed with many dyes. For example "Courtelle" can be dyed with basic dyes while "Acriln", "Zefran" can be dyed with acid dyes. An American acylic fibre can be dyed with direct dyes. This fibre is modified hy drafting radicals on the polymer to make it hydrophilic.

Acrylic fibres cannot be heat set like nylon or polyester. They are also not abrasion resistant. They are used mostly for knitwear and carpeting. The high bulked varieties are used in blankets, sweaters, fleeces and synthetic furs. Their special uses are factory clothing, filter cloths, sunblinds, umbrelias, tents, tarpaulins etc. Orlon curtains are fairly stable in sunshine.

Acrylic fibres when heated to high temperature get converted to carbon fibres of very high tensile strength. They are used for reinforcing plastics and laminates. Carbon fibre reinforced plastics have been used in racing car bodies, helicopter blades and hovercraft propellers.

"Dynel" is an important modacrylic fibre. It is a copolymer of 40% acrylonitrile and 60% vinyl chloride. The copolymer is dissolved in acetone and spun into a fibre by coagulating in water. The fibre is stretched to 14 times its original length. It has a pale cream colour which needs bleaching for whiteness. It is a thermoplastic above 80°C. It is resistant to inorganic chemicals, micro organisms and insects. It gets dissolved or softened by organic solvents. It is used in protective clothing for chemical plant workers, military and miners. It is used in cushions and pillows. It is used for military uniforms. **Doll's hair** are also made for **"Dynel"**.

"Verel" is another modacrylic fibre made by Eastern Chemicals from 60% acrylonitrile and 40% vinylidinechloride monomers. It can absorb more moisture and is comfortable to wear. It is used in carpeting and simulated furs.

10.8. Polyvinyl Fibres

Germans prepared polyvinyl fibre during Second World war known as "**Pe Ce**" for making clothes for air force. Polyvinylchloride was chlorinated to increase chlorine content to 64%. This material dissolved in acetone and was spun into fibres. In later processes chlorination was omitted and the fibre was spun from PVC dissolved in acetone and carbon disulphide. They are flexible at low temperature. These fibres are non-flammable, water-repellent and resistant to chemicals. They cannot be ironed. They are used in industrial filters, fishing nets, protective clothing, under wear and sweaters.

Polyvinyl-alcohol fibres are prepared by alkaline hydrolysis of polyvinly acetate and spinning of aqueous solution of polyvinyl alcohol in a coagulating bath containing sodium sulphate. The fibres are treated with formaldehyde and hot dried. These polyvinyl-alcohol fibres are used in fishing nets, linings of gloves, stockings, surgical threads and for reinforcemcent of tyres.

10.9. "Saran"

Saran is a vinylidine chloride based copolymer fibre containing at least 80% vinylidine chloride and 15-20% vinylchoride. They are melt spun. They are difficult to dye hence for coloured yarn pigment is added before spinning. The fibre is heat sensitive but non-flammable and resistant to water and stains. It is largels used in upholstery for cars, restaurants and ships. It is also used for lining chemical equipment.

Method of Manufacture

The raw materials used to produce vinylidene chloride are ethylene, which is derived by cracking natural petroleum, and chlorine, which is electrolyzed from seawater. Ethylene is treated with chlorine to produce trichlorethane, which in turn is treated with lime to produce unpolymerized vinylidene chloride. The vinylidene chloride is then copolymerized with a small quantity of vinyl chloride to produce a powdered resin. The resin is melted and excluded through a spinneret into a water-cooling bath. While still being cooled, it is drawn out 400 per cent to improve its crystallinity and molecular orientation, thereby increaing its strength. In a manner similar to nylon.

The fibre may be produced in different shapes. Some are round, some flat, others oval. One type, *Rovana*, appears to be flat. Under the microscope, it appears to be folded over on itself several times, giving added bulk with lighter weight than is generally available in saran fibre.

The natural colour of saran is pale gold or straw. It may be delustered to a translucent or to an opaque appearance by the addition of a delusterant such as titanium dioxide. The fibre may be solution-dyed, but the colours are not bright because the natural colour of saran tones down the hue.

A copolymer of vinylidene dicyanide and vinyl acetate was prepared and marketed as "*Darvan*". The two monomers alternate in the chain. It has good durability and elasticity and is resistant to weathering. Polystyrene fibres have been used as bristles for brushes.

10.10. Polyalkene Fibres

Polyethylene fibres are melt spun at 300°C and drawn to six times in length. The yarn has high tenacity and is abrasion resistance. Pigment can be added before spinning as the fibre can not be dyed. They are resistant to acids hence used for acids proof overalls, filter cloth etc. They are not suitable as texiles because of poor heat resistance.

Polyropylene is meet spun like polyethylene to get a strong fibre with higher melting point. It is used in rope making pile fabrics, carpets and sacking.

Method of Manufacture of Polypropylene Yarn

The basic procedure is to polymerize propylene gas with the aid of an appropriate metal compound, such as titanium chloride, in diluent. The polymer formed from the propylene is slurried in the diluent to decompose the catalyst, and is then filtered, purified, and reduced to powder form as polypropylene resin. The resin is melted and extruded through a spinneret is filament form. The filaments are then treated and drawn to get the proper crystallinity and molecular orientation in order to obtain such desired characteristics as strength, abrasion resistance, elasticity, resilience, and shrinkage.

The catalyst system is added to the diluent in the polymerizer, and a polymer is formed from propylene gas. The polymer slurried in the diluent is first treated to decompose the catalyst, then filtered and purified. After drying, it is pelletized to produced the finished **pro-fax polypropylene powder.**

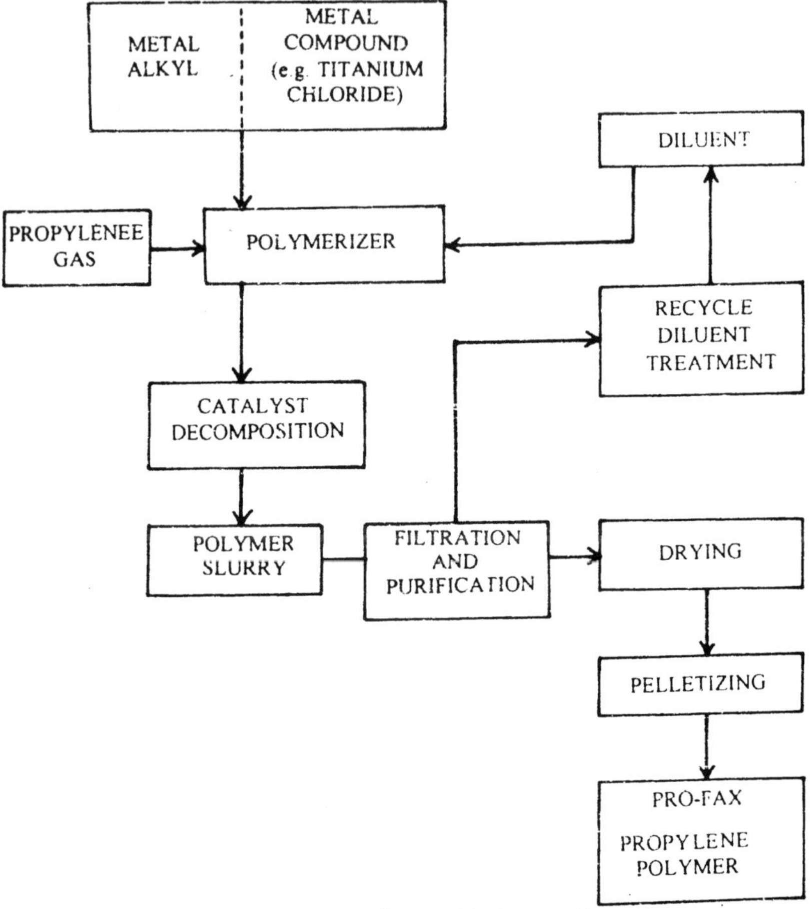

Fig. 10.10: Flow chart of Polypropylene Process.

Polytetrafluoroethylene fibres are by extruding a dispersion in an acid bath. The fibre is washed dried and heated to 385°C to form a homogeneous fibre. This is an expensive fibre resistant to heat and chemical attack. It is used as packing for pumps handling corrosive ... filter cloth, covers for ironing stand, protective clothing etc.

Polyethylene and polypropylene films and drawn and cut into ribbons. Ribbons are stretched to give weaving tapes. Woven polyproylene is used in carpet blacking, sacks, handbags, shoes, hats, suitcases, wall coverings etc. They are used in baler twine and ropes. Chopped polypropylene twine has been used for reinforcing non-load bearing concrete *e.g.* pipes, ducts, panels, etc. This type of concrete has greater bonding strength and impact strength. It holds together even after fracturing.

10.11. Polyurethane Fibres

These are synthetic fibres having elastic property. They are segmented polyurethanes prepared by copolymerising urethane with a polyol or polyester. The fibres are melt spun as multifilaments or monofilaments. Polyurethane segments are is a random state, they uncoil and gel straightened on stretching. On releasing stress they go back to original shape. The yarns may be used as such or covered with rayon or nylon. These fibres are marketed as **"Spandex"**, **"Lyera"** etc. These fibres can be dyed and have good resistance to chemicals and washings. They are used in foundation garments, belts, stockings, swimming costumes etc. Leather like synthetic material such as **"Corfam"**, **"Ortex"** have been made by impregnating synthetic fibre matrix with polyurethanes. This kind of material that allows passage of air are referred to as **Poromerics.**

Metallic Fibres

Fine metal foil, 0.025 mm in thickness can be sandwiched between layers of polyester or polyacetate film. The laminate is cut into narrow ribbons. These ribbons can be woven into textiles for decorative effects. These are made in a number of colours and can be washed. They are marketed under the name **"Lurex"**.

11

Polymer Analysis

11.1. Introduction

Commercially useful polymers are generally not made up solely of the macro-molecules itself. These samples also contain additives, filters residual monomers or catalyts that affect the performance of the polymers. In chapter eight we described some techniques for characterisation of polymers and some of these require making a solution of the sample. The size of polymeric molecules and reactions such as grafting and cross-linking tend to limit the solubility of many macro-molecules. Some polymers are only soluble at elevated temperatures, so we must understand the effect of temperature on the stability of these molecules to ensure that any solutions prepared at elevated temperatures truly represent the original sample.

Techniques such as viscometry, ultra-centrifugation, osmometry, end-group analysis and precipitation liquid chromatography, are used to characterise the molecular weight of polymeric systems. Polymers, particularly those prepared from emulsion technology, also contain particles of different sizes. Consequently, particle sizes techniques such as **light-obstruction** etc. are important part of polymer analysis.

The conventional spectroscopic techniques (such as I-R, Nmr, X-ray and mass spectrometry) described in Chapter 8 of this book are used to define the composition of complex polymeric systems. Pyrolysis techniques discussed in this chapter offer another technology for analysis of polymers. These pyrolysis techniques are beneficial for polymer analysis because they are applicable to solid in soluble samples, which eliminates the need to solubilize the macromolecules, and they are based on the controlled degradation of the large molecules to smaller molecules, which are amenable to conventional chromatographic and spectroscopic analysis. In addition to pyrolysis technology, other techniques such as differential scanning colorimetry (DSC), thermogravimetric analysis (TGA) etc. are used to define the state of a polymer system including properties such as glass transition temperature, melting point, degree of crystallity.

The end-use performance of a polymer depends on molecular weight, composition, particle size and the thermal properties. The analysis techniques presented in this chapter provided comprehensive characterisation of many synthetic polymers. Each technique may measure a specific property or the polymer but data from all, or at least many, of these techniques are required to fully characterise any polymeric system.

11.2. Molecular Weight Determinations

Size exclusion chromatography (SEC) is an important techniques for characterisation of macro-molecules. SEC is a high performance liquid chromatography (HPLC) techniques in which molecules are separated according to the differences in hydrodynamic volume. This separation mechanism is made possible by the packing material in the column. The packing material is made of porous spherical particles (3–20 μm). The retention in SEC is governed by the partitioning (or exchange) of the macromolecular solute molecules between the mobile phase (the eluent flowing through the column) and the stagnant liquid phase that occupies the interior of the pores. The range of macromolecular sizes that can be separated with a given column depends on the size (or *size distribution*) of the pores. The resulting chromatogram in an SEC experiment thus represent a **molecular size distribution.**

The relationship between molecular size and molecular weight depends on the conformation of the dissolved solute molecules. However, for any solute conformation (such as random coil, rigid rod, or hard sphere), *the molecular size increases with molecular weight.* The rate at which the molecular size increases with molecular weight varies for the different possible solute conformations. Most synthetic polymer samples can be categorized as random coil solutes. As such, the relationship between molecular size (radius of gyration, R_g) and molecular weight is $R_g \propto M^a$, where the exponent is a constant dependent on the solute composition, the temperature, and the solvent. Given that the molecular size is related to the molecular weight of the polymer, the molecular size distribution obtained in the SEC experiment can be converted into a molecular weight distribution (MWD) if we can establish a molecular weight versus retention volume calibration.

Experimental Details

The diagram in Fig. 11.1 shows the sequence of events that occur during an SEC experiment. Larger solutes elute from the column first; smaller solutes elute from the column later. Consider the analysis of a hypothetical mixture of large and small polymer solutes by SEC. The first frame in Fig. 11.1 shows the injection of the mixture as a narrow band at the head of the column. In the second frame, the development of size separation is illustrated. In this frame, the larger polymer solutes are clearly migrating through the column at a faster rate than the smaller solutes. The larger solutes are more excluded from the pores in the packing, and thus spend more time in the mobile phase than in the stagnant mobile phase contained within the pores. The smaller solutes can enter more of the pores containing the stagnant mobile phase. Due to this the smaller solutes spend more time in the column. In the third frame, the larger molecules elute from the column and are detected by a suitable detector, and in the last frame, the small molecules elute.

The sequence of events shown in Fig. 11.1 serves only to describe the mechanism of separation in SEC. Real samples of synthetic polymers generally contain a broad distribution of molecular sizes that cannot be separated into individual SEC peaks. Rather, an SEC chromatogram of a given polymer sample is typically a broad peak, with the earlier-eluting species representing the high-molecular-weight species and the late-eluting species representing the low-molecular-weight species.

In SEC, retention depends on the continuous exchange of solute molecules between the mobile phase and the stagnant mobile phase within the pores of the packing. This exchange is an equilibrium, so entropy-controlled processes and enthalpic processes such as adsorption are undesirable in SEC. Thus, the SEC retention volume

(V_r) is expressed as shown in Eq. (11.1)

$$V_r = V_0 + V_p K_{SEC} + V_s K_{LC} \quad ...(11.1)$$

where V_0 = the exclusion volume or interstitial volume between the packing particles (defined in Fig. 11.2), V_p = the pore volume (defined in Fig. 11.2), V_s = the stationary phase volume, K_{SEC} = the SEC solute distribution coefficient, and K_{LC} = the liquid chromatography solute distribution coefficient.

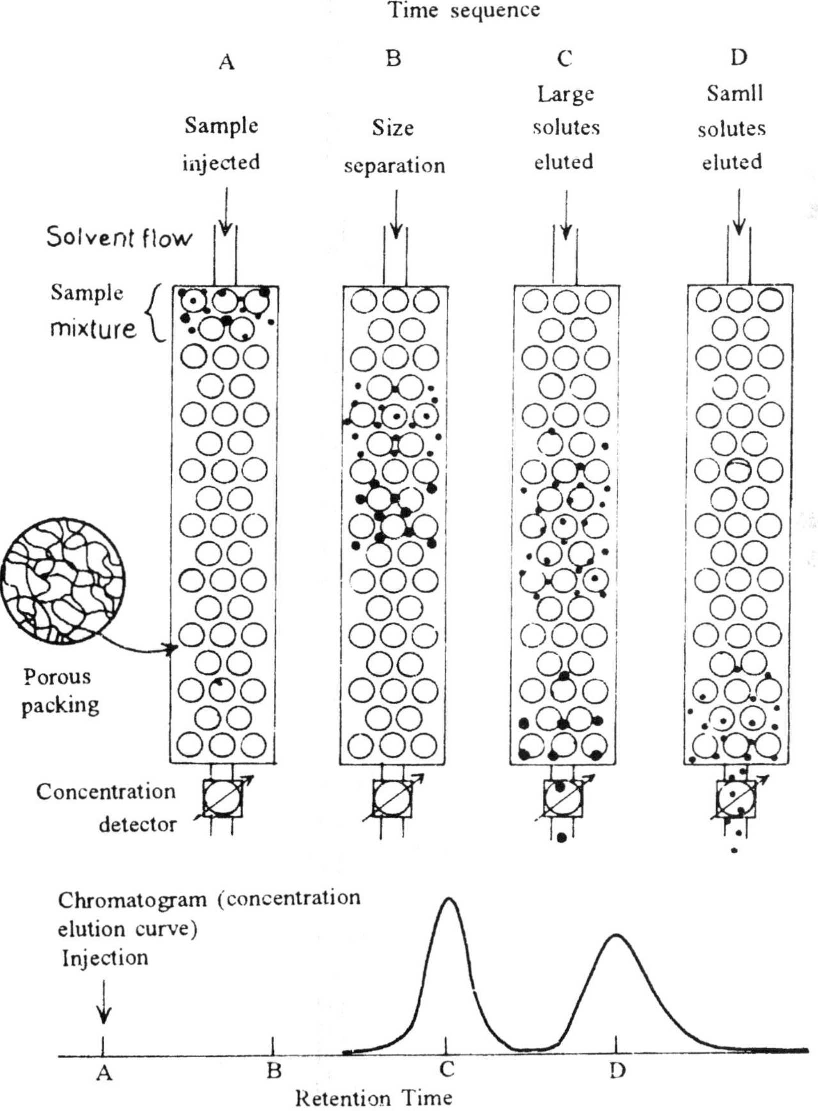

Fig. 11.1: Polymer separation in SEC.

Since ideal SEC retention is governed only by entropic contributions, the column packing material/eluent combination is chosen such that K_{LC} is minimized or ideally $K_{LC} = 0$.

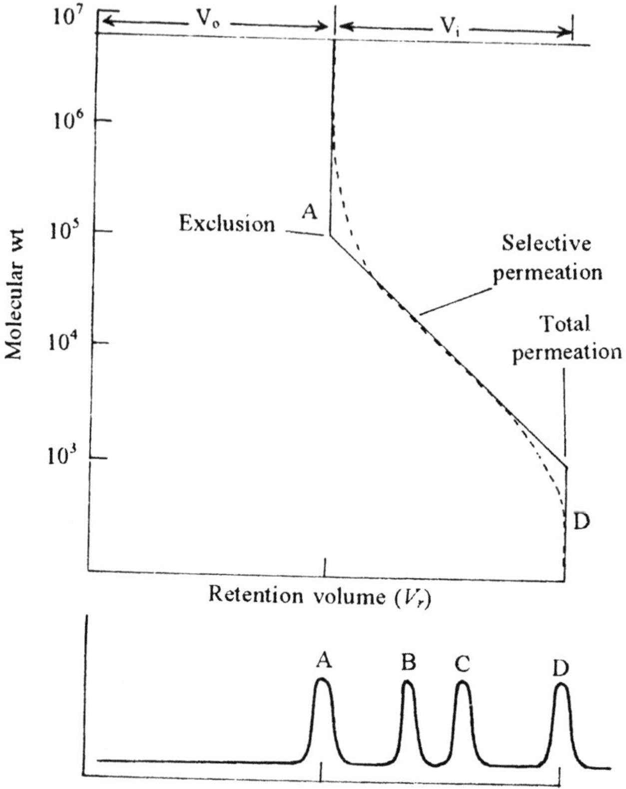

Fig. 11.2: Log molecular weight *versus* retention volume plot for a typical SEC experiment with four narrow fraction molecular weight standards.

The dependence of V_r on solute molecular weight represents the SEC calibration curve. An example calibration curve, generated from an injection of four narrow-fraction molecular weight standards, is shown in Fig. 11.2. In above example, the highest-molecular-weight species is totally excluded from the pores in the packing, so its retention volume is equal to the interstitial or exclusion volume, in which case K_{SEC} is zero. The lowest-molecular-weight standard permeates all of the pores within the SEC column, so its retention volume equals the sum of the interstitial volume and the pore volume. This volume is known as the total volume or total permeation limit. The value of K_{SEC} for species eluting at the total permeation limit is 1. The two intermediate-molecular-weight standards can permeate the pores to some extent, so they are separated according to their respective hydrodynamic volumes. The value of K_{SEC} for peaks eluting in this region is $0 < K_{SEC} < 1$.

The linear region in Fig. 11.2 defines the useful region for SEC separation of macromolecules. The molecular weight range covered by the linear region of the SEC calibration curve depends on the size of the pores in the packing material. For single pore size columns, the useful region for SEC covers approximately 2 decades in molecular weight. (Gels with pore sizes ranging from 106 to 50 Å are typically available in a wide variety of compositions). Most synthetic polymer samples have broad MWDs that span more than 2 decades in molecular weight. Individual pore size columns can be coupled to provide the required separation range, or mixed-bed columns (columns containing a mixture of pore sizes) can be used. Mixed-bed columns are designed to provide several orders of magnitude (4 to 5) in molecular weight separation range. The choice of column depends on the sample composition, solvent options, and molecular weight range. The goals are to have ideal SEC retention (that is, separation by size only) and optimal resolution. Resolution in SEC depends on the slope (and column dispersion) of the linear calibration region. The column should be chosen such that the linear range encompasses the entire MWD of the sample with the smallest (absolute value) slope available.

Uses of SCE

Size exclusion chromatography is widely used for determination of polymer molecular weight and is useful for determination of molecular weight and MWD, it is an important technique for helping in the establishment of structure/property relationships for polymers. Once molecular weight/property relationships are established, SEC can be used to ensure to production of polymers having desirable physical properties. SEC can also be used to help in optimization of process conditions for production of materials with desirable molecular weight. SEC may also be used for comparing the MWD of samples that may perform differently in end-use applications.

Synthetic polymers differ from small molecules in that they cannot be characterized by as single molecular weight. In a synthetic polymerization, a distribution of chain lengths (that is, molecular weights) is produced. This distribution can be described by any number of molecular weight averages. The most common molecular weight averages used in establishing molecular weight/property relationships are the number-average (M_n), weight-average (M_w), and z-average (M_z) molecular weights. These averages are defined by the following expressions:

$$M_m = \frac{\sum_i N_i M_i}{\sum_i N_i} = \frac{\sum_i W_i}{\sum_i W_i / M_i} = \frac{\sum_i h_i}{\sum_i h_i / M_i} \qquad ...(11.2)$$

$$M_w = \frac{\sum_i N_i M^2{}_i}{\sum_i N_i M_i} = \frac{\sum_i W_i M_i}{\sum_i M_i} = \frac{\sum_i h_i M_i}{\sum_i M_i} \qquad ...(11.3)$$

$$M_z = \frac{\sum_i N_i M^3{}_i}{\sum_i N_i M^2{}_i} = \frac{\sum_i W_i M^2{}_i}{\sum_i W_i M_i} = \frac{\sum_i h_i M_i^2}{\sum_i h_i M_i} \qquad ...(11.4)$$

Here N_i and W_i are the number and weight, respectively, of molecules having molecular weight M_i. The subscript i is an index representing all molecular weights present in the ensemble of chains. Each of the above equations contains three representations of the particular molecular weight average. The third representation in each case (farthest right) defines how one obtains these averages from SEC chromatograms. h_i is the height (from baseline) of the SEC curve at the ith elution increment and M_i is the molecular weight of species eluting at this increment. M_i is obtained via calibration with appropriate standards.

Conversion of the retention volume axis in SEC to a molecular weight axis (that is, calibration) can be done in a ways including peak position, universal calibration, broad standard (integral and linear), and determination of actual molecular weight. In the peak position method, a series of well-characterized, narrow-fraction molecular weight standards of known peak molecular weight (M_p) are injected onto the SEC system and the retention volumes determined. A plot of log M_p versus retention volume is constructed as shown in Fig. 11.3.

Depending on the type of SEC columns used, the calibration data points can be fit with either a linear or a third-order polynomial function. For determination of absolute molecular weights, the peak position calibration is of limited utility because only a few commercially available standards exist, and in order for the molecular weight calibration to be valid, the composition and topology of the standards and unknown must be the same.

We bound prepare and characterized calibration standards via preparative fractionation schemes and absolute molecular weight techniques (such as LALLS), but these efforts can be costly and time-consuming. In some instances, the accuracy of the calibration is not important because the researcher may be interested only in comparing two samples to determine whether there is a relative difference in molecular weight. In these cases, only apparent molecular weight information is obtained, and for comparison of sample MWDs, calibration may not even be necessary. However, in other cases, absolute molecular weight data may be necessary. In these cases, alternative calibration schemes exist.

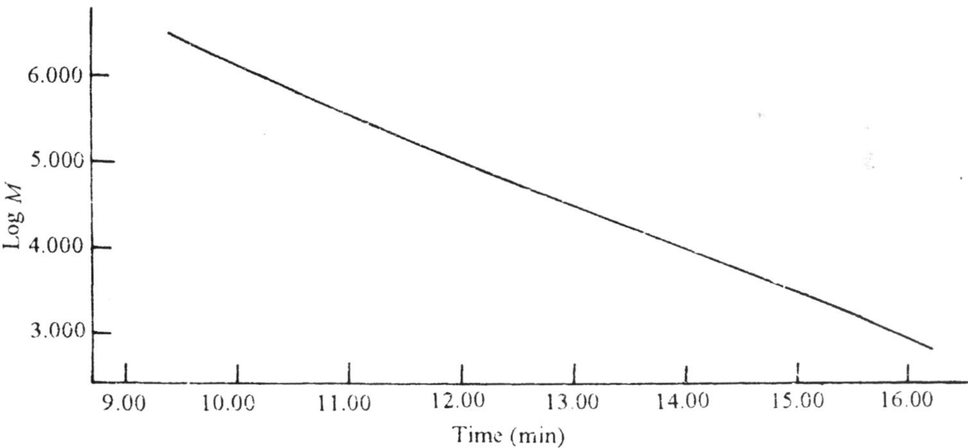

Fig. 11.3: SEC calibration curve for 17 narrow-fraction polystyrene molecular weight standards analysed using two mixed-gel SEC columns and tetrahydrofuran as the mobile phase. The line is a third-order polynomial fit of the data points.

Polynomial coefficients: Log $M = A + BT + CT^2 + DT^3$

A = 1.737168e + 001, B = – 1.911779e + 000, C = 1.043279e – 001, D = –2.578582e – 003

One such scheme is the **universal calibration principle** shown in Fig. 11.4. The product of intrinsic viscosity, $[\eta]$, and molecular weight is proportional to hydrodynamic volume, and in ideal SEC, molecules are separated according to hydrodynamic volume. In 1967, Benoit showed that the calibration for polymers of different types can be merged into a single line when plotted as log $[\eta]M$ *versus* retention volume, as opposed to the typical log M *versus* retention volume shown in Fig. 11.3. There are two ways to use the universal calibration principle to glean absolute molecular weight data. One way is to use an on-line viscometer to determine $[\eta]$ at each SEC retention increment. Well-characterized narrow-fraction molecular weight standards can be used to generate the universal calibration curve. The plot of log $[\eta]M$ *versus* retention volume and the measured $[\eta]$s for the sample allow one to calculate M at each retention volume increment.

Alternatively, the universal calibration principle and the Mark-Houwink equation may be combined to determine absolute molecular weights. From the universal calibration principle,

$$[\eta]_{i,A}M_{i,A} = [\eta]_{i,B}M_{i,B} \qquad \text{...(11.5)}$$

where the subscript i refers to the particular SEC retention time increment and the subscripts A and B refer to polymer of different composition. The Mark-Houwink equation is given below:

$$[\eta]= kM^{\alpha} \qquad \text{...(11.6)}$$

where k and α are constants that depend on polymer composition, temperature, and solvent. Combining Eqs. (11.5) and (11.6) yields

$$k_A M_{i,A}^{1+\alpha A} = k_B M_{i,B}^{1+\alpha B} \qquad \text{...(11.7)}$$

Rearrangement of Eq. (11.7) gives

$$M_{i,A} = (k_B/k_A)^{1/1+aA} M_{i,B}^{R} \qquad \text{...(11.8)}$$

where $R = (1 + \alpha_B)/(1 + \alpha_A)$.

Thus, with knoweldge of the Mark-Houwink parameters for the standards and the sample (in the SEC eluent), one can determine absolute sample molecular weights. Published values of Mark-Houwink parameters in common SEC solvents are available for many polymers. When published constants are not available, they may be measured via Eq. (11.1). However, the measurement of k and α require standards of known molecular weight.

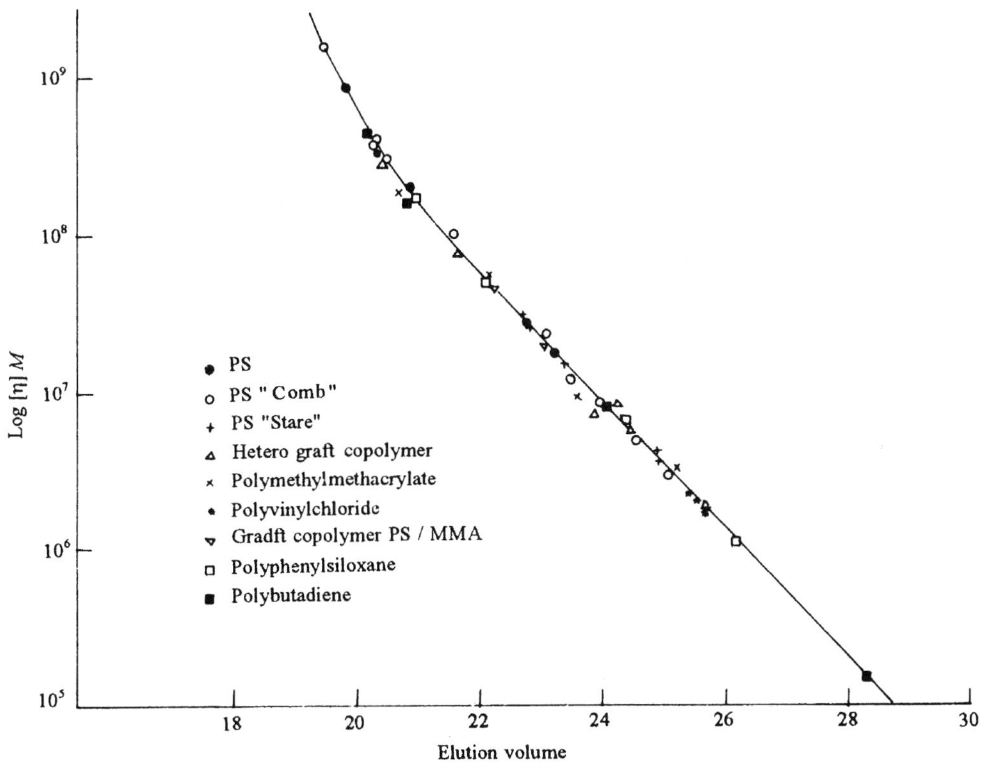

Fig. 11.4: Universal calibration plot.

A third type of calibration in SEC is known as broad-standard-calibration. Various types of broad-standard calibration schemes are used. The integral calibration scheme needs the complete MWD of a broad standard be known. Accurate characterization of a sample for the purposes of establishing it as a broad standard can be tedious and time-consuming. In many instances, a model such as the **Flory most-probable model** is used with the measured M_n or M_w to predict the complete MWD of a broad standard. Use of this approach requires that the polymers standard closely follow the theoretical model, which limits the applicability of this approach.

A second, more widely used method for broad-standard calibration is the **linear method.** This approach needs a broad standard of known M_n and M_w. The method consists of an iterative search for the coefficients of a linear calibration equation (log M versus retention volume), such that the equation yields the known values of M_w and M_n for the broad standard.

Finally, one may determine the absolute calibration by using an absolute molar mass sensitive detector, such as a LALLS detector. In this approach, the absolute M_w is determined at each SEC elution volume increment. The plot of the determined log M_w versus retention volume represents the calibration curve. The calibration curve is then applied to the concentration detector chromatogram for determination of molecular weight averages and MWD.

Analytical Information Provided by SEC

Qualitative

SEC is generally not used for qualitative analyses, because separation is governed by size, not composition. However, simple comparison of SEC chromatograms can provide qualitative information about difference in MWDs for unknowns having the same composition and topology.

Quantitative

SEC is used for determination of MWDs and molecular weight averages. MWD are generally calculated and presented in one of two ways. The cumulative weight fraction MWD of a typical polystyrene sample is shown in Fig. 11.5. The right-hand y-axis is the cumulative weight percentage of chains having molecular weight less than M_i. A second, more popular, way of presenting MWD data is the differential weight (d_w) fraction MWD, also shown for the same polystyrene sample in Fig. 11.5. Most commercially available SEC software packages proved these options for plotting MWD data.

Integrated SEC peak areas may also be used for determining the concentration of polymer in a matrix. For example, the concentration of polymer in a monomer sample may be determined via SEC. Generally, in these types of experiments, MWD information may not be needed. In these cases, a small pore size column is chosen such that the polymer is excluded (and thus well separated from the monomer peak). The excluded peak area is then used to calculate the concentration of polymer in the monomer sample. Determination of the concentration of polymer in other small molecule matrices may be carried out in a similar manner.

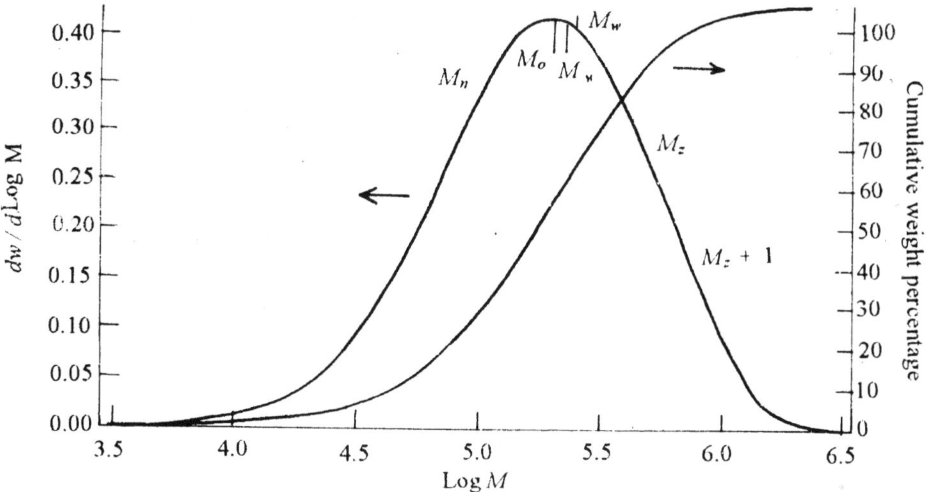

Fig. 11.5: Cumulative and differential weight fraction log MWDs for the same polystyrene sample. $M_p = 204685$, $M_n = 100988$, $M_w = 255949$, and $M_z = 482307$; polydispersity $(M_w/M_n) = 2.534$ with peak area = 906374.

Applications of SCE

The number of applications of SEC to the *characterization of synthetic polymers* and other macromolecules is enormous. Two examples are presented here to give the reader a small sampling of typical SEC applications.

1. Determination of the MWD and Molecular Weight Averages of a Polystyrene Sample

The absolute MWD of a polystyrene sample was determined by SEC using a peak position calibration curve generated from narrow-fraction polystyrene molecular weight standards. A solution of the polymer was prepared in tetrahydrofuran (THF) at a concentration of 2.0 mg/mL. The solution was filtered through a 0.2 μm filter before injection. Separation was performed on two mixed-gel SEC columns designed to provide a linear separation range for polystyrene from approximately 3×10^6 Da to 500 Da. THF at a flow rate of 1 mL/min was used as the eluent. A differential refractive index detector was used to detect the electing peaks. A total of 17 narrow fraction polystyrene molecular weight standards were injected to determine to log M_w versus retention time calibration via the peak position method. The calibration curve was fit to a third-order polynomial as shown in

Fig. 11.3. The resulting SEC chromatogram of the polystyrene sample is shown in Fig. 11.6, and the MWD is shown in Fig. 11.5 along with the calculated molecular weight averages.

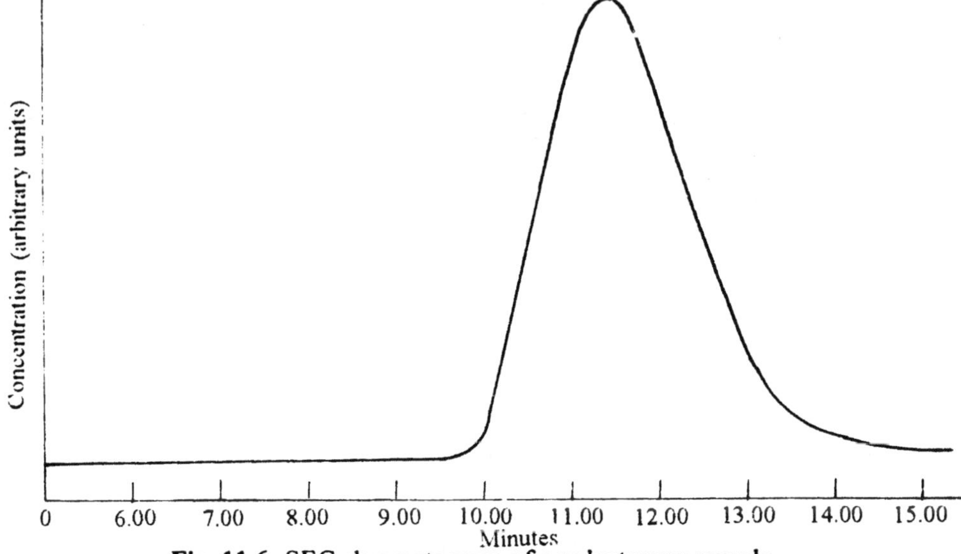

Fig. 11.6: SEC chromatogram of a polystyrene sample.

2. Comparison of the Apparent MWDs of Polyurethane Samples After Various Processing Steps

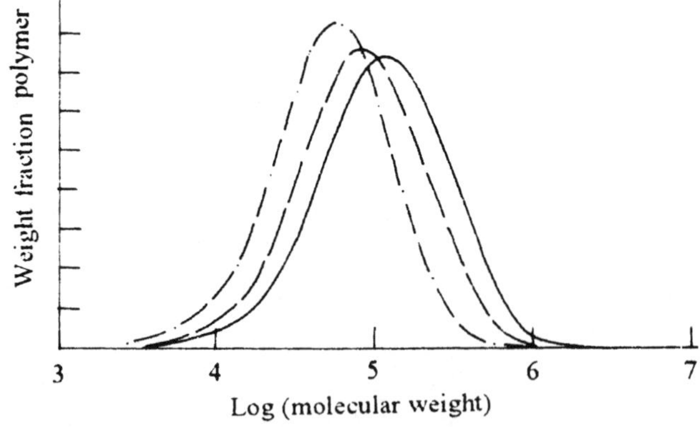

Fig. 11.7: PEO apparent MWDs of polyurethane samples after various thermal processing steps. Virgin = solid curve, compounded = dashed curve, injection-molded = dot-dashed curve.

Three polyurethane samples that had undergone various thermal cycles were analysed by SEC. The samples were virgin pellets, compounded pellets, and an injection-molded part. SEC was performed on the same column set and detector described in the previous example.

The eluent (and solvent for sample dissolution) was N,N, dimethyl-formamide containing 0.4% (*v.v*) lithium nitrate. Narrow-fraction polyethylene oxide (PEO) molecular weight standards were used for calibration. Thus, the results generated in this case are PEO apparent data. The resulting MWDs are overlaid in Fig. 11.7. Although the data are based on PEO standards, clearly there is a systematic decrease in molecular weight after each thermal processing step.

11.3. Low-Angle Laser Light Scattering (LALLS)

Introduction

Low-angle laser light scattering (LALLS) is used mainly to determine the *absolute weight average molecular weight* (M_w) *of homopolymers*. A LALLS experiment may be performed on a bulk solution wherein only M_w is obtained. A LALLS detector may also be used in conjunction with an SEC experiment wherein a measure of the absolute molecular weight distribution may be obtained.

Although SEC is the most common methods used to determine molecular weight, LALLs offers many advantages. SEC separates on the basis of hydrodynamic size. A calibration is necessary to correlate size with retention time. Unless appropriate molecular weight standards are available, the experiment yields molecular weight averages relative only to the particular standards used for calibration. Ideally, the standards should match the unknowns in terms of composition and topology (linear or branched, for example). Unfortunately, relevant standards are not generally available for most polymers. The use of SEC coupled to LALLS eliminates the need for calibration, as the LALLS detector provides absolute molecular weight data. Thus, the use of LALLS permits for accurate molecular weight measurements for homopolymers where suitable SEC standards are not available. In addition, LALLS may be used to distinguish between linear and branched samples which may have the same molecular size as determined by SEC, yet differ in absolute molecular weight.

The static or bulk LALLS experiment is useful for determining absolute weight average molecular weights of homopolymers, which are not well suited for SEC experiments.

Experimental Details

Determination of absolute molecular weight by light scattering involves measurement of the ratio of scattered light intensity at a particular angle, θ, to the intensity of incident light, per unit volume. This quantity is known as the **Rayleighl ratio,** R_θ, defined by the following equation:

$$R_\theta = \frac{f_\theta}{I_0 V} \quad \text{...(11.9)}$$

where f_θ = intensity of scattered light at angle θ, I_0 = incident intensity, and V = scattering volume.

Polymer solutions are considered to be optically heterogeneous due to local random fluctuations in density and refractive index. The fluctuation theory of light scattering may be used to related the Rayleigh ratio to physical characteristics of the polymer molecule causing the sacttering by the following equations:

$$\frac{kc}{R_\theta} = \frac{1}{M_w P_\theta} + \frac{2A_2c}{P_\theta} + \frac{3A_2c^2}{P_\theta} + \ldots \quad \text{...(11.10)}$$

$$\mathrm{k} = \left(\frac{2\mathrm{p}^2\mathrm{n}^2}{\lambda^2\mathrm{N}}\right)\left(\frac{\mathrm{dn}}{\mathrm{dc}^2}\right)(1+\cos^2\theta) \quad \text{...(11.11)}$$

where c = concentration of scattering species in g/mL, M_w = absolute weight average molecular weight of the scattering species, n = solution refractive index, λ = wavelength of incident light in vacuo, N = Avagadro's number, A_2, A_3, ... = virial coefficients, P_θ = form factor, which depends on size and shape of scattering species, dn/dc = specific refractive index increment, $R_\theta = R_{\theta,\text{solution}} - R_{\theta,\text{solvent}}$ where $R_{\theta,\text{solution}}$ and $R_{\theta,\text{solvent}}$ represent the Rayleigh factors for the solution and solvent, respectively.

The specific refractive index increment is measured in an independent experiment using a differential refractometer or it is measured on-line using some commercial instruments.

The form factor, P_θ, arises when the scattering molecules are large with respect to the wavelength of incident light, as is typically the case for polymers. For these larger particles, different parts of the molecule are exposed to incident light of different phase and magnitude. Thus, the scattered light will be of various phases and magnitudes, which results in interference of the scattered light coming from different parts of the polymer molecule. The net result is an angularly dependent scattered light intensity. Particle scattering functions have been derived for various particle shapes, as shown in

Fig. 11.8. (2) In all cases, P_θ approaches unity as the size parameter (which is directly proportional to the ratio of the particle diameter to the incident wavelength in solution) becomes small.

In general, P_θ is related to the square of the molecular radius of gyration (R_g). In the limiting case of small angle ($\theta \to 0$), the relationship may be expanded in a power series, keeping only the first-order terms as follows:

$$\lim_{\theta \to 0} P_\theta = \frac{1 - 16\pi^2}{3\lambda^2 \sin^2(\theta/2) R_g^2} \quad ...(11.12)$$

In this expression, P_θ reduces to unity as θ approaches zero.

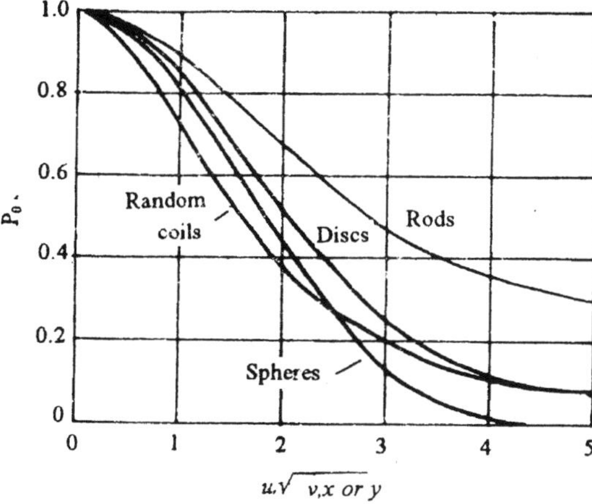

Fig. 11.8: Variation in the particle scattering function (P_θ) with the size parameter.

Uses

In a classic multiangle laser light scattering experiment, the excess Rayleigh ratio, R_θ, is determined at a number of angles for a series of polymer concentrations. A double extrapolation to zero concentration and zero angle is then performed in what is typically called a Zimm plot, as shown in Fig. (11.9). The extrapolation to zero angle corrects for the effect of finite particle size. The slope of the θ-dependence plot in this form can be shown to be equal to the quantity $16\pi^2 R_g^2/3\lambda^2$. Thus, an average radius of gyration is obtained in this type of experiment.

The second extrapolation, to zero concentration, corrects for the effect of finite concentration. The intercepts of both the zero concentration extrapolation and the zero angle extrapolation can be shown to be equal to $1/M_w$.

A LALLS experiment takes advantage of the fact that at small angle θ, the form factor P_θ approaches unity regardless of the molecular size or shape. Equation (11.13) reduces to the following in the limit of small angle:

$$\frac{kc}{R_\theta} = \frac{1}{M_w} + 2A_2c + \ldots \tag{11.13}$$

If R_θ for a series of sufficiently low concentration is measured such that higher-order concentration terms may be neglected, a plot of kc/R_θ versus concentration will be linear with an intercept of $1/M_w$ and a slope of $2A_2$.

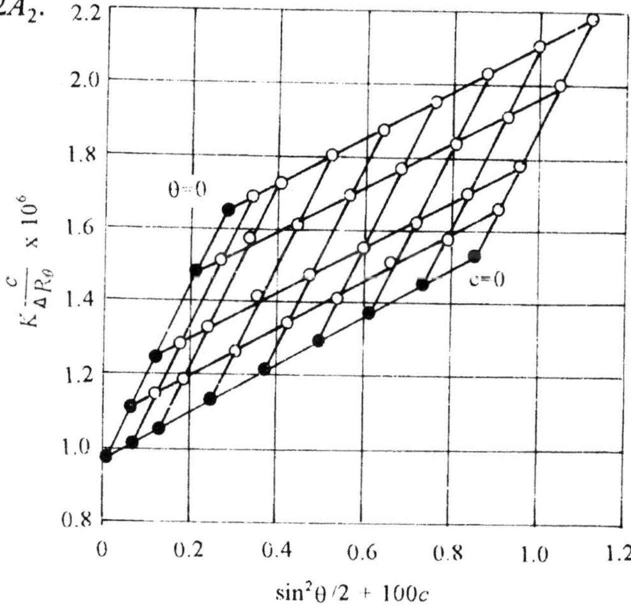

Fig. 11.9: Zimm plot from light scattering data for polystyrene in butanone.

In the SEC/LALLS experiment, the LALLS detector is used to measure R_θ and a differential refractive index (DRI) detector is typically used to measure concentration, c, at each retention time increment ($2A_2c$ is assumed to be negligible). The data are then plotted as log (M_w) versus retention volume over the range of the chromatogram. This curve or a fit to the curve may then be used as the SEC calibration curve and applied to the DRI chromatogram to generate an absolute molecular weight distribution.

A number of low-angle and multiangle light scattering instruments are commercially available with a variety of features. Some of the instruments may be used both in the static (or bulk) mode as well as an SEC detector. Others may be limited to one application.

Common components of light scattering instrumentation include a source, incident optics, sample cell, output collection optics, and a detector for the scattered light. Instrument design is crucial because the scattered light intensity is inherently small. Therefore, reflections and scattering from all interfaces must be minimized for optimum sensitivity.

Laser sources have now replaced mercury arc sources in currently available instruments. The increased available intensity of the lasers results in increased scattering intensities and, hence, increased sensitivity. Incident optics may include focusing lenses, polarizing lenses, and filters depending on the type of source used. Neutral density filters may also be necessary for measurement of the incident light intensity. The sample cell design varies greatly among commercial instrumentation. Cells designed for use in multiangle instruments are typically cylindrical cuvettes, which allow detection of scattered light at multiple angles from incident. LALLS cells may be rectangular, from which scattered light is collected via an annulus at a fixed angle. These cells are generally designed as flow-through cells, which may be coupled with an SEC experiment. Output collection optics may include lenses, polarizers, and apertures. Finally, a photomultipler tube, photodiode, or photodiode array may be used for detection of the scattered light.

Other features available in commercial instruments include temperature control and simultaneous differential refractive index (DRI) detection. Advantages to the DRI detection capability are that *dn/dc* can be measured on-line at the wavelength of the light scattering instrument and that no interdetector delay exists between the LALLS and the DRI signals in the coupled SEC/LALLS experiment.

Most currently available light scattering instrumentation is easily interfaced to data acquisition systems for data collection. Sophisticated data handling software is also available from various sources.

Analytical Information

Qualitative

LALLS is not generally used for qualitative analytical purposes.

Quantitative

The primary use of LALLS is quantitative determination of absolute molecular weights of homopolymers, as previously described. LALLS is extremely sensitive to high-molecular-weight fractions of polymers and, therefore, is most accurate for high-molecular-weight polymers. Accuracy decrease for low-molecular-weight polymers (less than 10,000 g/mol), but may still be acceptable if the experiments can be performed in a solvent where *dn/dc* is large.

Applications

1. Determination of Absolute Weight Average Molecular Weight, M_{w}, of Polystyrene by LALLS

A determination of absolute M_w was performed on a polystyrene sample without the use of molecular weight standards. The specific refractive index increment of the polystyrene sample in tetrahydrofuran (THF) was measured using a differential refractometer. Solutions of the sample were prepared in THF at concentrations ranging from 0.2 mg./mL to 2.5 mg/mL using volumetric glassware. The excess Rayleigh ratios of the solutions were measured using a LALLS photometer at an angle of 4.6°.

A plot of kc/R_θ versus concentration is shown in Fig. 11.10. As described earlier, the intercept of the plot yields $1/M_w$ and the slope of the plot is equal to $2A_2$.

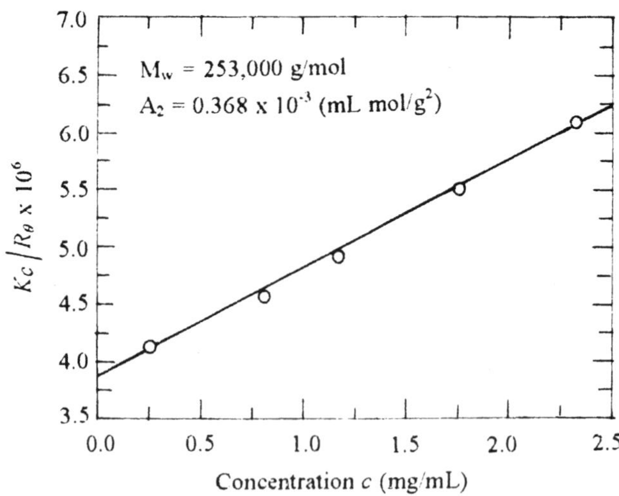

Fig. 11.10: Plot of Rayleigh ratio (kc/R_θ) versus concentration.

2. Determination of Absolute Molecular Weight Distribution of Polystyrene by SEC Coupled to LALLS

The absolute molecular weight distribution of a polystyrene

sample was determined with out the use of molecular weight standards using SEC coupled to LALLS. The specific refractive index increment of the polystyrene sample in tetrahydrofuran was measured using a differential refractometer. A solution of the sample THF was prepared at 2 mg/mL using volumetric glassware. A known quantity of the sample was injected onto the SEC system using DRI and LALLS detection with a known interdetector delay.

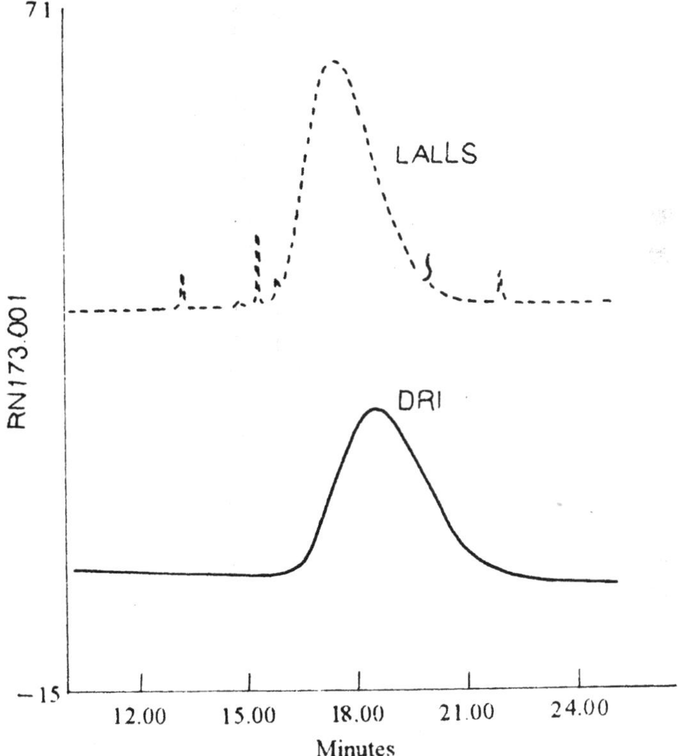

Fig.11.11: Comparison of LALLS and DRI chromatograms for a polystyrene sample.

Fig. 11.11 gives the DRI and LALLS chromatograms for the polystyrene sample. A plot of the logarithm of the absolute M_w of the eluting species as a function of retention time may be generated knowing the optical constants of the detector and the flow rate. The DRI response is used to measure to concentration as a function of retention time, which is necessary for the molecular weight calculation. A fit of a line to the plot was used as an SEC calibration curve to calculate the absolute molecular weight distribution of the polystyrene sample from the DRI chromatogram.

3. Determination of the Amount of Long-Chain Branching in Polystyrene by SEC Coupled to LALLS

SEC coupled to LALLS was used to quantify the amount of branching in branched polystyrene samples. The branched polystyrene was prepared using benzocyclobutenoyl peroxide (BCBPO) as a branching agent. The experimental procedure used was identical to that described in the previous application. Curvature in the plot of the logarithm of the absolute M_w versus retention time is attributed to branching in the sample. This is an effect of the SEC, which separates polymer chains on the basis of hydrodynamic size. For a given absolute molecular weight, a branched sample will have a smaller hydrodynamic radius than a linear sample. Therefore, the branched sample elutes from the SEC column later than a linear sample of the same molecular weight, resulting in curvature of the log (M_w) versus retention time plot.

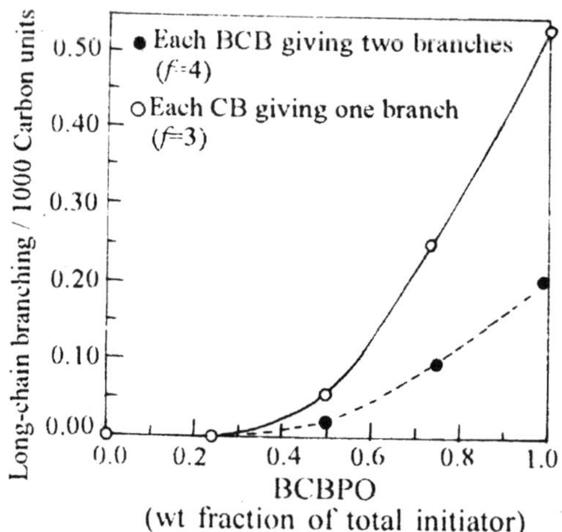

Fig. 11.12: Long-chain branching versus concentration of BCBPO.

The use of models developed by Zimm and Stockmayer may then be used to calculate the long-chain branching per 1000 carbon units of the polymer. Fig. 11.12 shows the calculated amounts of branching for various levels of BCBPO added during the polymerization.

11.4. Practice Size Measurement

Introduction

There are a number of analytical techniques available which

focus on the measurement of **particle size.** The first step in performing a particle size analysis is deciding which technique to use. The make such a decision, one must first determine what type of size information is required and what are the physical and chemical properties of the sample. The type of size information obtained can be as detailed as a complete size distribution or as simple as a mean size. Contained in a size distribution is information about the breadth of the distribution, the number of individual populations in the sample, and the percentage of particles greater than or less than a particular diameter. A mean diameter, on the other hand, is a way of describing particles size using a single number. Some commercially available particle size analyzers report only a mean diameter. Such analyzers may be capable of identifying differences in the particle size, but one must realize that any two samples having the same mean diameter may have completely different size distributions.

Physical properties like particle refractive index and density may be required for some particle size techniques. The refractive index of the particle and suspending medium is generally required for size measurement based on light scattering, whereas densities are required for techniques based on sedimentation and acoustics.

Examining the particles of interest under a microscope permits one to determine the general range of sizes and shapes present. Because most size measurement techniques have a finite range of applicability with respect to particle size, a preliminary examination under the microscope is quite useful. Nearly all particle size analysis techniques are sensitive to particle shape. In general, the more the shapes of the particles of interest deviate from spherical, the greater the error in the size assignment. Particles whose shapes deviate significantly from spherical, such as needles or rods, are best analysed using microscopy.

Knowing the chemical properties of the particulate also helps one determine what size analysis technique can be used. Particles are generally suspended in a fluid to facilitate the analysis. The fluid dilutes the particles so that they can be delivered to the analyzer at the appropriate concentration. When chosen properly, the suspending fluid helps dissipate any cohesive forces that tend to lead to particle agglomeration. Moreover, when chosen properly, the suspending fluid helps produce a suspension that is homogeneous in particle concentration and size, which can be delivered to the particle size instrumentation. The suspending fluid chosen must not dissolve the particles of interest and must be compatible with the instrument's materials of construction.

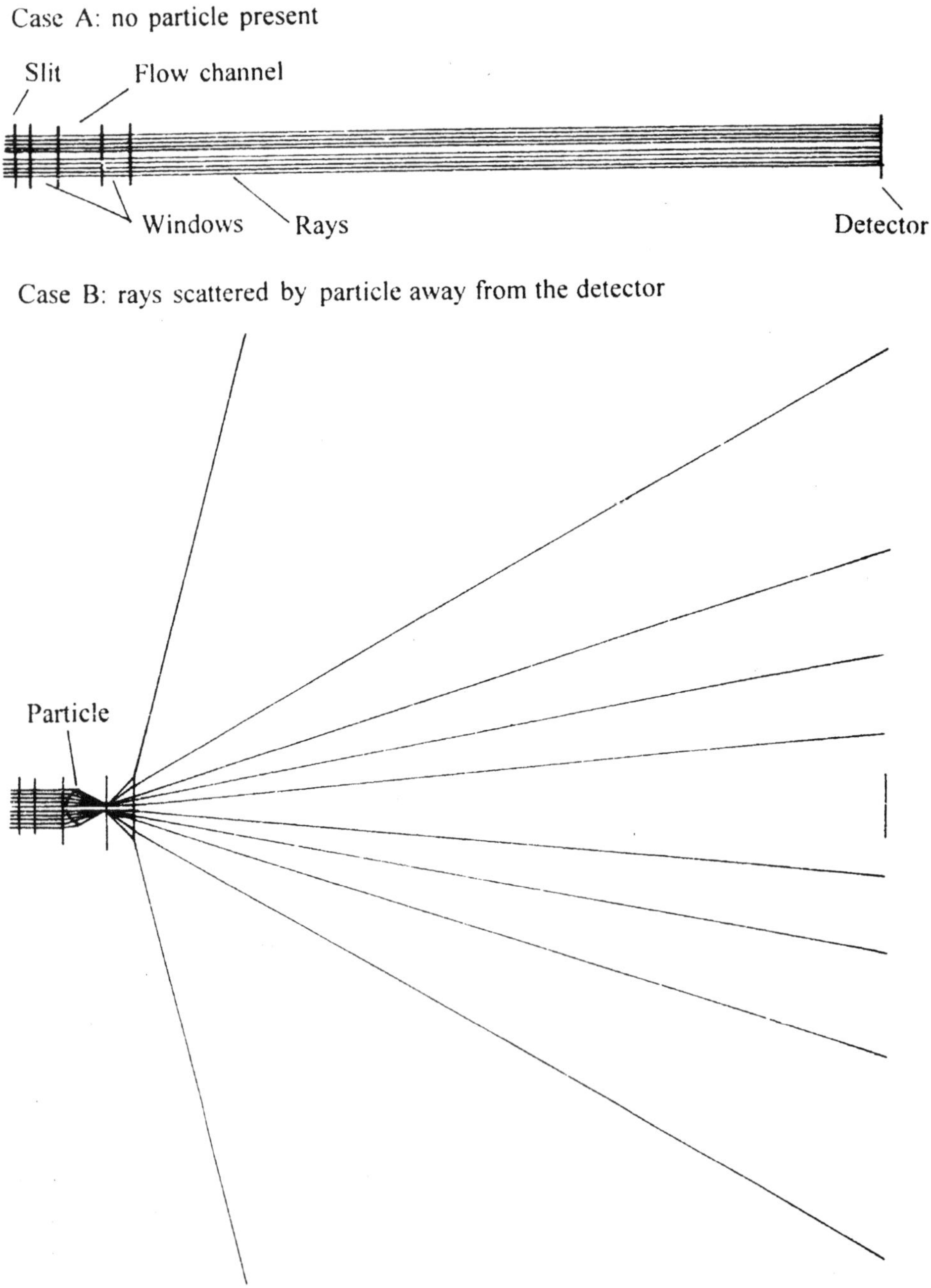

Case A: no particle present
Slit
Flow channel
Windows
Rays
Detector
Case B: rays scattered by particle away from the detector
Particle

Fig. 11.13: Ray trace for a light obstruction particle size experiment.

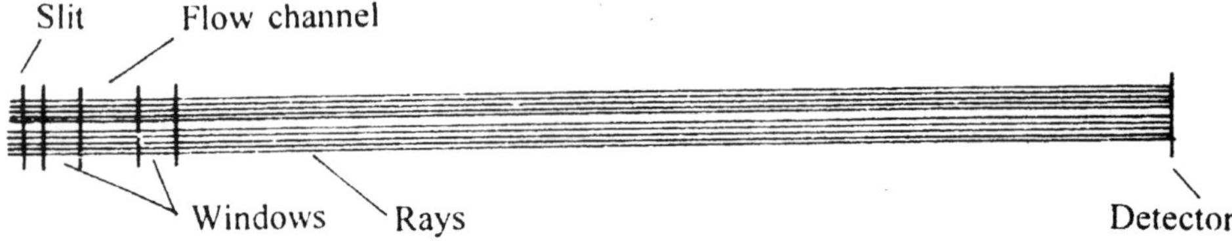

Light obscuration (LO) counting is a versatile light scattering technique because of the broad size range, approximately 1 to 2500 μm, and because of the flexibility in the choice of suspending fluid. Light obscuration is based on the fact that particles whose refractive index differs from that of the medium in which they are suspended scatter light away from a detector, as shown in Fig. 11.13. When a single particle is in the measurement volume at any given time, the amount of light scattered is proportional to the projected cross-sectional area of the particle.

Another common technique used in the particle size range of 1 to 2500 μm, known as **Fraunhofer diffraction,** suffers from a number of deficiencies. Fraunhofer diffraction has much lower size resolution than does light obscuration. The measurement technique for Fraunhofer diffraction is insensitive to systematic errors such as solubility of the sample in the suspending fluid or high particulate background in the fluid. Finally, Fraunhofer diffraction has been found to be unable to weight accurately the contributions of particles of widely differing sizes to the size distribution. The widespread use of Fraunhofer diffraction stems from the apparent ease of use and the reproducibility of the measurement.

Experimental Details

The light obscuration instrument consists of a sensor where the light scattering experiment occurs a pulse height analyzer, and generally a computer that handles all data storage and analysis. Some type of sample handling system is needed to pump the sample suspension through the sensor. The design of the sample handling system varies with each instrument, but can be as simple as a vacuum source with regulator, a large-volume vacuum flask to collect the fluid after it passes through the sensor, and a beaker for the sample suspension. Some attention must be given to the stirring of the sample suspension for particles that are dense or larger than 500 μm in diameter to ensure good mixing and representative sampling.

The pulses from the sensor are fed into a multichannel analyzer or similar electronic device that sorts the pulses based on the total change in light intensity as a consequence of scatter from the particles. To relate the pulse sizes to particle diameter, calibration using particles of known diameter is needed. The number of counts in each pulse size range is then communicated to a computer, where average diameters are calculated, graphic results are presented, and the data are stored.

For best, results, analysis parameters should be optimized as highlighted in the following paragraphs. Sensors are chosen according to the size of the material to be analysed. If the range of particle sizes in the material is not known, it can be determined by examining the material using a light microscope. The two principal factors to balance include avoiding a plugged sensor if particles are too large and having a small enough lower size limit to detect the majority of the particles. Common sensor ranges are 40 to 2000 μm, 10 to 500 μm, 5 to 250 μm, and 2 to 100 μm. Manufacturers' recommendations for solvent compatibility should be consulted when choosing a sensor.

The choice of a suspending fluid is crucial. The fluid should wet the particles of interest, allowing formation of a stable dispersion without causing dissolution. To check for dissolution of the material, the sample is prepared for analysis and a portion of the suspension is analysed. After 3 to 5 min of stirring, another portion is analyzed. Dissolution of the particles is indicated if the total particle count has decreased by more than 10%. The particle count may also be decreased by particle agglomeration or deposition on the sample suspension beaker.

When changing suspending fluids, care should be taken to wash out the old fluid. If the two fluids are not miscible, and intermediate solvent should be found that is soluble in both. Isopropanol is generally a good intermediate between most organic fluids and water.

It is imperative that the suspending fluid be free of contamination (particles) and bubbles. Bubbles will be detected by the sensor and included in the particle size analysis, so care should be taken to avoid creating bubbles when transferring fluid.

A 2- or 4-L samples suspension beaker should be used with the 500-μm or larger sensors. The beaker should contain a stirrer and baffle. The addition of a baffle has been found to be effective in eliminating the formation of a vortex. Bubbles will be formed from entrapped air whenever a vortex is present. The baffle also promotes mixing and homogeneity of the suspension. For smaller sensors, a 250–mL beaker will suffice and baffles are usually not necessary unless the particles are dense (greater than 3 g/mL).

The flowrate should be the same as that used to calibrate the sensor. The flow rate can be measured easily using a glass cylinder and a stopwatch or by measuring the weight loss per time. It is good practice to measure the flowrate frequently.

The background particulate level should be measured with a clean sample of fluid before sample analysis. The background level should be measured down to the lower size limit of the sensor. The background count rate should be no more than 2% of the expected count rate for the sample. If the background count rate is too high, an effort should be made to clean the beaker, baffle, and stir bar or the fluid should be allowed to stand for several minutes to allow bubbles to purge. If this does not reduce the background, the fluid must be filtered.

All glassware, stir bars, baffles, and other equipment must be kept free of particulate contamination. Multiple rinses with clean fluid are often necessary after cleaning the sample vessel. If a non-aqueous fluid is used, water must be removed from the sample handling system by drying or rinsing with a fluid that is compatible with the nonaqueous fluid and water. In addition, care should be taken to obtain a representative sample using techniques.

To be wetted by the suspending fluid (particularly water), some powders require a surfactant. It is important to ensure that the particles are wetted by the surfactant before adding them to a large volume of suspending fluid.

Dispersion of particles may be assisted by sonication. A stable dispersion will have a size distribution that is invariant to sonication. However, it is important to bear in mind the aim of the analysis when considering sonication. If the goal of the analysis is to determine the size of the ultimate particles, which are held strongly together, then sonication is appropriate. On the other hand, if the size distribution of the particles as they are, even if held lossely together, is needed, sonication is obviously inappropriate.

Dilute the sample such that the concentration is appropriate for the sensor. If the concentration is out of range, dilute or add material as appropriate. It often facilitates the measurement if the dilution steps are recorded so that they can be repeated.

The concentration of the sample should be high enough to count a statistically significant number of particles, but not so high as to create too much double counting or coincidence counting from having two or more particles in the sensing volume simultaneously.

The coincidence level can be calculated using the following equation

$$\text{Vdc} = \left(\frac{C * Sv * \sqrt{2}}{1 + C * Sv * R\sqrt{2}} \right) 100 \qquad ...(11.14)$$

where Vdc = the volume percentage of the measure size distribution due to coincidence counting, C = the number of particles per volume, and Sv = the volume of the sensing zone.

For samples with broad size distributions, the coincidence level should not exceed 10%. For samples with uniform size distributions, the effects of double counting will be observed above 2% coincidence level as a population of concentration-dependent particles in the size distribution at roughly $\sqrt{2}$ times greater diameter than the main mode of the size distribution.

A standard of known size should be measured with each series of samples. If the measured average diameter differs by more than 3% from the expected result, consider recalibration of the light obscuration system.

Calibration of sensors requires a series of standards which span the range of the sensor. Five standards, generally narrow-size-distribution styrene-divinylbenzene beds, that are supplied by a number of vendors, are generally adequate. Calibrations are generally stable for periods up to several months.

Each standard is analysed as a sample. The method by which the standard data are converted to a calibration depends on the instrumentation and software provided by the vendor. One procedure involves plotting cumulative counts versus pulse height in millivolts. The number median of the standard peak is chosen as the point halfway through the counts. The calibration is constructed by plotting the voltage that corresponds to the middle of the standards size distribution versus the diameter of the standard on a log-long plot, as shown in Fig. 11.14. The voltage and diameter pairs for the standards are then fit to the following expression:

$$\text{Diameter} = \left(\frac{\text{Voltage} - \text{Asymptote}}{\text{Intercept}} \right)^{1/\text{Slope}} \qquad ...(11.15)$$

where the asymptote, intercept, and slope are calculated using a least-squares fit of the calibration data. The slope is expected to have a value within 5% to 2 because the sensor measures to projected area of the particles. For spheres, the projected area is proportional to the square of the diameter.

It is also necessary to choose a lower size limit for each sensor. This lower limit must be greater than the noise limit for the sensor. The noise limit can be determined by having the sensor count without pumping fluid through the sensor. The voltage level at which the count rate is roughly one count per second can be used as a noise level. However, the calibration graph of voltage versus diameter should also be examined to ensure that the lower size limit is in the range where the logarithmic plot of voltage versus diameter is linear, as this plot tends to deviate from the expected slope of 2 as the lower limit of the sensor is approached.

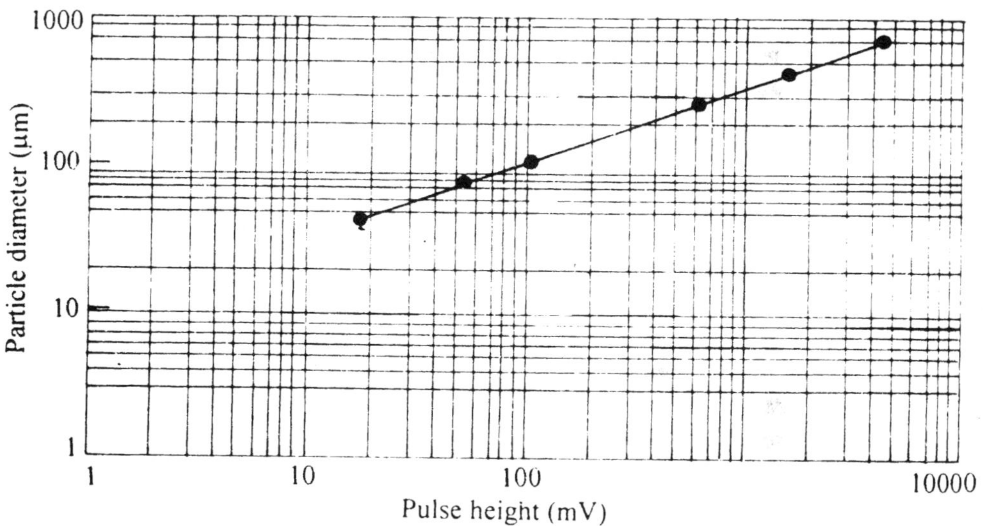

Fig. 11.14: Calibration plot (particle size versus pulse height) for a light obscuration particle size instrument.

An additional effect that should be considered when calibrating light obscuration sensors is the dependence of the light scattering on particle composition when the index of refraction of the particle is within 5% of that of the suspending fluid. If the contrast between the index of refraction of the particle and that of the fluid is not sufficiently large, the size of the particles will be underestimated. In that case, calibration with standards and suspending fluids having the same index of refraction ratio as the actual sample and suspending medium may suffice, provided that the sample has uniform composition.

Uses

The final result of particle size analyses using light obscuration counting instrumentation is size distribution that consists of number of particles versus diameter. From this result, a volume weighted size distribution can be calculated assuming spherical shape. Computer software can easily calculate a variety of averages from the counts versus diameter data. The volume or number median is often emphasized instead of the mean diameters because the median is a more robust measure of average size and it is not as strongly affected by a few outlying data points.

For spherical samples, the result measured by the light obscuration sensor will be accurate. However, for nonspherical particles, the accuracy of the size assignment depends on how closely the projected area is approximated by the square of the diameter. No useful size information can be measured from any instrumentation that uses an equivalent sphere assumption for particles such as rods and needles. Useful measurements can be made using light obscuration for particles that are irregular, but whose general shape does not deviate too strongly from that of a sphere. No quantitative guidelines are available, but in general if the aspect ratio of any two dimensions of the particle is less than 2, useful size information can be obtained by this instrumentation, which makes an equivalent sphere assumption.

Analytical Information

Qualitative

Light obscuration is generally used for quantitative particle size and concentration measurements. However, the output of a light obscuration experiment could be sent to a simple pulse height analyzer, allowing qualitative sample comparisons. Qualitative particle concentrations are determined from the total number of pulses registered by the pulse height analyzer, whereas qualitative information about the size distribution is obtained from the registered distribution of pulse heights.

Quantitative

Quantitative particle size analyses are obtained when there is a known relationship between particle size and pulse height. As described previously, such a relationship is obtained by measuring the distribution of pulse heights for several uniform particle size standards. Once a pulse height versus size relationship is obtained, the pulse height information is converted directly to a particle size. The signal processor places the individual size measurements into discrete size classes, thus forming as size distribution. The breadth and number of the size classes vary from system to system. Particle size distributions can typically be displayed in terms of particle, area, and volume.

A particle size distribution for a blend of uniform polystyrene standards is given in Fig. 11.15. The solid line shown on the distribution is the cumulative volume percent of the anlayzed particles. The individual bins represent the volume fraction within each individual size class. If the particles are spheres and the densities of all the analysed particles are the same, the volume fraction and weight fraction will be equivalent.

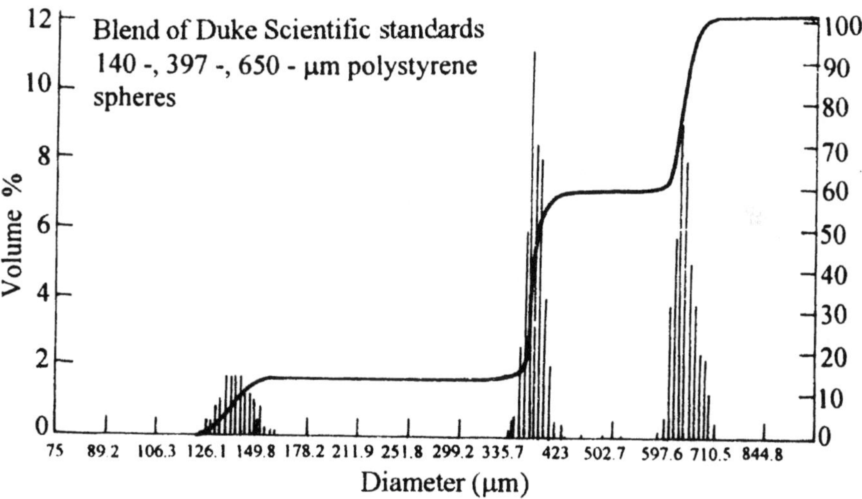

Fig. 11.15: Particle size distribution of a blend of uniform polystyrene standards.

For accurate size distribution measurements, one particle must be present in the measurement volume at any given time. As the concentration of particles in the suspending fluid increases, the probability of having more than one particle present in the measurement volume also increases. When multiple particles are present in the measurement volume, they appear as though they were one particle with a cross-sectioned area approximately equal to the sum of the individual cross-sectional areas. This double counting or coincident measurement generally leads to size distributions that are skewed to larger sizes. In general, the degree of coincidence should be kept below 5 to 10%, as discussed earlier in this chapter. The effects of coincidence can be readily observed with samples having uniform or narrow particles size distributions. At a coincidence level greater than approximately 1%, size distributions have a single population. At coincidence levels below approximately 2%, a distinct second population is observed in the measured size distribution. If further analysis of the new population is performed, one will see that the average size corresponds to a particle with a cross-sectional area that is twice as large as the population observed at low particle concentrations.

Applications

1. Characterizing the Effectiveness of a Screening Operation

Particle size measurements of a synthetic polymer resin were performed to determine the effectiveness of a sample fractionation procedure. Approximately 1 mL of a 1% aqueous solution of sodium lauryl sulfate (SLS) was added to 100 mg of a synthetic polymer resin. The SLS was used to facilitate dispersion of the resin into water by wetting the surface of the resin particles. The wet resin was then added to 4 L of distilled water, forming a dilute dispersion. Before the formation of a dispersion, the distilled water was filtered through a 0.1-μm filter. To create homogeneity, the dispersion was stirred using a magnetic stirrer. Baffles were added to the vessel that contained the dispersion to eliminate the vortex present and to promote mixing. A dip tube was added to the entrance of the LO sensor and a vacuum was applied to the exit of the sensor. The vacuum was adjusted so that the flowrate through the sensor was approximately 500 mL per minute. The sensor was calibrated with six different samples of uniform polystyrene shapers before the analysis. The analysis time was 3 min. Figure 11.16 gives the size distribution of the unfractionated resin, the original resin with the less-than-250-μm particles removed, and the original resin with the less-than-250-μm and greater-than-500-μm particles removed.

Analysis of the various fraction was consistent with the analysis procedure previously described.

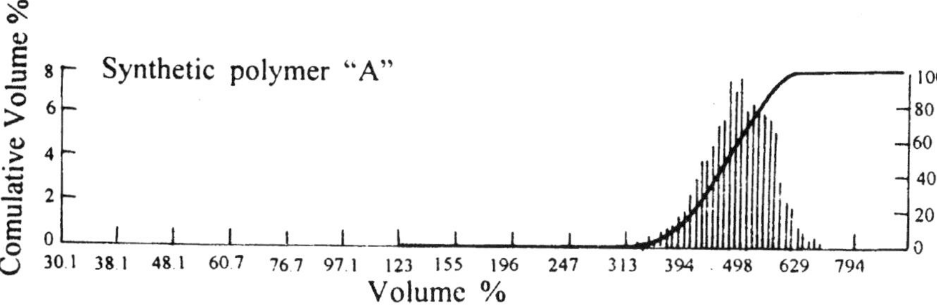

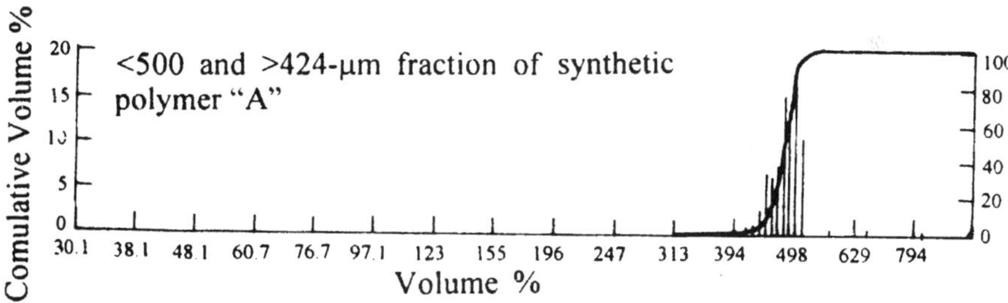

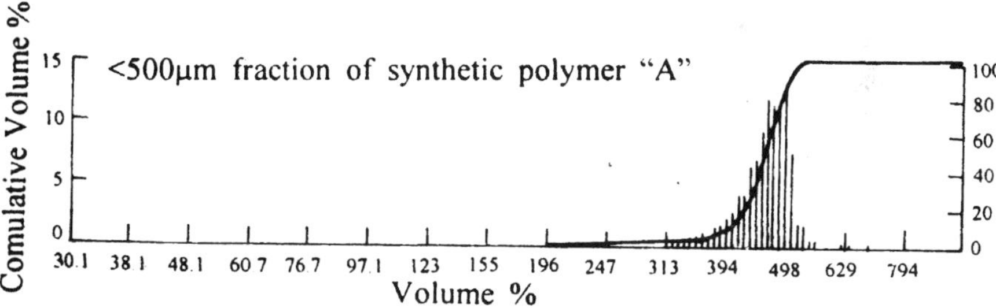

Fig. 11.16: Particle size distributions of a series of fractionated ion exchange resins.

2. Size Distribution Analysis of a Sample of Poly(Methyl Methacrylate) Beads

With an analysis procedure similar to the one described in the previous application, a sample of poly(methyl methacrylate) beads suspended in water were analysed using a light obscuration particle size analyzer. The particle size distribution of the beads is shown in Fig. 11.17.

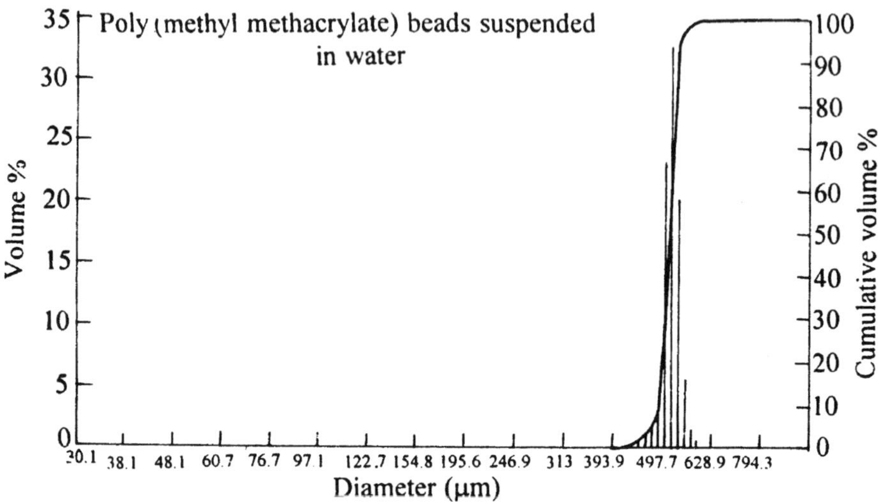

Fig. 11.17: Particle size distribution of a sample of poly(methyl methacrylate) beads.

11.5. Pyrolysis Measurement

Introduction

Pyrolysis refers the degradation of macromolecules by thermal energy alone. It is one of various commercially available degradative techniques that are used routinely for the characterization of synthetic polymers and other complex macromolecules. In the broadest sense, pyrolysis involves the application of thermal energy to a sample in the absence of oxygen. This results in breaking of molecular bonds and fragmentation into small molecular species that are related to the original sample composition. These small, characteristic molecules are used to qualitatively identify the structure of the original macromolecule and, with proper standardization, the characteristic small molecules can be used to determine quantitative information on the composition of the macromolecule.

Mechanisms for pyrolysis include *elimination of small neutral molecules, chain scission* or *depolymerization,* or *random cleavage.*

The mechanism of pyrolysis or themral degradation depends on the type of polymer or copolymer, whereas the types and *relative yields of various pyrolysis products* depend on variables including the *pyrolyzer type* and the *pyrolysis parameters.* To obtain reproducible pyrolysis data, these parameters must be optimized for the particular polymeric system and they must be carefully controlled. **Continuous-mode** and **pulse-mode pyrolyzers** are commercially available to control the necessary parameters required to give reproducible pyrolysis.

Since thermal degradation of complex macromolecules generally results in a complex mixture of smaller pyrolysis product molecules, gas chromatography provides easy separation of individual species. This coupling of the thermal degradation event (**pyrolysis**) with readily available separation science (**gas chromatography**) results in the technique known as **pyrolysis gas chromatography (PyGC)**. When the separation and detection system includes a gas chromatography mass spectrometer (**GC-MS**), the technique is known as **pyrolyis gas chromatography mass spectrometry (PyGC-MS)**. Thermal degradation of some macromolecules results in undesirable secondary reactions and some PyGC systems may result in discrimination or loss of degradation products during transfer from the pyrolyzer to the chromatographic system. When these problems occur or pyrolysis product separation is undesirable, a useful alternative involves pyrolysis within the mass spectrometer, giving rise to the technique called **pyrolysis mass spectrometry (PyMS)**.

Experimental Method

The polymer sample is rapidly heated in an oxygen-free atmosphere to a high temperature known as the *final pyrolysis temperature* (**FT**). Helium is generally the preferred atmosphere since its high thermal conductivity facilitates heat transfer from the sample and minimizes secondary degradation reactions, also helium is a common GC carrier gas. The time needed to raise the temperature of the sample from ambient (or another initial temperature) to this final pyrolysis temperature is the temperature rise time (TRT) and the total time required to raise the sample temperature and pyrolyze it at the final temperature is known as the *pyrolysis interval* or **total heating time (THT)**.

The relationship between these different temperatures and times depends on the tape of pyrolyzer. Historically, pyrolyzers can be classified as either *continuous-mode* or *pulse-mode systems. Furnace and microfurnace pyrolyzers,* that are preached to the desired final pyrolysis temperature before introduction of the sample, fall into the category of continuous-mode pyrolyzers. Fig. 11.18 shcws a schematic for a commercial microfurnace pyrolyzer.

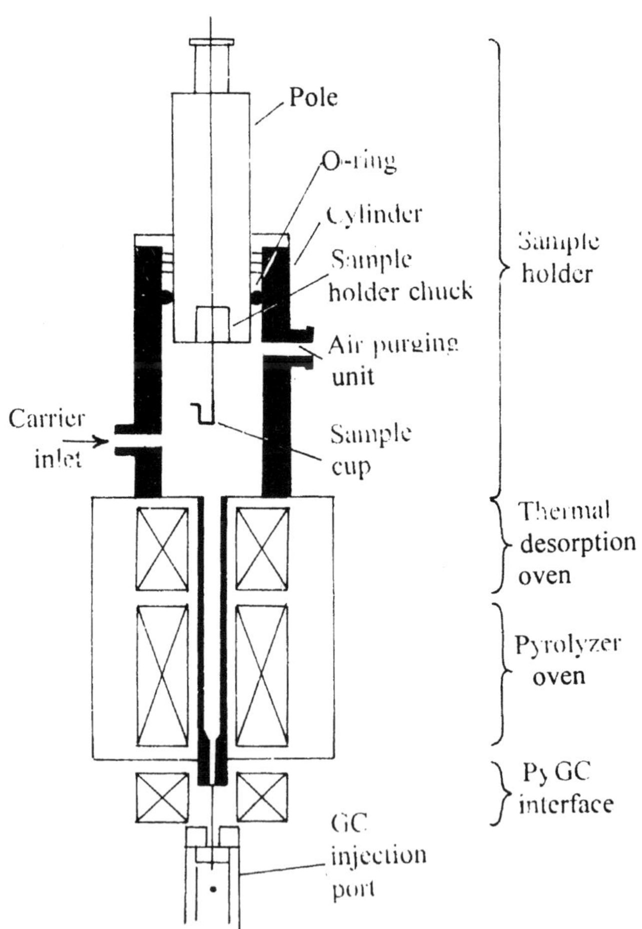

Fig. 11.18: Schematic of a two-stage microfurnace pyrolyzer.

Pulse-mode pyrolyzers include systems using resistively heated filaments, shown schematically in Fig. 11.19, or ferromagnetic metal wires inductively heated with radiofrequencies. The latter, shown schematically in Fig. 11.20, are called **Curie point (CP) pyrolyzers.** Although not commonly used for pyrolysis of polymers, **laser-**

pyrolysis systems are another class of pyrolyzer.

The reproducible yield of characteristic small molecule fragments from a polymer, known **pyrolysis products,** is also affected by the cleanliness of the sample holder and other sampling considerations. Pyrolysis using a contaminated surface can result in catalytic effects that can drastically alter the type and quantity of pyrolysis products. Heating Pt filaments in air at 1000°C is adequate to remove most organic residues, although some inorganic materials (like fillers) may remain behind even under these conditions. Cleaning Curie point wires or filaments in a flame, however, may form metal oxides that will affect subsequent pyrolysis results. For Curie point systems, the position of the wire in the induction coil also greatly affects the nature and the amount of pyrolysis products.

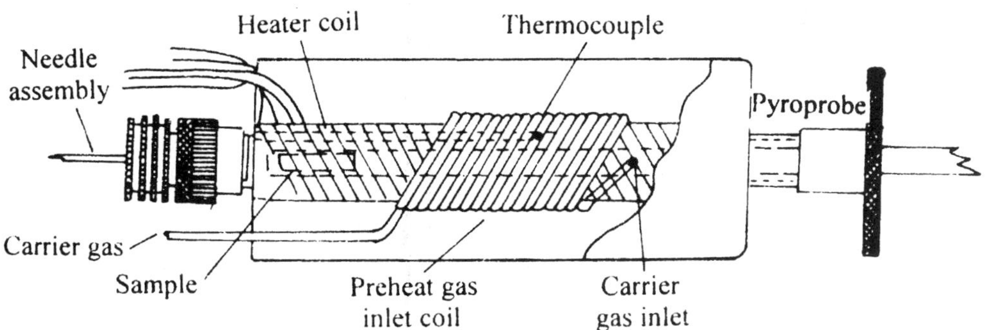

Fig. 11.19: Schematic of a resistively heated filament pyrolyzer.

Techniques of sample deposition can also affect the quality of pyrolysis results. For filament and Curie point pyrolyzers, use of uniform, thin polymer coatings gives the best results. Samples are normally deposited as a few microliters of a polymer solution or, in the case of Curie point wires, simply dip-coated from a polymer solution. With microfurnace pyrolyzers or filament systems terms using coiled filaments with quartz tubes should be carefully washed and heated before use. Sample sizes should be kept small to facilitate good heat transfer from the pyrolyzer to the sample.

Uses

Pyrolysis gas chromatographic system need interfacing the pyrolyzer with the injection port of the gas chromatograph. This is generally achieved using a simple needle through the septum of any commercial chromatogrpah. Other pyrolyzers need a separate interface or pyrolysis chamber, which is connected to the chromatographic injection port by a simple needle assembly. For any of these PyGC systems, this interface is critical to successful, reproducible pyrolysis experiments. **Cold spots** (placed where pyrolysis products can condense before injection into the chromatograph) or **overheated zones** that facilitate secondary thermal degradations will result in loss or undesirable differentiation of significant pyrolysis products. Therefore, care must be given to adequately heat needle assemblies and any transfer lines from the pyrolyzer to the gas chromatographic inlet. Thus, direct PyGC-MS experiments eliminate the requirement for these interfaces and the potential loss of reactive or polar pyrolysis products.

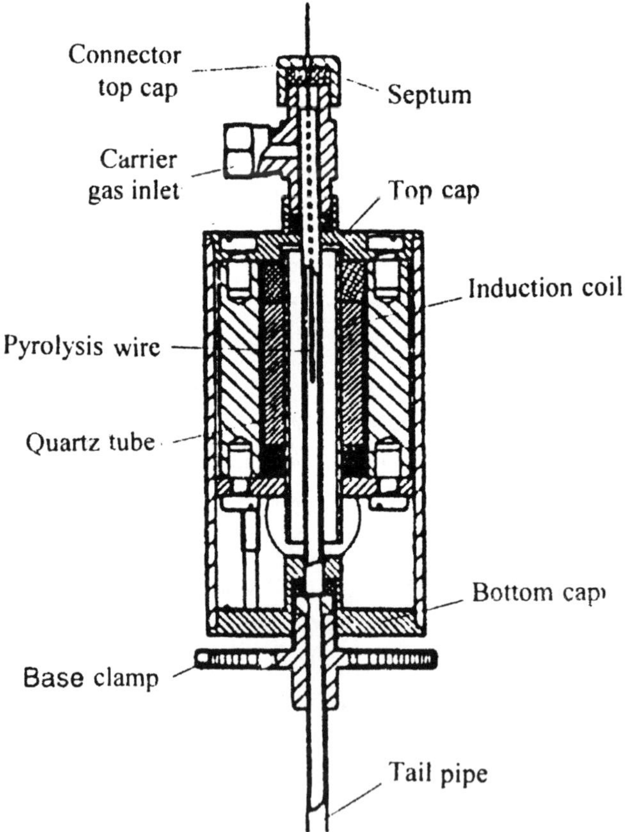

Fig. 11.20: Schematic of an inductively heated Curie point pyrolyzer.

The quality of the PyGC data depends on the quality of the chromatographic system. Generally, current PyGC systems use fused-silica capillary columns, that give a high degree of resolution to pyrolysis products. *Conventional thermal conductivity detectors* (**TCDs**) and *flame ionization detectors* (**FIDs**) are widely used to detect the separated pyrolysis products, although electron capture or flame photometric detectors can be used to obtain specificity for specific elements (such as Cl, S, and N) in the pyrolysis products.

Pyrolysis mass spectrometry (**PyMS**) systems eliminate some of the problems associated with the transfer of pyrolysis products from an external pyrolyzer to a gas chromatograph. Curie point systems are commercially available that can be interfaced with mass spectrometers so that pyrolysis takes place near the ion source. Transfer of temperature profiles from PyGC to PyMS may not be totally valid because the cooling processes are slower in the vacuum environment of a mass spectrometer. The direct insertion probe option for many spectrometers allows heating samples at temperatures up to 500°C directly in the ion source. Some instrument suppliers also provide pyrolysis systems that are used directly in the ion sources as a replacement for the conventional direct insertion probe.

Tunning of the pyrolysis temperature is important to obtain reproducible pyrolysis data using filament or pulse-type pyrolyzers. This standardization or tuning of the pyrolysis temperature is extremely important for kinetic studies or attempts to differentiate between homopolymer blends and copolymer based on changes in the programs. **Curie point wire temperature** is a well-defined physical parameter based on the metal alloy composition of the wire. The temperature repeatability for these Curie point wires is dictated by the quality of the metals used to manufacture the wires. For pulse-mode pyrolyzers, each filament probe is accompanied by a calibration table that correlates the controller set point with the resulting filament temperature as measured with an optical pyrometer. Assuming a clean filament probe, this calibration table serves as a guide to the user to define the controller settings needed to obtain a desired probe temperature. Obviously, contaminations that alter the thermal characteristics of a probe will invalidate the calibration table.

To compensate for changes in filament probe calibration (that is, actual temperature relative to the set temperature), Levy proposed the *model molecular thermometer approach for tuning pyrolyzers*. This approach was based on the pyrolysis of a styrene–isoprene (86/14%) copolymer whose fragmentation was highly sensitive to the actual pyrolysis temperature. Under a strict set of experimental conditions the ratio of major pyrolysis products (ratio of isoprene:dipentene) varied linearly with the pyrolysis temperature. By pyrolyzing this temperature-sensitive copolymer at different temperatures and examining the relative yields of isoprene and depending, an analyst can evaluate variations in the actual temperatures achieved by the pyrolyzer.

Automation of the pyrolysis process is achieved using systems like the Fischer AP-12 pyrolyzer, that is shown schematically in Fig. 11.21 . This unit, readily adapted to the injection port of a gas chromatograph, accommodates 12 Curie point wires, which are placed into separate induction fields. Using a series of valves and an accurate timing sequences, this unit permits unattended pyrolysis of the 12 samples with a high degree of accuracy and precision.

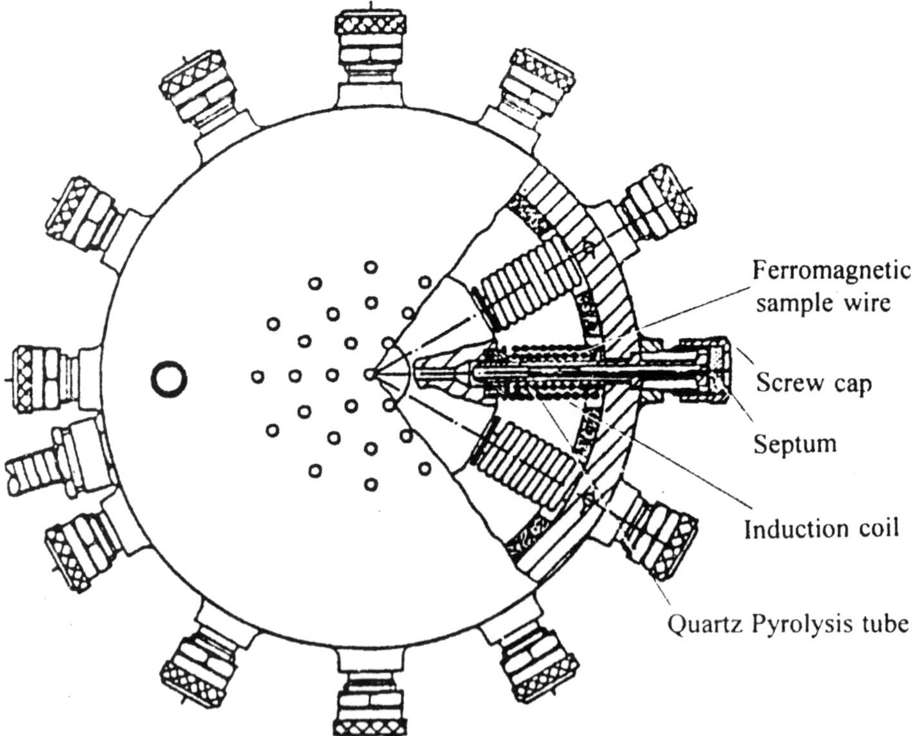

Fig. 11.21: Cross-section schematic of the Fischer automated Curie point pyrolyzer.

Older automated systems were based on commercially available solid sample injectors, but CDS Analytical now markets a special coiled filament pyrolysis insert to replace conventional gas chromatographic injection port liners. This cold filament accommodates automated injection of polymer solutions using commercial autosampler technology.

Analytical Information

Qualitative

Each type of polymeric material gives a unique set of pyrolysis products based on different mechanisms of thermal degradation. Polymer degradation mechanisms are generally classified as *depolymerization,* or *unzipping; elimination,* which features loss of a small, neutral molecule possibly followed by kick-back fission or cyclization; and *random cleavage,* which occurs when bond energies in a polymer are similar and rearrangement reactions are not possible. Table 11.1 lists examples that show these various degradation mechanisms for vinyl polymers and some condensation polymers. Gas chromatographic separation of these pyrolysis products results in a fingerprint or pyrogram that is considered to be unique for each type of polymer. For PyMS experiments, the mass spectrum is a composite of the mass spectra of the individual pyrolysis products.

Table 11.1: Examples of typical polymer degradation mechanisms.

Mechanism	*Polymer*	*Major Pyrolysis Products*
Depolymerization	Polystyrene	Styrene, dimers, trimers
	Poly(methyl methacrylate)	Methyl methacrylate
	Styrene-butadiene rubbers	Styrene, 1,3-butadiene, 4-vinylcyclohexene, oligomers
	Polyamide (Nylon 6)	Monomer(ε-caprolactam)
Elimination	Poly(vinly chloride)	HCl, benzene
	Poly(vinyl acetate)	Acetic acid
Random cleavage	Polyethylene	Homologous series of alkenes, alkanes, and dienes
	Polyurethanes	Disassociates, mono-ols, diols, and cyclic compounds

Qualitative identification of an unknown polymer by PyGC is accomplished by comparison of the sample pyrogram which reference chromatograms from pyrolysis of standard polymers. Tsuge and Ohtani prepared a compilation of programmes for polymer using methyl silicone fused silica capillary column technology. This and other publications are sources for reference pyrograms. If available, standard reference polymers can be pyrolyzed and characterized using the same system used for the unknown polymer.

Instead of using the whole pyrogram, some workers recommend use of several specific pyrolysis products as the basis for identification of unknowns by PyGC. Table 11.2 lists some of the useful comparisons.

Table 11.2: Individual pyrolysis products useful for identification of synthetic polymers.

Polymer	*Pyrolysis Products*
Polyisoprene	Isoprene, dipentene
Polybutadiene	1,3-Butadiene, 4-vinylcyclohexene
Butyl rubbers	Isobutylene
Nitrile rubbers	1,3-Butadiene, acrylonitrile
Polystyrene	Styrene, oligomers
Polyacrylates	Acrylate monomer
Polychloroprene	Chloroprene
Styrene-butadiene	Styrene,1,3-butadiene,4-vinylcyclohexene

Quantitative

Quantitative aspects of pyrolysis experiments include the *calculation of copolymer composition, microstructure measurements,* and the *determination of kinetic and degradation parameters.* The following paragraphs detail some of the important quantitative aspects of pyrolysis gas chromatography.

Composition calculations based on PyGC data need identification of one specific and unique pyrolysis product peak that is related to each monomeric component of a copolymer or homopolymer blend. In the example of an acrylonitrile–butadiene–styrene (ABS) compolymer pyroloysis yields characteristics peaks for the three monomers (acrylonitrile, 1,3–butadiene, and styrene). To quantify the normalized composition of a sample such as an ABS copolymer, a series of equations are written to define the relationship between the unique monomer peaks in the pyrogram and the concentration of these various moieties in the copolymer backbone.

$$K1 = \frac{P_s}{P_a} \div \frac{A_s}{A_a} \qquad K2 = \frac{P_b}{P_a} \div \frac{A_b}{A_a} \qquad ...(11.16)$$

$$P_s = K1\,P_a \frac{A_s}{A_a} \qquad P_b = K2\,P_a \frac{A_b}{A_a} \qquad ...(11.17)$$

$$P_a + P_b + P_s = 100\% \qquad ...(11.18)$$

where P_a, P_b, P_s = weight % monomer units in a copolymer of acrylonitrile, butadiene, and styreen and A_a, A_b, A_s = monomer peak area responses from the pyrogram for the respective monomers.

Copolymer standards of known composition are pyrolyzed. The area responses for the monomer peaks and the known composition data are substituted into Eqs. (11.16) and these equations are solved to determine the value of constants $K1$ and $K2$. Once the values for these constants are known, these same equations, rearranged as shown in Eqs. (11.17) and (11.18), and the respective peak areas from the unknown sample pyrogram are used to calculate concentrations P_a, P_b, *and* P_s for the unknown.

Standardization for quantitative analyses base on PyGC data requires careful choice of reference polymers. Pyrolysis of homopolymer blends results in product peaks that are unique for each component of the blend. Pyrolysis of a copolymer containing the same monomeric moieties, however, results in a pyrogram containing peaks for hybrid oligomers in addition to these unique monomeric species. Yields of unique monomeric pyrolysis products can be significantly different for blends and copolymers that have the same nominal composition because monomer yield from a homopolymer blend may be independent of polymer composition. As a result, use of blended standards to calculate the composition of copolymers may lead to erroneous results.

Fig. 11.22 compares 1,3-butadiene yields from pyrolysis of a series of polystyrene-polybutadiene blends and rubber-modified impact polystyrene samples with the same nominal compositions and similar rubber microstructure. Note that the difference in 1,3-butadiene yield between the blend and the true copolymer increases with increasing nominal concentration of polybutadiene. The slopes of these correlation curves are considerably different for samples containing higher levels of polybutadiene. Based on a 100-μg sample, a 1,3-butadiene peak height was equivalent to 6% polybutadiene in a blend, but the same peak height was equivalent to 6.7% polymerized butadiene in a styrene-butadiene copolymer.

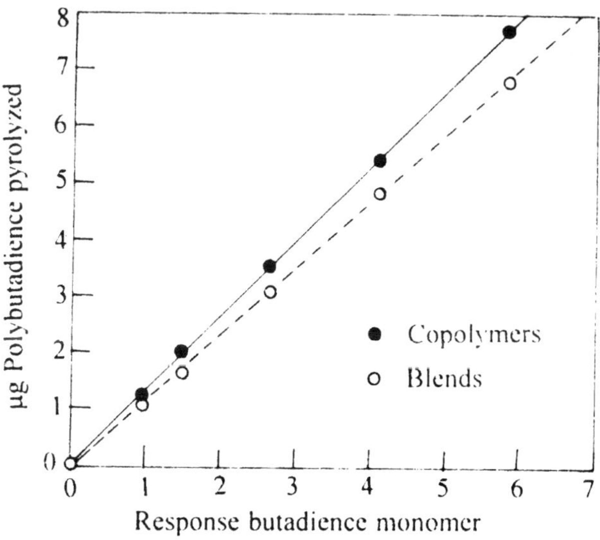

Fig. 11.22: Comparison of 1,3-butadiene yields from pyrolysis of copolymers and homopolymer blends with polystyrene.

Using an internal standard may improve the precision when PyGC data are used to calculate the concentration of a single polymeric component of a blend or copolymer (such as polybutadiene in a polystyrene sample). To be useful as an internal standard for PyGC experiments, a polymer must degrade (fragment) quantitatively into a single pyrolysis product and the peak for this product must occur at a retention time free of interference by peaks for products from pyrolysis of the sample. For example, poly(ethyl methacrylate) pyrolyzes to give predominantly ethyl methacrylate monomer, so this polymer was used as an internal standard in the PyGC examination of other acrylic polymers.

Microstructure measurements from pyrolysis data take advantage of the dependence of product yields from pyrolysis of copolymers on the effects of boundaries between the different monomeric units. For example, the pyrolysis of vinyl chloride (V)–vinylidene chloride (D) copolymers follows a dehydrochlorination–cyclization mechanism. Copolymer molecules initially lose HCl, then the resulting six-membered alkene chains cyclize to form chlorinated benzenes. The yields of monochloro benzene, m-dichlorobenzene, and 1,3,5-trichlorobenzene are related to the relative number of various VVD, VDV, VDD triads in the copolymer, as shown in Table 11.3.

Table 11.3: Triad sources for products from pyrolysis of vinyl chloride (V)-vinylidene chloride (D) copolymers.

Pyrolysis Product	*Sequence Designation*	*Contributing Triad Sequences*
Benzene	V3	VVV
Chlorobenzene	V2	VDV,DVV,VVD
m-Dichlorobenzene	V1	DVD,DDD,VDD
1,3,5-Trichlorobenzene		DDD
Vinylidene chloride		DDD

Pyrolysis followed by high-resolution capillary gas chromatography has been used to study cross-linking in styrene–divinylbenzene (DVB) copolymers. Linear correlations were obtained between the normalized yields of DVB isomers (counts per microgram) and the known DVB content of styrene–DVB copolymers. Similar correlations were observed using the isomers of ethylvinylbenzene.

Kinetic calculations are made from data obtained by pyrolyzing samples over a range of temperatures using different periods of time. From these pyrolysis data, it is possible to calculate the *isothermal pyrolysis rate for the polymer at different temperatures*, then the Arrhenius equation is used to determine the activation energy for the degradation. With pulse-mode pyrolyzers, a sequential pyrolysis approach was used by Andersson and Ericsson to obtain these data. The same polymer sample was pyrolyzed for a fixed time at some present temperature. The product yield decreased for each subsequent pyrolysis and the process was repeated until no additional product peaks were observed. Data from these experiments were used to calculate the rate constant for formation of the various pyrolysis products. Repeating the experiment at other temperatures gave data useful in calculation of Arrhenius parameters.

Applications

1. Identification of Components in a Polymer Blend

Fig. 11.23 compares the pyrograms using a 60-m fused silica capillary column (J&W DBWAX; 0.25 μm film) from pyrolysis (CDS Pyroprobe set temperature = 700°C) of 300 μg of an unknown polymer blend and comparable comparable portions of two reference polymers. Highlighted peaks in the ABS copolymer pyrogram include acrylonitrile (AN), 1,3-butadiene (B), and styrene (S), whereas the major peak from pyrolysis of nylon 6 polyamide was ε-caprolactam (C). Examination of the unknown blend pyrogram shows the presence of these peaks attributable to the pyrolysis of the ABS copolymer and nylon 6 components of the blend.

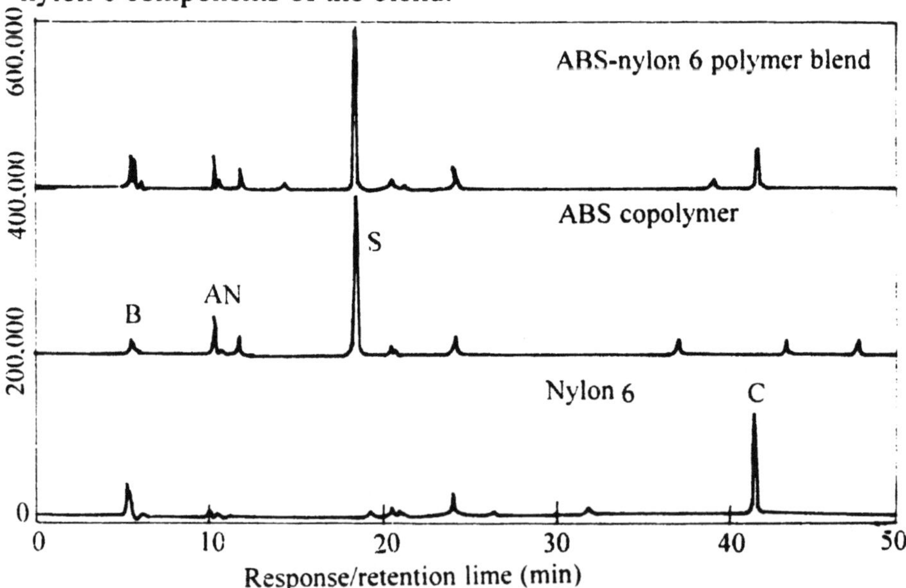

Fig. 11.23: Pyrograms of a polymer blend (ABS–nylon 6) and the respective polymer standards.

2. Identification of Poly(methyl methacrylate) Contamination in a Styrene Copolymer

The increased selectivity and sensitivity of the PyGC-MS technique are demonstrated in this example. Fig. 11.24 compares the partial (0- to 5-min region) ion current pyrograms for a contaminated copolymer sample and a sample of the virgin copolymer. From mass spectral data, the peak at 2.5 min in the contaminated sample pyrogram (M/Z = 41, 39, 69, 100, 99) was identified as methyl methacrylate (MMA) monomer from the depolymerization or unzipping of poly(methyl methacrylate). Pyrolysis of a series of blends containing poly(methyl methacrylate) and the virgin copolymer resulted in a curve correlating the yield of MMA monomer with he concentration of poly(methyl methacrylate) in the blend. The level of PMMA contamination in the sample was estimated using this correlation curve and the MMA peak response in the contaminated sample pyrogram.

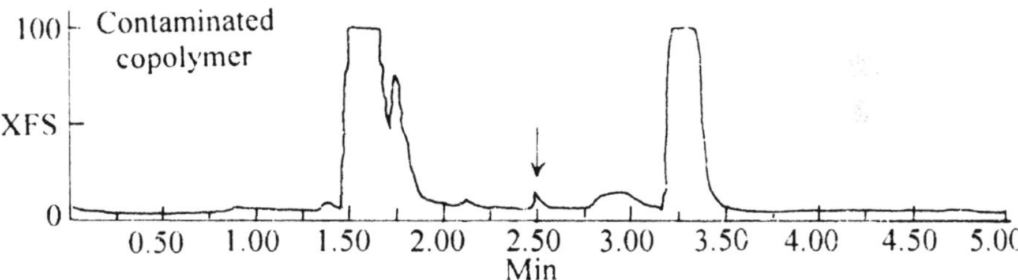

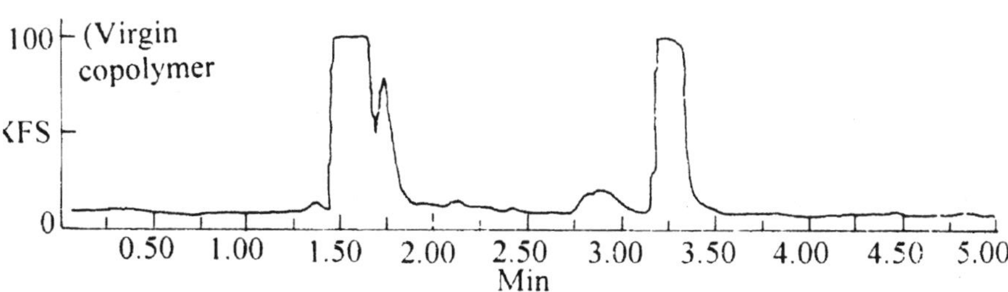

Fig. 11.24: Ion current pyrograms of a copolymer sample contaminated with poly(methyl methacrylate and a virgin copolymer standard.

3. Determination of the Vinyl Chloride Triad Sequence Distribution of Vinyl Chloride–Vinylidene Chloride Copolymers

The pyrogram (50°C) of a vinyl chloride–vinylidene chloride copolymer, presented in Fig. 11.25 shows resolution of vinylidene chloride, benzene, and chlorinated benzene isomers, which are related to the various triad sequences as summarized in Table 11.3. Chromatography of a synthetic mixture containing benzene, vinylidene chloride, chlorobenzene, m-dichlorobenzene, and 1,3,5-trichlorobenzene allows calculation of molar response values for these pyrolysis products relative to benzene. Division of the peak areas in a sample pyrogram by these relative molar response values gives the respective molar areas which are used in Eq. (11.19) to calculate the mole fraction vinyl chloride (*V*) in the copolymer.

$$V = \frac{B + (2/3)C + (1/3)D}{\Sigma A + B + C + D + E} \qquad \text{...(11.19)}$$

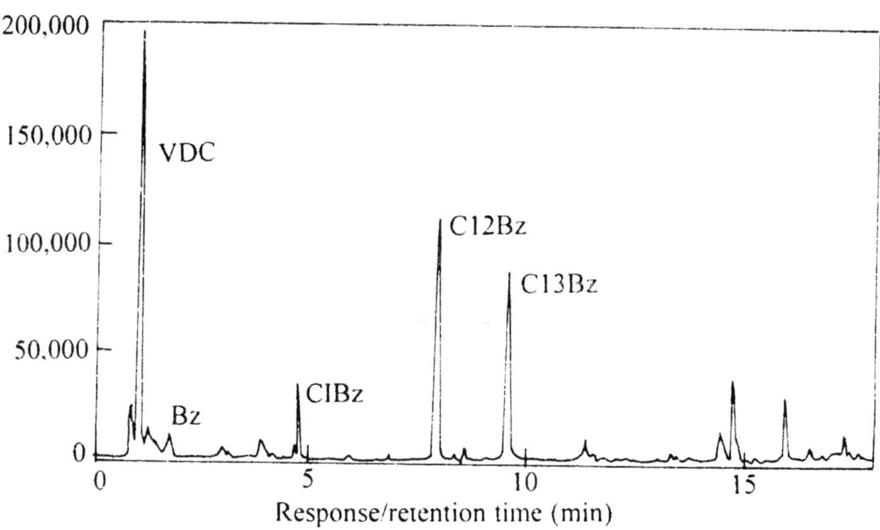

Fig. 11.25: Pyrogram for a vinyl chloride–vinylidene chloride copolymer.

In Eq. (11.19), A = 1/3 molar area of vinylidene chloride peak, B = molar area of benzene peak, C = molar area of chlorobenzene peak, D = molar area of m-dichlorobenzen peak, and E = molar area of 1,3,5-trichlorobenzene peak.

Equation (11.20) and the respective molecular weights for vinyl chloride (62.5) and vinylidene chloride (96.95) are used to convert from mole fraction vinyl chloride to the corresponding weight percent vinyl chloride in the copolymer sample.

$$\text{Wt\% Vinyl chloride} = 6250V/62.5V + 96.95\,(1 - V) \qquad \text{...(11.20)}$$

Finally, the relative vinyl chloride (VC) percentages in the triads described in Table 11.3 are calculated using the molar area responses, the total VC concentration, and Eqs. (11.21), (11.22), and (11.23).

$$V3 = (B)\%\text{VC}/B + 2/3C + 1/3D \quad ...(11.21)$$
$$V2 = (2/3\text{C})\%\text{VC}/B + 2/3C + 1/3D \quad ...(11.22)$$
$$V1 = (1/3D)\%\text{VC}/B + 2/3C + 1/3D \quad ...(11.23)$$

10.6. Thermal Analysis Techniques

Thermal analysis for polymers is a series of techniques, including calorimetry, TGA, TMA, Dynamical mechanical analysis (DMA), and dielectric analysis, which are used to characterise different aspects of the polymer systems. Thermal analysis methods often cannot be used for identification purposes by themselves without data from some complementary techniques. For example, a polymer with a melting point of about 220°C can be either nylon 6 or poly(butylene terephthalate). If an amine functional group appears in the IR spectrum of the same material would be a nylon, but may not be specific as to which type. Using both pieces of data, thermal and IR, the material could be identified as nylon 6, demonstrating the complementary nature of thermal analysis.

Thermal analysis most useful for **quantitative analysis** and **studying the state of a polymer.** Thermal methods are good qualitative tools because they are used to rapidly determine whether a polymer is the same or different from a standard, or to determine the quantity of a component after it has already been established that the component is present in the system. Characterizing the state of a system can include determining physical properties, like *heat capacities* and *linear and volumetric expansion coefficients,* as also reaction, including *degradation reactions* and *kinetics.*

Following a brief description on the theory of each technique, several examples of the more usual applications are given to show how these techniques can be applied to polymer characterization.

Experimental Methods

1. DSC

Differential scanning calorimetry (DSC) is the most commonly used calorimetric method because it is rapid, easy to operate, and readily available. In DSC, a sample contained in an appropriate holder and an empty reference holder are placed in the DSC cell in the sample and reference positions. Power is applied to heaters in the system to either ramp the temperature of the DSC at a specified rate (such as 10°C/min) or to hold the DSC isothermally at a given temperature. The DSC instrument *measures the difference in the heat flow between the sample and the reference.*

Differential scanning calorimeters are classified as either a *power-compensated DSC* or *a heat-flux DSC*. In the power-compensated DSC, the temperature of the sample and reference are measured and compared as shown in Fig. 11.26. Depending on the difference in observed temperature, more or less power is applied to the sample to adjust the sample temperature so that it is the same as the reference temperature, and the instrument records this change in power. In a heat-flux DSC, a single heating element is used to heat both the sample and the reference, which are positioned on a thermocouple as shown in Fig. 11.27. The difference in temperature is then measured and used to determine the heat flow.

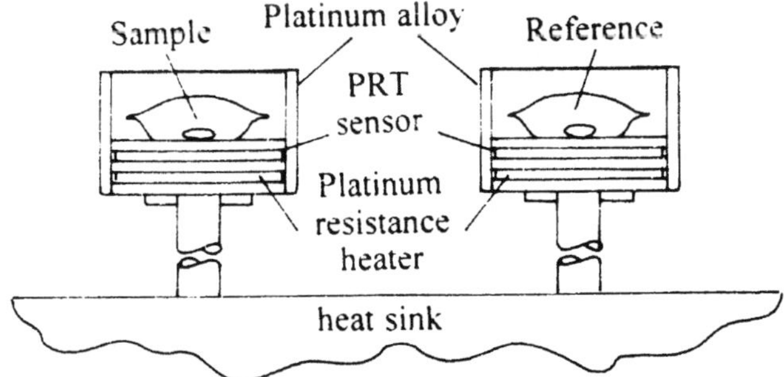

Fig. 11.26: Schematic of a power-compensated DSC. The computer generates the program temperature (T_p) signal and the platinum resistance thermometer (PRT) sensor detects a small error signal between T_p and the sample temperature. The resistance heater provides power to each heater to keep both sample holders on program. The difference in power required between the sample and reference is amplified and recorded.

A DSC experiment involved encapsulating a small portion (0.5 to 100 mg. generally 2 to 10 mg) of the sample into a sample pan. Sample pans are generally made from aluminium, although pans made from gold, stainless steel, and in some cases even sealed glass ampoules can be used for DSC. A gas purge is applied when the sample is placed into the DSC cell. In different experiments, performing the study in an inert atmosphere such as nitrogen is desirable; in other cases the effects of oxidation may be studied by using an air or oxygen atmosphere. Most commercial DSCs are able to run at low temperatures, ranging to about −150°C, although in practice special care needed to obtain good data below about−100°C. Most commercial DSCs have an upper temperature limit of about 700°C, which is more than adequate for most polymer studies. There are instruments that will go to 1700°C, but these are used mainly in the study of inorganic materials.

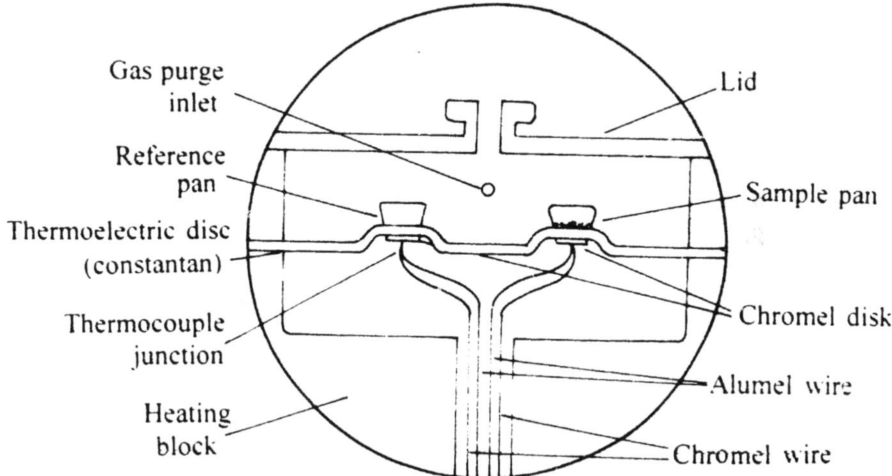

Fig. 11.27: Schematic of a heat-flux DSC.

Once the sample is the DSC cell with the desired purge, the temperature is adjust to 10 to 40°C below where the first data point is desired because it can take 1 to 2 min for the DSC to come to an equilibrium heating rate. Until this equilibrium rate is achieved, heat flow data represent equilibration of the cell, not the sample.

Most commercial DSCs make use of a computer interface to set up temperature programs, control the DSC during the heating program, and collect data.

As part of the computer control package, most instrument also are provide with software packages for the analysis of the collected data. The software helps to determine melting points and glass transition temperatures and to integrate peaks to quantify the heat from exotherms or endotherms observed in the DSC scan.

DSCs are calibrated for both heat and temperature and standards exist for both temperature and enthalpy calibrations. For enthalpic calibrations, the melting of a known quantity of a material with a well-characterized heat of fusion (indium is probably most often used) is measured and a correction factor determined. Temperature calibrations are done by measuring the melting point for well-characterized materials and comparing the observed melting point to the literature value. Temperature calibrations should be performed over a range wide enough to include the temperature range being studied for a particular sample.

2. TGA

TGA is used to follow the change in mass of a sample as it is heated or held isothermally at a specific temperature. For TGA, the balance mechanism is placed outside the furnace area to maintain good mass sensitivity. Manufacturers use either a horizontal or vertical configuration for the relationship of the balance to the furnace, with wires or rods connecting the sample and the balance, to allow mass determination. A typical TGA schematic is given in Fig. 11.28. Both configuration are subject to drift with increasing temperature. While the balance is outside of the furnace area, temperature effects remain on the ability to measure weight. Once common problem is buoyancy effects as the sample is heated, particularly for instruments where the sample hangs vertically from the balance. The horizontal furnace design was developed to overcome this problem; however, in the case the material used to connect the sample to the balance, generally a quartz rod, expands upon heating and thus the force (mass times distance) measured by the balance changes with changes in the length of the rod. The TGA instrument interprets this as a change in mass.

Operating temperatures for TGA typically range from about room temperature to 1000°C, however some instrument have a maximum temperature of 2400°C. The 1000°C temperature range permits the use of quartz tubes for containing the experiment. TGA must use high-temperature inert sample holders, and quartz, ceramic materials, and platinum are the most commonly use materials of construction. In a TGA experiment, a gas is purged through the system to provide a suitable atmosphere for the measurement and remove decomposition products from the sample chamber. Purge gases for TGA experiments are generally nitrogen (inert) or air (oxidizing) although other atmosphere are also used. Care must be used when a reducing atmosphere is desired because the use of hydrogen as a purge gas can create a serious hazard. Decomposition products for polymer systems often are *condensable tars,* which must be trapped or flushed from the system. Coating parts of the sample holder system with these tars results in recording an inaccurate weight loss. Consequently, TGA experiments must be performed in well-ventilated areas, like a hood, so as avoid buildup of degradation gases in the area.

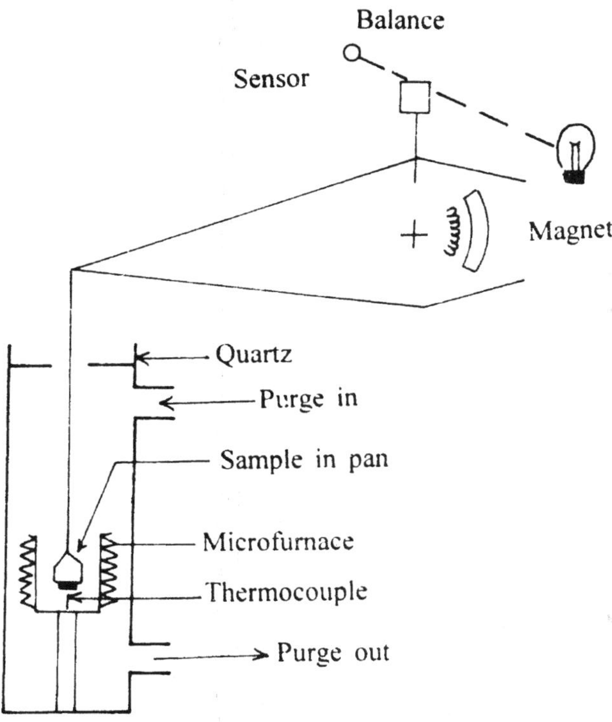

Fig. 11.28: Schematic of a typical TGA balance and heater assembly. As the sample weight changes, the balance tips, resulting in a change in the light hitting the sensor. The sensor output is amplified and provides current to the coil, which restores the beam position. This restoring current is the ordinate (y) and the thermocouple signal is the abscissa (x) of a conventional TGA plot.

For a TGA experiment, sample mass ranges from about 2 mg to as much as 3 g in some systems, with typical sample masses being 5 to 150 mg. Many typical TGA system balance are sensitive to better than 5 μm (not taking into account buoyancy problems), giving TGA a sensitivity that can each 10 ppm.

Temperature calibration in TGA experiments is done at a series of temperatures using either melting point standards or the Curie temperature of ferromagnetic materials.

3. TMA

Thermomechanical analysis (TMA) and thermodilitometry are used to characterize volumetric changes in polymer systems. These changes are generally used to characterize the changes in the physical state of a polymer and are critical for many applications of polymer systems. A commercial TMA, shown in Fig. 11.29 consists of a dimensionally stable sample holder (quartz is typical), a measuring probe (again with good dimensional stability), a device to measure the deflection of the probe, a linear variable differential transformer (LVDT), and a temperature control unit. TMA experiments can be

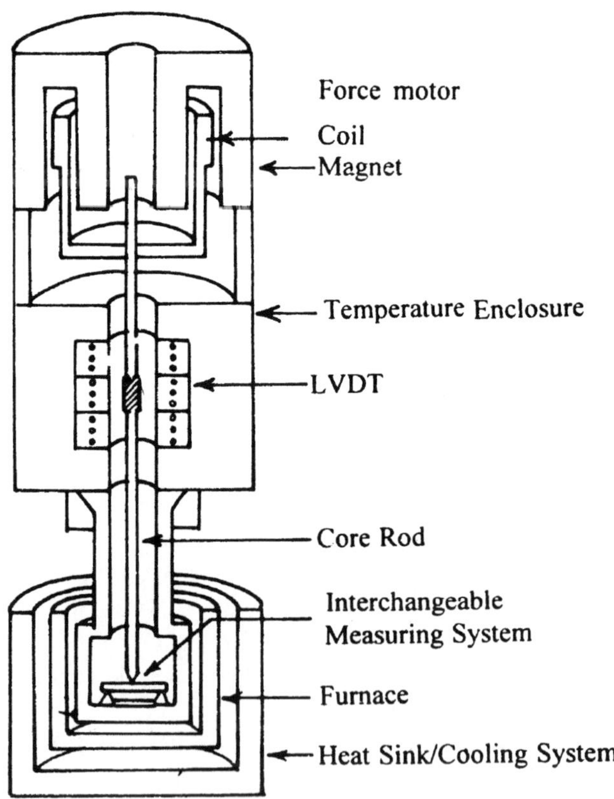

Fig. 11.29: Schematic of a typical TMA assembly. Current is supplied to the force coil to support the weight of the core rod and probe. Additional current is provided to apply the desired force on the sample. The LVDT position transducer provides a voltage proportional to the deflection from center null. As the furnace heats, the sample expands or softens and the analyzer output is the sample height versus sample temperature.

run from about—100°C to 700°C; however, the upper use temperature is generally much lower and is dictated by the thermal stability of the polymer sample. Because it is critical to maintain good contact between the probe and the sample, a load is generally applied to the system. The load is typically a force of a few grams, although newer TMAs have the ability to electronically set the force to range from 0.1 to about 100 Newtons. Probes for TMA instruments can range from a fine-tripped quartz rod for measuring penetration of the sample to various large, flat surfaces intended to measure the expansion of the sample.

Although the instrument is designed and constructed with materials of low thermal expansion, obtaining accurate results needs calibrating the instrument for temperature and expansion. This calibration must be done under the load to be used for testing the polymer. Temperature calibration is done using various metals as melting point standards, with indium, tin and zinc being typical metals. Aluminium is used if a sample of the melting standard is sandwiched between two higher-melting metal discs. In this case, a deflection of the TMA probe is observed when the melting standard melts and the temperature at which this occurs corresponds to the melting point of the standard.

Dimensional calibration of the LVDT is done by running a sample (such as aluminium) of known expansivity. In this method, the initial length of the sample is determined at room temperature and then the change in length of the sample is determined with changes in the temperature. As the temperature is changed, a finite time is needed for the sample to come to the equilibrium length for that new temperature. This needs using slow temperature ramp rates (less than 5°C/min) or temperature step isothermal methods. The linear coefficient of thermal expansion (LCTE) is defined by Eq. (11.24).

$$\text{LCTE} = (1/L)\,(\delta L/\delta T) \quad \text{...(11.24)}$$

where L = length of the sample and T = temperature (°K). The calculated LCTE is then compared to the standard value for the standard and a correction factor is applied to the data.

Once the instrument is calibrated, a TMA experiment is run in a manner similar to the calibration experiment. The sample is placed onto the sample holder, the temperature is adjusted to the starting temperature, and the temperature is ramped or stepped over the desired temperature range. As with other modern thermal analysis instruments, the entire experiment, sometimes including the initial measurement of the sample dimension, is computer controlled and data analysis software permits for easy calculation of the LCTE or deflection temperature.

4. DMS

Dynamic mechanical spectroscopy (DMS) applies a physical force to one portion of a test specimen and measures the response of the sample at some distance from the specimen. A test specimen mounted in the dynamic mechanical spectrometer is subjected to a sinusoidal mechanical deformation of a preset force. Commonly, the frequency (typically 0.001 to 10 Hz) is varied but the applied force remains constant during any given experiment. Various types of deformation can be used, including torsional, elongational, and bending motions. DMS experiments are generally performed over a range of temperature (typically –100°C to 500°C). The frequency shift and the force transmitted through the sample are measured and used to determine the storage and loss moduli. The ratio of the loss modulus to storage modulus remains nearly constant, or changes only slightly in a regular fashion, over temperature ranges that allow no new relaxations. At temperatures where new molecular motions like glass transition temperature (T_g) begin, there is an increase in the loss modulus and generally a decrease in the storage modulus so that there is a maximum in their ratio. This maximum shows the temperature range over which the molecular motion begins to occur and it is used to assign the **temperature of the transition.**

The loss modulus curve in Fig. 11.30 illustrates another point important in polymer analysis. If the melting point of a pure inorganic material or small organic molecule (such as indium metal to benzoic acid) is determined, a single sharp melting peak is observed generally over a small temperature range of 1 to 3°C. In characterizing polymer systems, however, that is generally not the case. Most polymers consist of a molecular weight distribution and many polymers also have a variety of defects (like branches). Therefore, a thermal transition, such as T_g, does not occur at a single temperature but over a range of temperatures. For DMS testing, the use of a variety of frequencies helps to determine the distribution of properties in a sample because a response to a deformation is observed only when the relaxation time for the motion under study is the same as the deformation frequency. If DMS were used to study the T_g relaxation in a polymer, a different value for the loss modulus peak maximum would be observed for the different frequencies used in the test.

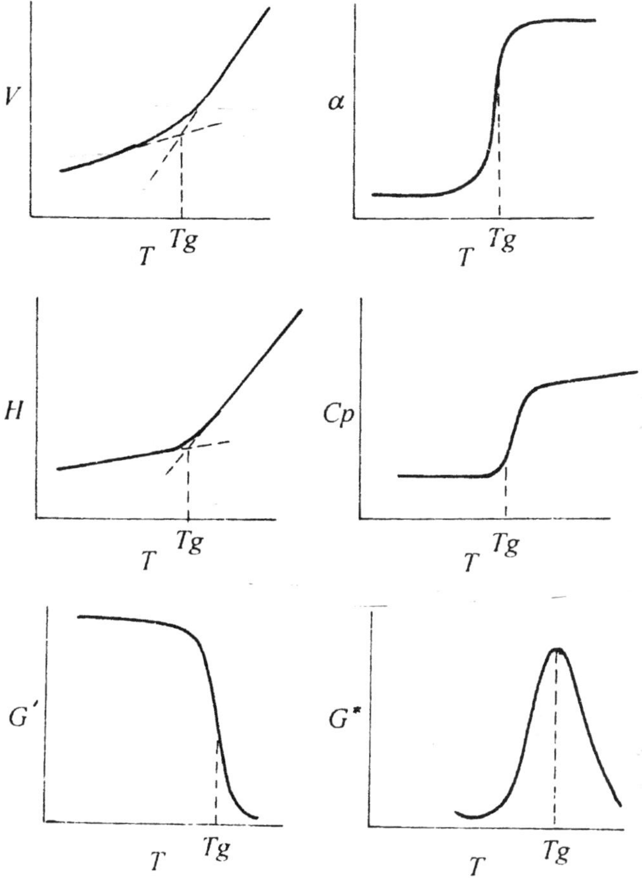

Fig. 11.30: The glass transition temperature as observed by changes in volume and expansion (*V* and α), heat flow and heat capacity changes (*H* and C_p), and the shear modulus and loss modulus (G' and G*).

5. Dielectric

In a dielectric experiment, a test specimen is fixed between capacitor electrodes (plates). A sinusoidal voltage oscillation (ranging from about 0.001 to 10^7 Hz) at a specified voltage is applied across a pair of electrodes. The shift in phase and amplitude of the electric current, relative to the applied field, is measured as the field passes through the polymer. As in the case of mechanical testing, the measured values are used to determine the stored energy (capacitance) and lost energy (resistance) for the sample. These values are commonly reported as the *dielectric constant* or *permitivity* and the *dielectric loss value* or *dissipation factor*.

With the similarities between DMS and dielectric spectroscopy it becomes difficult to choose one method and not the other. The choice depends on the type of transition being studied because some transitions are more sensitive to mechanical deformation and others to electrical probing. Fig. 11.31 illustrates this point by comparing the dielectric and mechanical tan δ (dielectric loss factor/dielectric

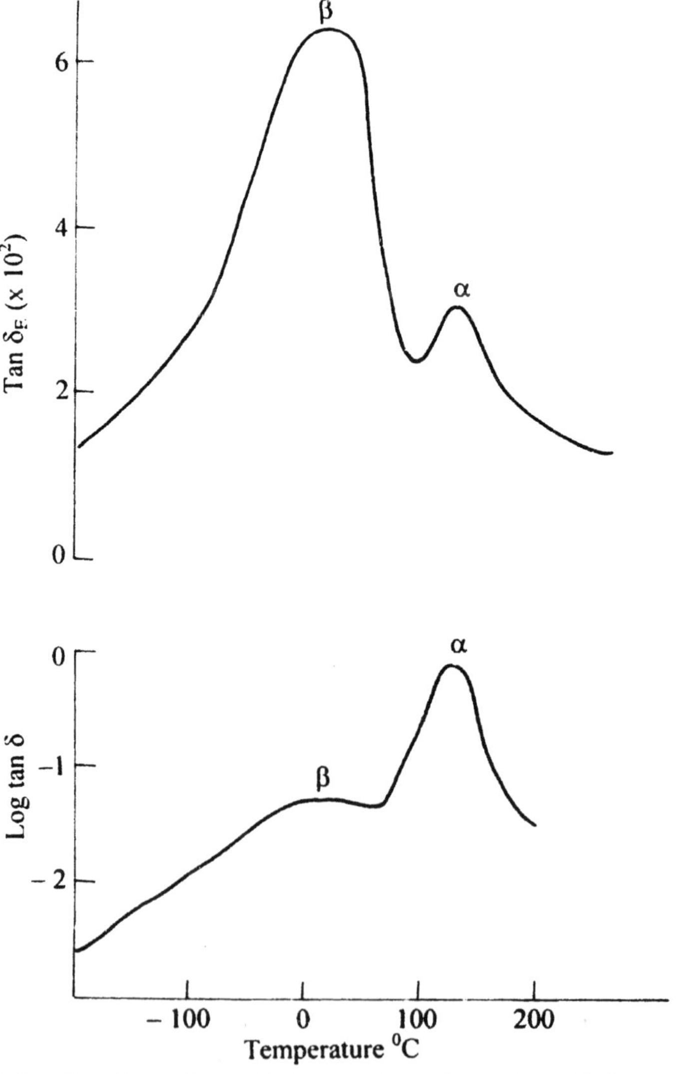

Fig. 11.31: The dielectric tan δ (top) and mechanical tan δ (bottom) for poly (methyl methacrylate).

constant, loss modulus/storage modulus) for poly(methyl methacrylate) (PMMA). For PMMA, the α transition is T_g and the β transition is the rotation of the methacrylate group around the main polymer chain. Because T_g is a main chain motion that affects modulus and polymer stiffness, this motion is most intense for a mechanical deformation. The rotation of the methacrylate group, on the other hand, is not a main chain motion and it only weakly affects the stiffness of the polymer, and is therefore a weak mechanical transition. The methacrylate group is a highly polar segment and it gives a more intense dielectric response. This demonstrates that comparison of mechanical and dielectric responses, combined with a knowledge of the polymer structure, will give added insight to the relaxation temperature and what motions are occurring at these temperatures.

Uses

Calorimetry is one of the most commonly applied techniques for polymer characterization. In polymer systems it is possible, using different techniques, to observe a number of transitions. However, two transitions that involve main chain motions, and therefore are of most importance, are the **glass transition** and the **melting of semicrystalline polymers.**

At the glass transition temperature (T_g), the physical properties of a polymer change dramatically. At this temperature, where large-scale molecular motions involving more than 20 atoms of the main chain are permitted, the polymer goes from a glassy state to a rubbery state. The physical are represented in Fig. 11.30. *At the glass transition, there is a change in the rate of volume expansion and a change in the total heat content.* For the expansion coefficient and heat capacity, this change at T_g leads to step *changes in the linear coefficient of thermal expansion* (α) and the *heat capacity* (c_p). This change in heat capacity is readily measured **calorimetrically.** This step in heat capacity can be directly related to the difference in heat capacity for the glassy and liquid (rubbery) states of the polymer. For a pure high-molecular-weight polymer, both the temperature where as glass transition occurs and the specific heat capacity change (ΔC_p/gram of polymer) are constants; however, variations take place when studying low-molecular-weight polymers or polymers with additives. Calorimetric methods are generally the easiest method for determining T_g.

The **melting point** (T_m) and the **extent of crystallinity** are two other important thermal transitions of polymers. Not all polymers can crystallize and a variety of factors affect the ability of a polymer to crystallize. Probably the most important factor affecting whether a polymer can crystallize is **polymer tacticity.** *Atactic polystyrene cannot crystallize, but both isotatic and syndiotactic polystyrene can crystallize,* although they have different melting points and heats of fusion. This being the case, the melting point and the heat of fusion can be used to determine the presence and the quantity of a polymer present in blend. The heat of fusion can be used to determine the extent of crystallinity in a polymer sample, and because the extent of crystallinity can have a dramatic effect on polymer properties, this can be a critical measurement for characterizing the polymer.

Thermogravimetric analysis (TGA) is used to study the weight loss characteristics of a polymer when the sample is heated over a temperature range or held isothermally at a temperature. TGA is generally used to determine the temperature for the onset of degradation, which becomes the upper use temperature for a polymer. TGA is also used to characterize polymer systems and the TGA decomposition curve is used for identification purposes. Additionally, TGA is used to quantify the presence of volatile components (such as adsorbed water in a hygroscopic polymer) that may be present in a sample.

When studying degradation, the analyst must be certain to understand the meaning of the term *degradation* and know that degradation as measured by different techniques can mean different things. For degradation phenomena to be observed by a TGA experiment needs the generation of a molecular fragment that is small enough to be volatile at the experimental temperature. If a polymer chain with 1000 carbon atoms in the backbone "degrades" into two polymer chains, each with 500 carbon atoms in the backbone, it is likely that no product that would be volatile at 300°C would be made. However, because the molecular weight was reduced by a factor of two, changes in other important properties (such as modulus) may have occurred and these are best examined using other techniques.

Thermogravimetric analysis requires a sensitive balance and a well-controlled furnace or other suitable heating system to maintain a specified temperature or temperature ramp program. Data collected during a TGA experiment includes time, temperature, and sample weight. Commercially available TGA are interfaced with a computer that can be used to set up the experiment, control the instrument during the experiment, collect data, and analyze the data after the experiment.

Thermogravimetric analysis can be linked to other methods to permit for even greater characterization of a sample. Methods like TGA Fourier transform infrared, TGA mass spectrometry, or TGA gas chromatography mass spectroscopy are used to characterize the degradation products given off by the sample. In a multicomponent sample, these data are used to determine which part of the sample is degrading over a given temperature range, resulting in some evaluation of the degradation mechanism.

Thermomechanical analysis (TMA) and **thermodilitometry** are the methods used to characterize *volumetric changes in polymer systems.* As noted in Fig. 11.30, the glass transition temperature can be characterized as a change in volumetric expansion. Some authors describe TMA as the measurement of the temperature of deformation under a static load, whereas thermodilitometry determines the change in dimensions under a static load. The same instrument is used for both measurements and is usually called TMA.

Dynamic mechanical spectrosocpy (DMS or DMA) and **dielectric spectroscopy** are used to determined changes in modulus, the last example defining T_g. Of the three different manifestations of the glass transition temperature shown in Fig. 11.30, changes in enthalpy/heat capacity and volume/expansion were already discussed. Polymers, being long-chain molecules, have a variety of possible molecular motions that, upon heating from a low enough temperature, are allowed at specific temperatures. For some motions, like T_g or the melting point, there are major changes in bulk polymer properties, such as modulus or toughness. The temperature at which these and other molecular motions are permitted is also useful in identifying and defining a polymer. In some cases, this temperature information helps to define the occurrence of less dramatic, although important, changes in polymer properties, such as the *change from brittle to ductile fracture.* These sub–T_g relaxations generally involve the motion of molecular fragments that are only a few atoms in length, and are significantly weaker in intensity than the T_g motion. This makes it difficult to detect these transitions, and in fact they often cannot be detected using DSC or TMA but are detectable using DMS or dielectric spectroscopy.

Dielectric spectroscopy is much like DMS in that a sinusoidal oscillation is applied to a test specimen, resulting in a measurable response. In the case of dielectric spectroscopy, however, an electrical forces in used instead of a mechanical force. Polymer chains that contain dipoles, like acrylates or halogenated polymers, are studied by dielectric spectroscopy. Although we would not expect a nonploar polymer like polyethylene to have a significant dielectric response, the presence of small levels of impurities, like oxidized polymer fragments, give enough of a dielectric response that dielectric measurements on these polymers often are possible.

Analytical Information

Qualitative

Although *glass transition temperature* and *melting point* are useful for classifying the polymers present in a sample, thermal methods generally cannot be used for absolute identification of materials present in a polymer system. Thermal methods, particularly DSC and TGA, are commonly used in quality control applications to verify that a given lot of material is the same as a control sample of that chemical.

Quantitative

Once the presence of a given material has been demonstrated by other analytical techniques, thermal methods are used for quantitative analysis [such as the determination of vinyl acetate content of poly(ethylene-co-vinyl acetate]. The use of DSC to determine heats of fusion and the extent of crystallization for crystalline materials have been discussed in the Applications section. Levels of inorganic fillers are also determined by thermogravimetric analysis.

Applications

1. Determination of Glass Transition

From a thermodynamic and mechanical standpoint, the glass transition temperature is one of the most important parameters for characterizing a polymer system. Therefore, the determination of the glass transition temperature is one of the first analyses to be done on a polymer system. The glass transition temperature can be determined by various techniques, for example TMA, DMA, and DSC. These three methods may be used to characterize a cured epoxy system.

Fig. 11.32 illustrates the type of data obtained for each of the three methods for an epoxy/woven glass composite. At T_g, the TMA curve drops sharply as the material softens. The extrapolated onset on this curve is used for defining the glass transition temperature as 122.9°C. In the DSC curve (solid line), a step change in the heat capacity is observed in the same region. The mid-point of the step transition, 121.3°C for this sample, is taken as T_g. The other three curves in the figure are various representations of the DMA data for this sample.

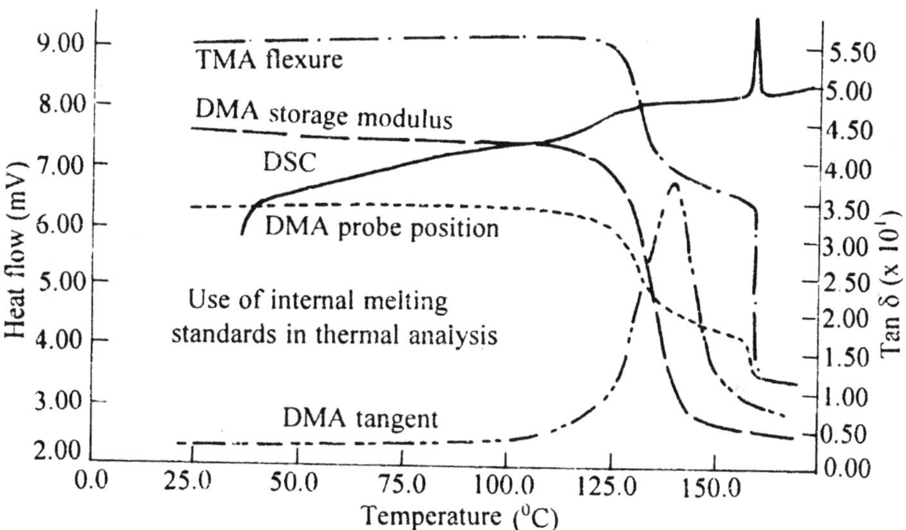

Fig. 11.32: The glass transition temperature for an epoxy/woven glass composite as determined by TMA, DMA and DSC.

As already mentioned, glass transitions occur over a temperature range, and there are a number of options for where to choose the glass transition temperature. Here, the extrapolated onset of the storage modulus at 0.1 Hz gave a value of 122.1°C which was in good agreement with values obtained by TMA and DSC. For this sample, these DMA conditions give the best agreement with other methods. For other materials, some other temperature, possibly corresponding to the *loss modulus, tangent delta extrapolated onset temperature,* or *peak temperature* might give better agreement with other methods. It is important to remember that it is not critical that all methods give the same value of T_g. Because the different methods can represent different deformation conditions, it is generally valuable to use a test method that most closely represents the conditions under which the polymer will be used. For strictly analytical purposes, the determination of T_g by DSC using published parameters is the method of choice, because this test is more rapid than TMA or DMA.

2. Heat Capacity Determinations

The basis for all calorimetric measurements is the determination of heat capacities. In the DSC scan of a polymer system, in the absence of any other transition, the DSC curve represents the change in heat capacity of the sample over the experimental temperature range. Even at the glass transition temperature of a polymer, the step observed in the DSC curve is related to the change in heat capacity when the polymer goes through the transition from the glassy to the liquid state.

The determination of heat capacities is performed using DSC methods. In this method, a DSC scan of an empty pan is recorded followed by the DSC scan of a well-characterized heat capacity standard, such as sapphire. Finally, the sample of interest is run in the same DSC pan. The data for the empty pan is subtracted from the curve data for the reference standard, and the DSC is calibrated for heat capacity. (Calibration of the DSC for heat flow, with indium for example, is not the same as calibrating for heat capacity determinations). The baseline is then substracted from the sample run and the heat capacity of the unknown is determined using the heat capacity calibration.

The reported determination of the heat capacity of various nylon polymers is a good example of how to determine heat capacities. Of particular interest in this work was the relationship between the molecular structure of the nylons studied, that is, the number of methylene groups and the number of amide groups. An equation was developed to calculate the heat capacity at any temperature between the polymer T_g and about 600°K from the chemical structure. To be analytically useful, this type of data also helps to understand the relative molecular motions in the polymer liquid.

3. Characterization of Polymer Blends

As seen in Fig. 11.32, a shift in the heat capacity curve appears at T_g in a DSC experiment, a peak in the loss modulus or tan δ appears in DMA experiments, and a change in the volumetric expansion occurs for TMA experiments. When more than one polymer is present, such as in a blend, we observe more than one T_g. One of the most commonly used materials is *rubber modified high-impact polystyrene* (HIPS). This material contains two polymeric phases, often a polybutadiene phase with a T_g near –85°C and a polystyrene phase with a T_g near 104°C. When scanned from a low temperature (about –120°C) to 200°C, two steps in the heat capacity are observed for DSC measurements and two maxima in tan δ are observed for DMA experiments.

This can be useful for studying blends of similar polymers such as the characterization example of styrene–butadiene latex blends. Because both components are composed of styrene and butadiene, a blend of a 75% styrene and 25% butadiene latex with one containing 25% styrene and 75% butadiene would look like a single polymer based on data from spectroscopic methods such as infrared (IR). In a DSC scan, however, different T_gs would be observed for each of the styrene-butadine compositions, confirming the existence of two different compositions in the blend.

4. Compositional Effects on the Glass Transition

Homopolymers and blends of homopolymers, such as the high-impact polystyrene example discussed above, show a T_g for each of the polymers present in the blend. In the case of copolymers, where two or more monomers are blended together and reacted in an essentially random fashion, the T_g is intermediate between the T_gs of the homopolymers for each of the monomers used. Various relationships for calculating the change in T_g with composition have been developed and a simple relationship is given in Eq. (11.25)

$$1/T_g = W^a/T^a_g + W^b/T^b_g \qquad ...(11.25)$$

where W = the weight fraction of monomers a and b and T_g = respective T_g for homopolymers of monomers a and b.

Fig. 11.33 shows the change in T_g for various copolymers of styrene with other monomer units. Therefore, for any copolymer, at any mole fraction of styrene, only a single T_g value exists. If the T_g of a copolymer is determined and the T_g composition curve exists, then the composition of the copolymer can be approximated from the T_g. This holds only when the copolymers being studied are prepared in a random fashion.

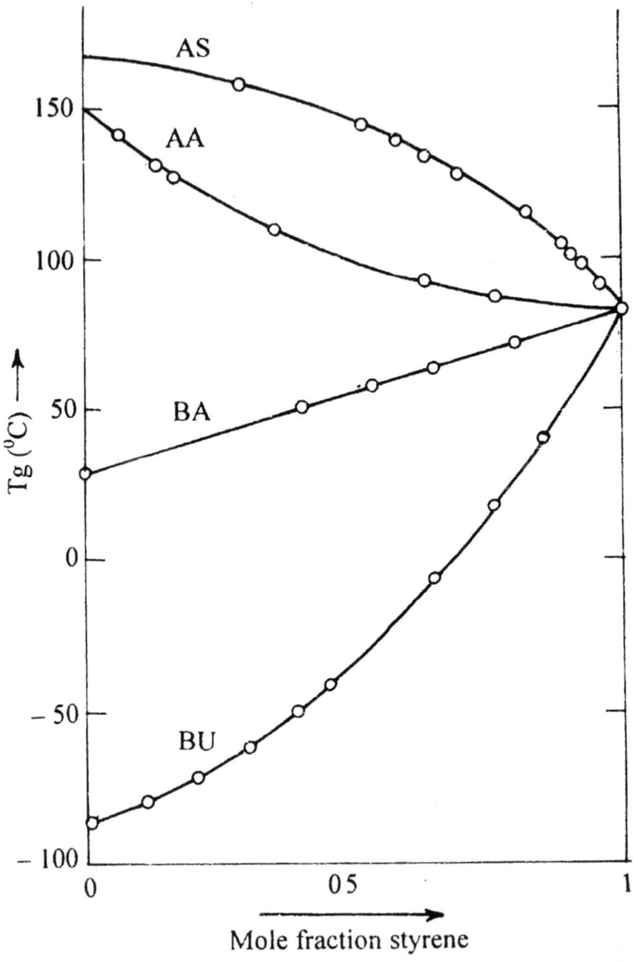

Fig. 11.33: Glass transition temperature at various styrene mole fractions for copolymers of styrene with acrylic acid (AS), acrylamide (AA), t-butyl acrylate (BA), and butadiene (BU).

As polymers are beign developed to have more specialized properties, it is becoming more common to see terpolymer systems containing three or more monomer types. Although the equation for calculating T_g can be expanded, the relationship between T_g and composition becomes more complicated, and therefore of less analytical value.

5. Additive Analysis

Differential scanning calorimetry (DSC) is used to study additives added to polymeric systems to increased flowability in the melt, to reinforce the toughness of samples, or to improve the oxidative stability or reduce the flammability of polymer samples.

One example of DSC use for studying oxidative stability and the effect of adding antioxidants involves characterization of the oxidative stability of polyethylene. The oxidation of nearly any polymer is an exothermic event, so when the polymer begins to oxidize and exotherm is observed. The presence of a material to inhibit the oxidation of the polymer, an antioxidant, significantly reduces the rate of oxidation of the polymer until the antioxidant is consumed, and the exotherm is then observed.

A scanning method and isothermal method are used to study the oxidation of polymers. In the scanning methods, the sample is heated at a specified rate in an air atmosphere to a high enough temperature that the exotherm of oxidation is observed. The extrapolated onset temperature is then determined. By using this method with polymers containing different antioxidants or different levels of the same antioxidant, it is possible to determine the relative oxidative stability of polymers in this manner.

In the second method, an **isothermal test**, the sample is rapidly heated to the temperature of interest, which was determined from previous range-finding experiments. The time the sample is held in the oxidative atmosphere at temperature is used to determine the time the polymer system can spend at that temperature in air before the onset of oxidation. *This points out the difference between the scanning and the isothermal test.* The scanning test is a faster test for screening and is good for relative information (determining whether antioxidant 1 is better than antioxidant 2). The isothermal test takes longer but gives a more accurate picture of the lifetime of a polymer under use conditions. The time to the onset of the oxidation peak is a kinetic parameter, so if a series of tests at different temperatures are run on the same sample, the Arrhenius kinetics can be determined, and the time to oxidation at lower temperatures, where longer times are observed, can be calculated if there is no change in mechanism over the range where the extrapolation is made.

Thermogravimetric analysis (TGA) is used to determine inert (inorganic) additives and organic additives such as pigments. Glass fibres are sometimes added to polymer to increase their rigidity above T_g. Glass fabric is sometimes used as the support for epoxy resins in larger cured parts such as aircraft parts and computer boards. In an oxidative atmosphere, the polymer will degrade but the glass will not. When a sample of a glass-filled polymer is heated to about 900°C in an air atmosphere, the polymer will be totally oxidized and the weight remaining will be that of the glass fibre from the composite formulation.

With inert atmospheres (such as nitrogen and helium), a polymer will unzip or depolymerize (that is, return to monomer form) when heated to a sufficiently high temperature. Polyethylene and polystyrene are examples of this type of polymer. Some organic additives such as carbon black are of a more complex structure and they do not undergo any weight loss when heated in the same inert atmosphere to temperatures as high as 600°C. However, these materials will oxidize in air at these temperatures. TGA methods for the *determination of carbon black in polyethylene involve rapidly heating the polymer sample in an inert atmosphere* (over 100°C/min) to about 600°C and holding the sample isothermally at that temperature for some period of time. The atmosphere is then changed to air and the sample is heated to about 1000°C. The weight loss observed after switching from inert to oxidative atmosphere gives the level of carbon black in the system.

6. Crystallinity and Crystallization Rate

In crystalline polymers, the level of **crystallinity** can be one of the most important factors for determining properties of the polymer systems including *strength, ultimate elongation,* and *barrier properties.* Thus, a method for determining the level of cystallinity is important for studying polymer systems. There are a variety of methods, including *IR spectroscopy, density, X-ray diffraction,* and *calorimetric methods,*. In many cases, calorimetric methods are the easiest method for determining crystallinity. Moreover, calorimetric methods are not sensitive to orientation of the polymer crystal structure, and this can be an advantage for looking at total crystallinity. This is disadvantage, however, if we are interested in studying the effects of orientation, then IR spectroscopy or X-ray diffraction would be preferable methods.

Use of calorimetric methods for determining the level of crystallinity is based on a knowledge of the heat of fusion for the sample. Inorganics and small molecule organic materials have well-characterized heats of fusion, which can be used to determined the quantity of the material present in a system. Polymers also can have well-characterized heats of fusion, and by measuring the heat of fusion the level of crystallinity can be determined. **Dole** gives a derivation of the expression in Eq. (11.26) for determination of the level of crystallinity in a polymer.

$$\text{Percent crystallinity} = \Delta H^{\text{sample}}/\Delta H^{\circ} \qquad ..(11.26)$$

where ΔH^{sample} = measured heat of fusion for the sample of interest and ΔH° = heat of fusion for the fully crystalline sample.

It is important to distinguish the two meanings of ΔH°. In one case ΔH° can be the *heat of fusion for a sample that is* 100% *crystalline*. It is virtually impossible to make a sample that is 100% crystalline, so it can be difficult to determine the value for the 100% crystalline sample. Values of the heat of fusion for polymers are tabulated or can be determined. Two methods that are used to determine the heat of fusion include **X-ray** and **heat capacity measurements.** In the X-ray method, the level of crystallinity is determined by X ray for several samples and the heat of fusion is determined for the same samples by DSC. The heat measured is then divided by the fraction crystallinity, as determined by X ray and ΔH° is then determined.

A second method involves the use of both heat of fusion measurements and the heat capacity change at T_g. As already discussed, the heat capacity change at T_g for a polymer is used for quantitation of the amount of polymer present. In the present application this can be related to the amount of polymer present in the glassy phase. To determine the heat of fusion for a 100% crystalline polymer, both the heat of fusion and change in heat capacity at T_g are measured for samples of the polymer that have different levels of crystallinity (that is, different ΔH values). A plot of ΔH versus ΔC_p should give a straight line, and the intercept where ΔC_p is zero is the heat of fusion for the 100%–crystalline polymer.

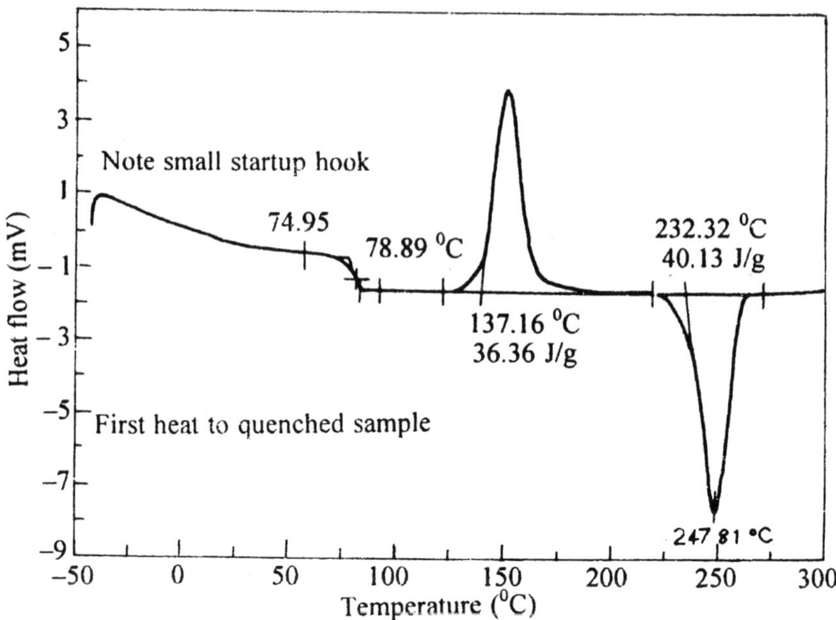

Fig. 11.34: DSC scan for a sample of poly(ethylene terephthalate) that was quenched from the melt. T_g is observed at 78.89°C. The onset of crystallization is observed at about 137°C and the onest of melting is observed at about 232°C. The melting endotherm (40.13 joules/gram) is slightly large than the crystallizing exotherm (36.36 joules/gram), indicating that the sample was not fully amorphous when quenched.

A number of authors report using DSC measurements to characterize the crystallinity of polymer systems, including measuring the level of crystallinity and the relationship to branching. In this work, the change in crystallinity level and melting point were studied and found to decrease as a function of the number of branch points in the polymer chain. Another study also reported the effect of branching on the DSC trace and the extent of branching characterized by DSC traces.

Another aspect of crystallinity involves studying the *rate of crystallization*, or *crystal growth*, in a polymer system. Calorimetry is also used for studying crystallization kinetics. Crystallization is an exothermic event, as shown in Fig. 11.34, and calorimetry is used to follow the rate of heat release from the system. Differential scanning calorimetry was applied to the study of the crystallization of polymer blends as well as homopolymers.

7. Reaction Kinetic Studies

Like the crystallization process, many chemical reactions are exothermic, and as such are readily followed by calorimetric methods. In using calorimetry to study reaction kinetics, the important parameters include the determination of the amount of the rate (δ_{time}) of heat released during the course of the reaction. From this knowledge the extent of conversion versus time can be determined.

In *studying the kinetic of polymerization,* various complicating factors should be considered. When studying the polymerization of monomers, the volatility of the monomer generally creates a problem in determining accurate heats because the mass can change during the course of the experiment. Some instrument manufactures market a high-pressure DSC pan to overcome this volatility problem and a sealed glass capillary method is described in the literature. When studying *curing reactions for epoxy systems, volatility is not normally a concern.*

In the *study of polymerization reactions,* the physical changes that occur during the reaction can also affect the measured kinetics. A material below its glass transition temperature is a glass, and as such has limited mobility. If reaction kinetics for a system are studied at temperatures below the maximum T_g for the final polymer, which is mostly the case for *thermoset systems,* then the reaction will take place only to an overall extent of reaction such that the T_g of the product is essentially the same as the reaction temperature. If the sample is then heated to a higher temperature (above the current T_g of the sample), the reaction will continue until the reaction completes or the T_g of the sample reaches the current reaction temperature. As a reaction nears completion, the viscosity of the system can limit the kinetics, so the rate of reaction becomes diffusion controlled rather than kinetically controlled. This phenomenon is commonly observed in the study of cure kinetics for thermoset systems.

Polymerization kinetic are studied both in a scanning and an isothermal mode. Reaction kinetics are generally determined from a scanning DSC curve, using the method of Borchardt and Daniels.

Isothermal DSC experiments were reported for studying the cure of various epoxy systems. In these the extent of conversion was followed in two ways. The heat of reaction (or the rate of heat release) was determined at various times, and these measurements were used to determine the kinetics. Additionally, kinetics are determined by holding the sample isothermally at the reaction temperature for some time interval, and then cooling the sample and determining the remaining heat (extent) of reaction. In this case, the extent of reaction is 1 minus the unreacted reaction. The rate of conversion is determined by doing this at several isothermal times.

8. Polymer Degradation

The degradation of polymers can be important for determining the maximum use temperature of polymer systems. As previously discussed, the term *degradation* can be somewhat ambigures, so methods for measuring degrodation can lead to different results. When a polymer endothermic or exotherimc energy is related with the process. Measuring the energy release gives a measure of the extent of degradation. Studying the same process by TGA may not give the same results because TGA requires that the degradation product formed be volatile at the test temperature. In deciding what test to use to follow degradation, we must determine which methods will most closely represent the end-use conditions and the conditions that would constitute material failure.

In addition to being used to study the oxidative degradation of polymers, calorimetry is used to study the oxidation of polymers without additives.

Thermogravimetric analysis is often used for studying degradation. Two aspects of degradation that are studied are thermal degradation and oxidative degradation of polymers. Thermal degradation is important because degradation that occurs during processing (such as extrusion) may be largely thermal and not oxidative. A classic example of thermal degradation is the study of halogenated aliphatic polymers. In studying the degradation of halogenated polymers, such as poly(vinylidene chloride), TGA is often the preferred methods. These polymers often degrade by the elimination of a hydrohalogen molecule, so the weight loss is directly related to the rate of polymer degradation.

Oxidative degradation is also studied by TGA. In these studies, the weight loss profiles are different from those for thermal degradation studies, but they may be more applicable to performance at high temperature encountered during the use of the part or in some processing applications (such as thermoforming).

For either the thermal or oxidative degradation of polymers, understanding the degradation pathway is important for designing end-use applications for the polymer and this information may enhance the ability to stabilize the polymer. In addition to pyrolysis methods, various combined techniques such as TGA infrared, TGA mass spectrometry, or TGA gas chromatography mass spectrometry have been developed for studying polymer degradation. These methods allow real-time analysis of the degradation products of the sample to be characterized. This information aids in building a model for the degradation mechanism for polymer systems under a variety to end-use applications.

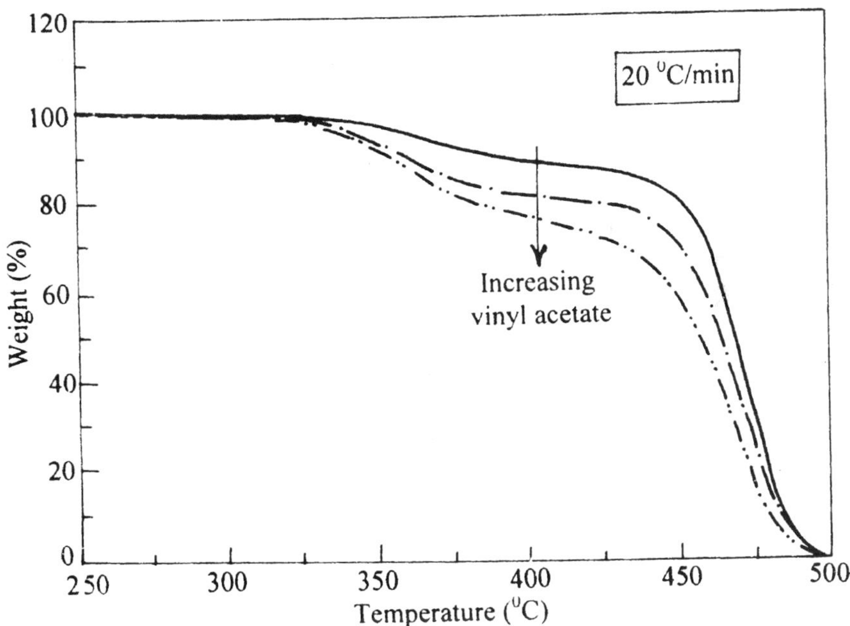

Fig. 11.35: Determination of vinyl acetate levels in poly(ethylene-co-vinyl acetate) by TGA. As the vinyl acatate level increases, the weight loss corresponding to the elimination of acetic acid increases, giving a quanti:ative measure of the vinyl acetate level.

In some cases, TGA degradation studies are useful for quantitative analysis. Poly(ethylene-co-vinyl acetate) is made by reacting ethylene and vinyl acetate. When heated, a polymer containing vinyl acetate eliminates acetic acid, producing an ethylene linkage in the polymer. Because this is a stoichiometric elimination, each mole of acetic acid lost corresponds to one mole of vinyl acetate in the copolymer. Fig. 11.35 shows the weight loss for several copolymers with different vinyl acetate levels, and demonstrates how TGA is used for quantitative analysis.